International Code of Nomenclature for algae, fungi, and plants
(Madrid Code)

I0085391

Regnum Vegetabile
Volume 162

REGNUM VEGETABILE is the book series of the International Association for Plant Taxonomy and is devoted to systematic and evolutionary biology with emphasis on plants, algae, and fungi. Preference is given to works of a broad scope that are of general importance for taxonomists.

All manuscripts intended for publication in REGNUM VEGETABILE are submitted to the Editor-in-chief. Authors interested in publishing in the series are requested to send an outline of their book, including a brief description of the content, to the Editor-in-chief. Proposals for new book projects will be scrutinized by the editorial advisory board.

Editor-in-chief: Laurence J. Dorr, National Museum of Natural History, Smithsonian Institution, Washington, D.C., U.S.A., E-mail: DORRL@si.edu

Production Editor: Franz Stadler, Institute of Botany, Slovak Academy of Sciences, Bratislava, Slovak Republic and Vienna, Austria, E-mail: production@iapt-taxon.org

Editorial advisory board: Domingos Cardoso (Universidade Federal da Bahia, Brazil), Xue-Jun Ge (South China Botanical Garden, Guangzhou, China), Vinita Gowda (Indian Institute of Science Education and Research Bhopal, India), Muthama Muasya (University of Cape Town, South Africa), Helga Ochoterena (Universidad Nacional Autónoma de México, México), and Nicholas Turland (Botanischer Garten und Botanisches Museum Berlin, Freie Universität Berlin, Germany).

International Code of Nomenclature
for
algae, fungi, and plants
(Madrid Code)

Accepted by the Twentieth International
Botanical Congress, Madrid, Spain, July 2024

Prepared and edited by the Editorial Committee:

Nicholas J. Turland, Chair
John H. Wiersema, Secretary
Fred R. Barrie
Kanchi N. Gandhi
Julia Gravendyck
Werner Greuter
David L. Hawksworth
Patrick S. Herendeen
Ronell R. Klopper
Sandra Knapp
Wolf-Henning Kusber
De-Zhu Li
Tom W. May
Anna M. Monro
Jefferson Prado
Michelle J. Price
Gideon F. Smith
Juan Carlos Zamora Señoret

The University of Chicago Press
Chicago and London

The University of Chicago Press, Chicago 60637

The University of Chicago Press, Ltd., London

© 2025 by International Association for Plant Taxonomy

All rights reserved. No part of this book may be used or reproduced in any manner whatsoever without written permission, except in the case of brief quotations in critical articles and reviews. For more information, contact the University of Chicago Press, 1427 E. 60th St., Chicago, IL 60637.

Published 2025

Printed in the United States of America

34 33 32 31 30 29 28 27 26 25   1 2 3 4 5

ISSN: 0080-0694

ISBN-13: 978-0-226-83946-2 (cloth)

ISBN-13: 978-0-226-84199-1 (paper)

ISBN-13: 978-0-226-83947-9 (e-book)

DOI: https://doi.org/10.7208/chicago/9780226839479.001.0001

Library of Congress Control Number: 2024948960

♾ This paper meets the requirements of ANSI/NISO Z39.48-1992 (Permanence of Paper).

How to cite this *Code*: Turland, N.J., Wiersema, J.H., Barrie, F.R., Gandhi, K.N., Gravendyck, J., Greuter, W., Hawksworth, D.L., Herendeen, P.S., Klopper, R.R., Knapp, S., Kusber, W.-H., Li, D.-Z., May, T.W., Monro, A.M., Prado, J., Price, M.J., Smith, G.F. & Zamora Señoret, J.C. 2025. *International Code of Nomenclature for algae, fungi, and plants (Madrid Code).* Regnum Vegetabile 162. Chicago: University of Chicago Press.

Important note concerning names of organisms treated as fungi: After publication of this *Code,* the content of its Chapter F (names of organisms treated as fungi) could be amended by the International Mycological Congress scheduled for 2027. In case such amendments are made, users of Chapter F should consult the online version of the *Madrid Code.*

EDITORIAL COMMITTEE OF THE *MADRID CODE*

Nicholas J. Turland, *Botanischer Garten und Botanisches Museum Berlin, Freie Universität Berlin, Königin-Luise-Str. 6–8, 14195 Berlin, Germany; n.turland@bo.berlin* (Chair)

John H. Wiersema, *Department of Botany, NMNH - MRC 166, Smithsonian Institution, P.O. Box 37012, Washington, D.C. 20013, U.S.A.; wiersemaj@si.edu* (Secretary)

Fred R. Barrie, *Missouri Botanical Garden, 4344 Shaw Blvd., St. Louis, Missouri 63110, U.S.A.;* and *Gantz Family Collections Center, The Field Museum, 1400 S. Lake Shore Drive, Chicago, Illinois 60605, U.S.A.; fbarrie@fieldmuseum.org*

Kanchi N. Gandhi, *Harvard University Herbaria, 22 Divinity Avenue, Cambridge, Massachusetts 02138, U.S.A.; gandhi@oeb.harvard.edu*

Julia Gravendyck, *Plant Biodiversity Section, Bonn Institute of Organismic Biology (BIOB), University of Bonn, Meckenheimer Allee 170, 53115 Bonn, Germany; gravendyck@uni-bonn.de*

Werner Greuter, *Botanischer Garten und Botanisches Museum Berlin, Freie Universität Berlin, Königin-Luise-Str. 6–8, 14195 Berlin, Germany;* and *Herbarium Greuter, c/o Orto Botanico, Via Lincoln 14, 90133 Palermo, Italy; w.greuter@bo.berlin*

David L. Hawksworth, *Comparative Plant and Fungal Biology, Royal Botanic Gardens, Kew, Richmond, Surrey TW9 3AE, U.K.;* and *Natural History Museum, Cromwell Road, London SW7 5BD, U.K.;* and *Jilin Agricultural University, Changchun, Jilin 130118, China; d.hawksworth@nhm.ac.uk*

Patrick S. Herendeen, *Chicago Botanic Garden, 1000 Lake Cook Road, Glencoe, Illinois 60022, U.S.A.; pherendeen@chicagobotanic.org*

Ronell R. Klopper, *Foundational Research & Services Directorate, Foundational Biodiversity Sciences Division, South African National Biodiversity Institute, 2 Cussonia Avenue, Brummeria 0184, Pretoria, South Africa;* and *H.G.W.J. Schweickerdt Herbarium, Department of Plant and Soil Sciences, University of Pretoria, Lynnwood Road, Hatfield 0002, Pretoria, South Africa; r.klopper@sanbi.org.za*

Sandra Knapp, *Natural History Museum, Cromwell Road, London SW7 5BD, U.K.; s.knapp@nhm.ac.uk*

Wolf-Henning Kusber, *Botanischer Garten und Botanisches Museum Berlin, Freie Universität Berlin, Königin-Luise-Str. 6–8, 14195 Berlin, Germany; w.h.kusber@bo.berlin*

De-Zhu Li, *Kunming Institute of Botany, Chinese Academy of Sciences, 132 Lanhei Road, Heilongtan, Kunming, Yunnan 650201, China; dzl@mail.kib.ac.cn*

Tom W. May, *Royal Botanic Gardens Victoria, 100 Birdwood Avenue, Melbourne, Victoria 3004, Australia; tom.may@rbg.vic.gov.au*

Anna M. Monro, *Australian National Herbarium, Centre for Australian National Biodiversity Research, GPO Box 1700, Canberra ACT 2601, Australia; anna.monro@dcceew.gov.au*

Jefferson Prado, *Instituto de Pesquisas Ambientais, Av. Miguel Estéfano 3687, CEP 04301-012, São Paulo, SP, Brazil; jprado.01@uol.com.br*

Michelle J. Price, *Conservatoire et Jardin botaniques de Genève, Chemin de l'Impératrice 1, 1292 Chambésy, Geneva, Switzerland; michelle.price@geneve.ch*

Gideon F. Smith, *Ria Olivier Herbarium, Department of Botany, P.O. Box 77000, Nelson Mandela University, Gqeberha, 6031 South Africa; smithgideon1@gmail.com*

Juan Carlos Zamora Señoret, *Conservatoire et Jardin botaniques de Genève, Chemin de l'Impératrice 1, 1292 Chambésy, Geneva, Switzerland; juan-carlos.zamora@geneve.ch*

CONTENTS

Contents

Contents

PREFACE

The rules that govern the scientific naming of algae, fungi, and plants are revised at the Nomenclature Section of an International Botanical Congress (IBC). This edition of the *International Code of Nomenclature for algae, fungi, and plants* embodies the decisions of the XX IBC, which took place in Madrid, Spain in July 2024. This *Madrid Code* supersedes the *Shenzhen Code* (Turland & al. in Regnum Veg. 159. 2018), published seven years ago after the XIX IBC in Shenzhen, China; like its six predecessors, it is written entirely in (British) English. The *Shenzhen Code* was translated into Chinese, French, Japanese, Portuguese, and Spanish, and it is likely that the *Madrid Code,* too, will become available in several languages. In questions about the meaning of provisions in translated editions of this *Code,* the English edition is definitive.

AMENDING THE *CODE* – FROM SHENZHEN TO MADRID

Altogether, 433 numbered proposals to amend the *Shenzhen Code* were published in *Taxon,* the journal of the International Association for Plant Taxonomy (IAPT), between 31 July 2020 and 27 October 2023. A "Synopsis of proposals", with comments by the Rapporteur-général and Vice-rapporteur, was published on 21 February 2024 (Turland & Wiersema in Taxon 73: 325–404. 2024) and served as the basis for the preliminary guiding vote ("mail vote") cast by members of the IAPT, authors of the proposals, and members of the Permanent Nomenclature Committees, as specified in Division III of the *Shenzhen Code.* Tabulation of the mail vote was handled at the IAPT Secretariat in Bratislava, Slovakia by Eva Kráľovičová and Matúš Kempa. The results were published on 4 July 2024 ahead of the Nomenclature Section (Turland & al. in Taxon 73: 1096–1109. 2024).

The Nomenclature Section met from Monday to Friday, 15–19 July 2024 in the conference hall of the central campus main building of the Consejo Superior de Investigaciones Científicas (CSIC) at Calle de Serrano 117, Madrid 28006, Spain (and was followed, from 21–27 July, by the main part of the IBC at the North Convention Centre of IFEMA Madrid). At the Section, 173 registered members were in physical attendance, carrying 433 institutional votes (from 192 institutions) in addition to one personal vote each, making a total of 606 possible votes. The Section was also livestreamed on the internet, permitting observation of the proceedings but not interactive commenting or voting, with a total of 219 unique users accessing the livestream at least once. The Section officers, previously appointed in conformity with Division III of the *Shenzhen Code,* were Sandra (Sandy) Knapp (President), Nicholas (Nick) Turland (Rapporteur-général), John Wiersema (Vice-rapporteur), and Inés Álvarez and Anna Monro (Recorders). The Vice-presidents were Werner Greuter, David Mabberley, Gideon Smith, Carmen Ulloa Ulloa, and Karen Wilson. The discussions of the Section were conducted in English.

The Nomenclature Section was entitled to define its own procedural rules within the limits set by Division III of the *Shenzhen Code.* These procedures are detailed in the Report of Congress action mentioned in the next paragraph. Of the 433 published proposals to amend the *Shenzhen Code,* 134 were accepted and 22 were referred to the Editorial Committee; an additional seven were accepted from among 14 new proposals made from the floor of the Section.

The rules of the *Madrid Code* became effective immediately upon acceptance of the resolution, moved on behalf of the Section at the closing plenary session of the XX IBC on 27 July 2024, that the decisions and appointments of the Nomenclature Section be approved. The "Report of Congress action on nomenclature proposals", detailing the results of the 433 published proposals and 14 floor proposals, the membership of the Permanent Nomenclature Committees (except those for fungi, see below), the establishment of four Special-purpose Committees, and the election of the Rapporteur-général for the XXI IBC (Nicholas Turland), was published on 23 September 2024 (Turland & al. in Taxon 73: 1308–1323. 2024). The full, day-to-day proceedings of the Section will be a separate publication, planned for late 2025 or 2026, following the format of previous Nomenclature Section reports (see, for example, Lindon & al. in PhytoKeys 150: 1–276. 2020 [Shenzhen]; Flann & al. in PhytoKeys 41: 1–289. 2014 [Melbourne]; Flann & al. in PhytoKeys 45: 1–341. 2015 [Vienna]).

The Nomenclature Section also elected the Editorial Committee of the *Madrid Code.* In accordance with Div. III Prov. 7.4, the Nominating Committee proposed members of the Section who were physically present there (with one exception) to serve on the Editorial

Committee, with the Rapporteur-général and Vice-rapporteur serving as the Chair and Secretary, respectively. The Editorial Committee was increased in size from the previous 16 to the present 18 members to ensure representation from each continent, to include expertise in the main groups of organisms covered by the *Code* (vascular plants, bryophytes, fungi, and algae, both extant and fossil), and to improve gender balance (there are now five women on the Committee, compared with three previously).

The Editorial Committee has a mandate (Div. III Prov. 7.12) to incorporate into the new *Code* the changes agreed by the Section, to clarify any ambiguous wording so long as the meaning is not changed, to ensure consistency and optimal placement of provisions while retaining the present numbering as far as possible, to add, amend, or delete Examples to best illustrate the provisions, and to revise the Glossary to explain the terms used in the *Code*.

A first draft of the main body of the *Madrid Code,* incorporating the changes decided by the Section, was prepared during September 2024 by members of the Editorial Committee, as follows: Fred Barrie (Art. 7 and 10), Julia Gravendyck (Art. 8 and 9), Patrick Herendeen (Art. 11), Ronell Klopper (Art. 13–24), Sandra Knapp (Art. 60), Wolf-Henning Kusber (Art. 30–38), Anna Monro (Art. 40–49), Jefferson Prado (Glossary), Gideon Smith (Art. 61, 62, and Chapter H), Nicholas Turland (Division III), and Juan Carlos Zamora Señoret (Preamble, Rec. 5A, Art. 6 and 51–56). Chapter F was updated by the Editorial Committee for Fungi (see below). During October 2024, in response to a request made at the Section, the Glossary was thoroughly reviewed by Jefferson Prado and Michelle Price, who added new entries and amended existing entries. During October and November 2024, the first draft was updated according to edits and comments subsequently received from the members. In addition, Kanchi Gandhi compiled numerous new Examples and amendments to existing Examples. The resulting revised draft was used at the Editorial Committee meeting as the basis for discussion.

The full Editorial Committee met from 25–29 November 2024 at the Botanischer Garten und Botanisches Museum Berlin, Germany, for five days of hard work: scrutinizing the entire *Code,* reviewing not only the changes made in Madrid, but also reviewing the existing wording as well as potential new Examples and Glossary items referred to the Committee or suggested by Gandhi, Prado, and Price. It was an intense but highly productive week.

On 5 December 2024, a semi-final draft of the *Madrid Code* was completed and distributed to all Editorial Committee members for proofreading. After making final adjustments and corrections, the finished text was sent to Franz Stadler, the Production Editor of *Regnum Vegetabile,* on 31 December 2024 to begin the formatting and page layout. The Index of scientific names and the Subject index, compiled by Anna Monro, followed on 31 January 2025. After formatting, page layout, and final proofreading, the *Madrid Code* was sent to the University of Chicago Press on 3 April 2025 for publication.

AMENDING CHAPTER F – FROM SAN JUAN TO MAASTRICHT

Because both the XII International Mycological Congress (IMC12), scheduled for 2022, and the XX IBC, scheduled for 2023, were delayed until 2024, the *Maastricht Chapter F* is published here in the *Madrid Code* instead of separately, as was the case for its immediate predecessor the *San Juan Chapter F* (May & al. in IMA Fungus 10(21). 2019).

Seven proposals to amend the *San Juan Chapter F* were published in *IMA Fungus,* the journal of the International Mycological Association (IMA), on 14 August 2024 (May & Hawksworth in IMA Fungus 15(25). 2024). A "Synopsis of proposals", with comments by the Secretary and Deputy Secretary of the Fungal Nomenclature Session, was published on 16 August 2024 (May & Bensch in IMA Fungus 15(26). 2024). Drafts of the proposals and Synopsis, made available via the IMA website in July 2024, served as the basis for the mail vote and, together with a draft of the results of the mail vote, were made available at the Nomenclature Session.

The Nomenclature Session met on Wednesday, 15 August 2024 in Auditorium 1 of the MECC Maastricht conference centre, Maastricht, The Netherlands as part of IMC12, which took place on 11–15 August 2024. The maximum number of persons in attendance was 322, each registered for at least that day of the IMC and carrying one personal vote (there are no institutional votes at an IMC). The Session officers were Amy Rossman (Chair), Tom May (Secretary), Konstanze Bensch (Deputy Secretary), and Ewald Groenewald and Jos Houbraken

(Recorders). The Deputy Chairs were Catherine Aime, Lei Cai, Pedro Crous, Irina Druzhinina, and David Hawksworth. Nicholas Turland, as Rapporteur-général elected for the XXI IBC, attended as a non-voting advisor to the Session. The discussions of the Session were conducted in English.

Of the seven published proposals to amend the *San Juan Chapter F*, three were accepted; an additional three were accepted from among eight new proposals made from the floor of the Session.

The rules of the *Maastricht Chapter F* became effective immediately upon acceptance of the resolution, moved on behalf of the Session at the closing plenary session of IMC12 on 15 August 2024, that the decisions and appointments of the Nomenclature Session be approved. The "Report of Congress action on nomenclature proposals relating to fungi", detailing the results of the seven published proposals and eight floor proposals, the membership of the Nomenclature Committee for Fungi and Editorial Committee for Fungi, the establishment of a Special-purpose Committee, and the election of the Secretary for the Nomenclature Session of IMC13 (Tom May), was published on 21 November 2024 (May & al. in IMA Fungus 15(36). 2024). The full proceedings of the Session will form a separate publication.

The Editorial Committee for Fungi, created at the XX IBC in Madrid as a new Permanent Nomenclature Committee, consists of the following members elected at IMC12: Tom May (Chair), Konstanze Bensch (Secretary), David Hawksworth, Juan Carlos Zamora Señoret, Nicholas Turland (ex officio, Chair of Editorial Committee), and John Wiersema (ex officio, Secretary of Editorial Committee).

The Editorial Committee for Fungi prepared a first draft of the *Maastricht Chapter F* by incorporating the changes decided by the Session. It was sent to the Rapporteur-général, who added it to the first draft of the *Madrid Code* in preparation for the Editorial Committee meeting (see above). The Editorial Committee for Fungi continued to work with the Editorial Committee of the *Madrid Code* (the two Committees have five members in common) for the remainder of 2024.

AMENDMENTS TO THE *CODE* RESULTING FROM PROPOSALS ACCEPTED IN MADRID AND MAASTRICHT

Major changes

Several changes have been made to the definitions of "specimen" and "gathering" with respect to fossil-taxa. A new **Art. 8.6** rules that for fossils a specimen is an individual of a fossil-species or infraspecific fossil-taxon selected from a sample of sediment or rock, which may contain multiple individuals (i.e. specimens) of the same and other fossil-taxa. Each individual (specimen) is treated as a separate gathering. Accordingly, **Art. 8.2 footnote** adds to the definition of "gathering" that, for most fossils, a sample is not presumed to be of a single fossil-taxon but is a set of gatherings, and **Art. 8.3 footnote** adds to the definition of "duplicate" that, for most fossils, there are no duplicates. A new **Art. 8 Note 2** clarifies that, for the purpose of typification, both the "part" and "counterpart", where a rock is split to reveal a fossil organ on both parts, comprise the same specimen, not separate specimens.

Finding type specimens again can be a problem with fossil-taxa, and this is addressed by the new **Art. 40.8**, which requires the protologue to clearly indicate the position of the holotype specimen within the rock, sediment, or preparation (for names published on or after 1 January 2026).

New **Rec. 8A.5** recommends that, if a type specimen is prepared on a microscope slide, the position of that specimen on the slide should be indicated by an England Finder reference (or an equivalent) to facilitate finding it again. **Rec. 8A.6** recommends that, for palaeopalynological samples, a part of the sample from which the type was selected be deposited along with the type.

Art. 9.4 on original material has been rewritten, and one important change is that illustrations are no longer original material for names of fossil-taxa. Hence if the nomenclatural type is lost or destroyed, and if there are no specimens among the remaining original material, then there is no original material and a neotype may be designated. Under the *Shenzhen Code,* an illustration that was original material could not be designated as a lectotype because it would be contrary to Art. 8.5, and the existence of original material precluded designation of a neotype.

Concerning priority of names of fossil-taxa (diatoms excepted) in relation to names of non-fossil taxa, two provisions in **Art. 11** have been rewritten. The previous Art. 11.8 has become **Art. 11.7** ruling that when the names of a non-fossil taxon and a fossil-taxon are treated as synonyms, the correct name of the non-fossil taxon must be accepted, even if it is later. The previous Art. 11.7 has been reformulated as **Art. 11.8** concerning "dual nomenclature", which accommodates "taxonomic equivalence" between a fossil-taxon and a non-fossil taxon when the names of the two taxa are not considered to be synonyms. The previous Art. 11 Note 5 on later homonyms being illegitimate whether the type is fossil or non-fossil has been moved to **Art. 53 Note 3**.

Art. 20.2 has been amended to end the problem of identifying generic names that coincide with Latin technical terms in use in morphology at the time of publication. Previous cases where this provision was applied are to be resolved through binding decisions, as the new **Art. 20 Note 1** explains; the related **Art. 20.4(b)** and **Rec. 20A.1(j)** are also new.

Art. 40 has been thoroughly restructured, **Art. 40.4** now explicitly permits the incorrect use of the term "lectotypus" or "neotypus" (or equivalents) in the protologue of the name of a new taxon to be corrected under Art. 9.10 to "holotype" to resolve the uncertain status of some names that had been treated as not validly published because the correct term "typus" or "holotypus" (or equivalents) had not been used. **Art. 9 Note 9** has also been amended to accord with the new Art. 40.4. **Art. 40.8** is also new and is discussed above, under names of fossil-taxa.

Art. 42 has been augmented with several new provisions (**Art. 42.3, 42.5, 42.6, 42.7, Notes 1, 3**, and **4, Rec. 42A.1** and **42A.2**) to establish a mechanism to enable the future voluntary registration of nomenclatural novelties and type designations for algae and plants.

New **Art. 51.2**, together with an amendment to **Art. 56.1**, permits a legitimate name of a new taxon or a replacement name published on or after 1 January 2026 to be rejected if it, or its epithet, is derogatory to a group of people. In addition, the new **Rec. 51A.1** advises authors to avoid publishing names of new taxa or replacement names that could be viewed as inappropriate, disagreeable, offensive, or unacceptable.

New **Art. 61.6** rules that epithets with the root *caf[f][e]r-*, considered as highly offensive especially in Africa, are not permitted under this *Code* and are to be treated as orthographical variants to be replaced by epithets with the root *af[e]r-*. For example, the epithet *'caffra'* is replaced by *afra*.

Div. III Prov. 3, concerning the allocation of institutional votes for a Nomenclature Section, has been amended so that each listed institution now has one vote, instead of one to seven votes as previously, in an effort to reduce geographical imbalance in the exercising of institutional votes. An institution applying for a vote for the first time (i.e. not on the list of institutions entitled to vote) should show that it is registered in an online, open-access international or regional index of herbaria, collections, or institutions.

Div. III Prov. 5.1 and **5.2** have been amended to put an end to almost 20 years of confusion at the Nomenclature Section when the recommendations of the General Committee are subject to a decision. A simple majority (more than 50%) of votes cast is now required to accept these recommendations. Previously, a qualified majority (at least 60%) of votes cast was required to reject them.

Div. III Prov. 7.1(c), **7.5**, and **7.13** add a new Permanent Nomenclature Committee, the Editorial Committee for Fungi, responsible for preparing Chapter F of the *Code*, which contains the provisions solely related to names of organisms treated as fungi.

Div. III Prov. 7.10(g)(2) and **8.13(e)(2)** permit Special-purpose Committees, with a specific mandate, to be appointed by the General Committee between IBCs or by the Nomenclature Committee for Fungi between IMCs for reporting back to the next IBC or IMC.

Other changes

Numerous other, smaller changes to the *Code* were made in Madrid. The following list is not intended to cover every change, but it discusses the more important items.

Art. 6.1 footnote, in defining "illustration", establishes that photographs of habitat are not illustrations for the purpose of typification. **Art. 9 Note 3** explains that, because such photographs are not illustrations, they cannot be original material or types.

New **Art. 7 Note 4** clarifies that the effective typification of a name automatically establishes the same typification for all names sharing the same basionym or replaced synonym and for that basionym or replaced synonym. The same holds for the typification of an autonym and the name from which it is derived.

Rec. 7A.2 is new and recommends that duplicates of type material be conserved in different herbaria, collections, or institutions, preferably in different areas of the world. New **Rec. 7A.3** encourages deposition of type material in herbaria, collections, or institutions in the country or countries of origin of the newly described taxon.

Art. 8.1 has been amended to clarify that the nomenclatural type of a name of a species or infraspecific taxon, which determines the application of the name (Art. 7.1 and 7.2), is a holotype, lectotype, neotype, or conserved type.

Art. 9.1 has been amended and **Art. 9.2** and **Note 1** are new. They provide a more precise definition of "holotype" and explain the circumstances under which a holotype can exist.

Art. 9.4 on original material now reflects a more hierarchical arrangement of elements in parallel with the order in Art. 9.1–9.7 and Art. 9.12. Clause (e) of Art. 9.4 establishes that specimens and published or unpublished illustrations (excluding illustrations of fossil-taxa, as discussed above) are original material when they were associated with the taxon by, and were available to, either the publishing author(s) or the author(s) of the validating descriptive matter.

Art. 9.24 (previously Art. 9.2) has been expanded to now allow corrections of errors made in the designation of a lectotype, neotype, or epitype. It was moved to Art. 9.24 so as to avoid renumbering the whole of Art. 9 pending a thorough review by the Special-purpose Committee on Types and Typification established in Madrid.

New **Art. 10.9** provides for the automatic typification of a generic name published without species names by the type of the first name of a species assigned to that genus and validly published solely by reference to its description or diagnosis. Other cases of automatic typification are newly noted in Art. 7.3, 7.4, 7 Note 4, and 10.10.

Art. 14.14 no longer permits the proposing of changes to the places of publication of family names in App. IIB, thereby eliminating potential disruption to nomenclature and pointless editorial and committee work.

A new added clause (3) to **Art. 14 Note 4(c)** establishes the date of conservation of some names of fossil-taxa.

Previous guidance on the formation of generic names to avoid names not readily adaptable to the Latin language and that are difficult to pronounce in Latin has been deleted from **Rec. 20A.1**. A similar clause concerning specific epithets has also been deleted from **Rec. 23A.3**. These amendments were made with the aim of expanding the diversity of new generic names and epithets.

Art. 23.2 now limits, for names published on or after 1 January 2026, the number of characters allowed in a specific epithet, from at least two to not more than 30.

Art. 23.5 now accounts for the independent relationship between the gender of the generic name and two groups of epithets: adverbs, and nouns and their accompanying adjectives in the genitive case.

Art. 23.6 introduces a new nomenclatural act (requiring effective publication) that provides for a choice between two different grammatical categories that would be possible for certain epithets.

Art. 23.7(a) now prohibits as species names designations consisting of a generic name and an epithet in the form of a phrase in the ablative case (e.g. *Solanum "fructu-tecto"*).

Art. 30.4 clarifies that electronic material that has been effectively published remains as it was when effectively published even if later altered or retracted.

Adjustments to **Art. 33.1** and **Art. 38.14** eliminate a conflict between these two Articles for names published between 1953 and 1972, inclusive, where the requirement for a description or diagnosis had been previously fulfilled.

Art. 34 Note 1 defines works considered for suppression as being separately published books or numbered parts or supplements of a journal while **Rec. 34A.2** cautions against proposing individual journal papers or series of papers for suppression, thereby discouraging proposals seeking to suppress minor works.

Art. 36.3 has been amended and redundant wording deleted so that alternative names are now defined as "two or more different names based on the same type accepted simultaneously for the same taxon by at least one author in common in the same publication". This rule does not apply to the same combination simultaneously used at different ranks (either for infraspecific taxa or for subdivisions of a genus), and this exclusion has been extended to suprageneric names formed from the same generic name that are simultaneously used at different ranks.

New **Art. 37.5** rules that statements on ranks associated with individual infraspecific names can be used to assign rank throughout a whole publication provided that they do not result in misplaced terms contrary to Art. 5 and when no general statement on the different infraspecific ranks used in that publication is made.

New **Art. 38.4** provides the definition of "description" that the *Code* previously lacked. Because it also rules that a description need not be diagnostic, Art. 38 Note 2 in the *Shenzhen Code* became redundant and has been deleted.

Art. 41 Note 2 provides guidance on how to cite page numbers from publications that lack them. This new Note suggests four options while pointing out that a DOI or URL is not by itself sufficient to indicate a page. **Rec. 41A.2** has been amended to recommend enclosing page numbers (or an indication of their absence, e.g. "without page number") in square brackets when they are cited according to Art. 41 Note 2.

A new opening sentence in **Art. 48.1** contrasts a misapplication of an existing name, when the type has not been excluded, with creation of a later homonym, when the type has been excluded. Further clarification on what constitutes exclusion or inclusion of the type of a name appears in **Art. 48.2**, with exclusion of the name itself now also to be considered. New **Art. 48 Note 3** indicates that inclusion of an apparent basionym with an expression of doubt, or only in part, does not by itself constitute exclusion of its type.

Two new Notes address the implications of nothogeneric names not having types (see Art. H.9 Note 1). **Art. 52 Note 5** explains that nothogeneric names, because they have no types, do not cause nomenclatural superfluity and **Art. 53 Note 2** points out that nothogeneric names, even though they have no types, can be homonyms (of each other or of non-hybrid generic names).

Art. 38.5 and **Art. 53.4** now rule that binding decisions on valid publication and homonymy, respectively, take retroactive effect upon publication of the General Committee's report containing the decisions, but this is subject to ratification by a later IBC. This now permits binding decisions to be added to App. VI and VII between IBCs, thereby streamlining the decision-making process.

An addition to the final clause of **Art. 60.8** precludes changes to terminations that conform to classical Latin adjectival usage other than that dealt with by Rec. 60C.1, e.g. *Cephalotaxus harringtonia* (Knight ex J. Forbes) K. Koch is not to be changed to *C. 'harringtonii'*.

New **Art. 60 Note 3** clarifies that epithets derived from personal names with a well-established latinized form may be formed according to either Rec 60C.1 or Art. 60.8, hence *martini* and *martinii* both commemorating Martin, which has the well-established latinized form Martinus, are both correct and are not to be changed.

New **Art. 60.9** rules that an epithet (or final portion thereof) formed from abbreviation of one or more personal names is considered to have been composed arbitrarily (which is permitted by Art. 23.2) and is not to be changed, e.g. under Art. 60.8.

Art. 60.12 has been amended to hopefully provide more clarity on the use or non-use of hyphens in epithets, i.e. when must a hyphen (present at valid publication) be deleted and when may a hyphen be added after valid publication. A set of four conditions, (a) (1 and 2) and (b) (1 and 2), is provided, and a hyphen is permitted only when at least one condition of (a) and at least one condition of (b) are met. However, this still does not result in a single outcome in every case. When, for example, a name is published with an epithet consisting of two words separated by a space, Art. 60.12 permits a hyphen to be inserted (replacing the space), while Art. 23.1 requires either uniting the two words (by deleting the space) or inserting a hyphen (replacing the space). On the other hand, when a name is published with an epithet containing neither a hyphen nor a space, a hyphen may not be inserted.

Art. F.2.1 and **F.7.1** have been revised to more clearly set out the steps involved in the procedures for lists of protected and/or rejected names of organisms treated as fungi. **Art. F.2 Note 1** makes it clear that, when preparing lists of names for protection, included names may be proposed with or without the listing of synonyms, because Art. F.2.1 rules that protected names are treated as conserved against any competing synonyms.

New **Art. F.3.5** rules that an earlier homonym of a sanctioned name remains unavailable if the sanctioned name is rejected under Art. 56 or F.7.

An addition was made to **Art. F.5 Note 3** to clarify that an identifier is not required when proposing a conserved type of a name of an organism treated as a fungus.

Two new Recommendations concerning types that are living cultures of organisms treated as fungi were added at the end of Chapter F. **Rec. F.11A.1** augments Rec. 8B.1 by recommending that when ex-type cultures are deposited in institutional culture or genetic resource collections, those collections should be public. **Rec. F.11A.2** recommends utilizing the oldest progeny of an ex-type culture when it is permitted to designate a neotype to replace a nomenclatural type that has been lost or destroyed.

Art. H.5.2 has been amended with respect to naming a nothotaxon with parent taxa at unequal ranks (species and infraspecific taxa). The appropriate rank of the nothotaxon is the lowest of these ranks, unless the nothotaxon is the only one known for hybrids between the species to which the parental taxa of the nothotaxon belong.

A new **Div. III Prov. 4 Rec. 2** recommends that individuals or groups should be able to observe the Nomenclature Section of an IBC online and that the Organizing Committee of the IBC in consultation with the Bureau of Nomenclature should be responsible for implementing this. Exactly this happened at the XX IBC in Madrid in July 2024.

EDITORIAL AMENDMENTS TO THE *CODE*

Several improvements have been made to the organization and presentation of the *Code's* provisions. In addition to the usual titles of Divisions, Chapters, and Sections, the *Madrid Code* now includes titles of individual Articles. These titles were inspired by those used in the online version of the *Shenzhen Code,* with adjustments to make them more self-contained, and were the joint effort of Julia Gravendyck and Werner Greuter. They are now also mirrored in an expanded Contents, allowing users to more easily find what they want.

For Articles that group multiple items together, often as conditions with any one or all required, the items have been organized into lettered and sometimes numbered clauses, as was already done for some provisions in the *Shenzhen Code*. This format makes the Articles more digestible and facilitates citation of individual clauses. When an Example is relevant to a particular lettered clause, the respective letter is cited in parentheses, usually at the beginning of the Example, e.g. "(c)" in Art. 9 Ex. 19. Numbered clauses are sometimes also cited, and a condensed format, e.g. "(a1, b2)", is used to avoid repetitive parentheses in Art. 60 Ex. 48–51.

Translations of words, phrases, and quotations in languages other than English (e.g. Chinese, Dutch, French, German, Latin, Russian, or Spanish) have now been provided wherever these appear, thanks to the efforts of Tom May with Latin advice from Werner Greuter.

In Division III, the Editorial Committee noticed that Prov. 7.10, which claimed to list the functions of the General Committee, did not in fact provide a comprehensive list, omitting several functions mentioned in Art. 42 and elsewhere in Div. III. Therefore, Prov. 7.10 has been expanded to enumerate all the functions of the General Committee compiled editorially from content already in the *Code*. Mirroring Prov. 7.10, a new Prov. 8.13 has been added to enumerate all the functions of the Nomenclature Committee for Fungi compiled editorially from content already in Chapter F and Div. III. A small, editorial addition to Prov. 7.15 comes from Art. 42.2 specifying that the Registration Committee considers applications for recognition as nomenclatural repositories for organisms other than those treated as fungi.

The Editorial Committee of the *Shenzhen Code* added bibliographic references to the Examples, as explained in the Preface of the *Shenzhen Code* (p. xxii). Over the last seven years, a number of errors have been discovered, in particular by Paul van Rijckevorsel, and these have been corrected. Where the same reference (same author(s) and publication) appears more than once within an Example, the abbreviation "l.c." (loco citato, in the place cited) is used to avoid undue repetition.

Throughout the *Madrid Code*, the abbreviation "nom. sanct." (nomen sanctionatum) has been added wherever sanctioned names of organisms treated as fungi have been cited (replacing the previous convention of citing author : sanctioning author).

A number of linguistic improvements have been implemented as well, such as using more modern and plain English where the meaning is not changed and avoiding ambiguous words, e.g. altering: "bear" to "have"; "collection" to "specimen" or "gathering" or "material collected" (unless a herbarium, collection, or institution is meant); "common" to "frequent" (unless used in the sense of shared, e.g. "in common"); "commonly" to "often"; "despite the fact that" to "even though"; "effected" to "achieved" (where appropriate, as requested by the Section in Madrid); "failed to" to "did not"; "in order to" to "to" (where appropriate); "ineffective" to "not effective"; "irrespective" to "regardless"; "location" to "locality" (where geographical) or "position" (e.g. on a microscope slide); "male" and "female" to "staminate" and "pistillate" (where describing angiosperms, not people or hybrid parents); "name of a person" to "personal name"; "notwithstanding" to "despite"; "ought to" to "should" (where appropriate); "parent" to "parental" (where used as an adjective); "prior" to "earlier" (where used as an adjective); and "prior to" to "before" (where appropriate). The phrases "for the purpose of" and "for the purposes of" were reviewed for the appropriate usage of singular or plural. A few words or phrases have been deleted, such as "pleonasm" (Rec. 23A.3(e)) and "bona fide" (Rec. 7A.1), although "infringements" (Art. H.10.1) remain for someone to propose a change at a future IBC.

LATIN TERMS AND THEIR ABBREVIATIONS
USED IN THE *CODE*

A number of Latin terms and/or their abbreviations appear throughout the *Code* that may not be readily understood, especially by first-time users, so these have been compiled, in particular by Tom May and Werner Greuter, into a separate listing as part of the front matter. Terms are listed alphabetically together with the abbreviations, if they exist, and their meaning is provided in English.

THE GLOSSARY

The Glossary has retained its basic structure but has been revised and updated, as mentioned above, by Jefferson Prado and Michelle Price. New entries in the Glossary are as follows: "adopted name", "circumscription", "conserved type", "dual nomenclature", "nomenclatural repository", "original author(s)", "original type", "personal name", "phrase name", "registration", "taxonomic equivalence", "type citation", "type indication", and "typify". Some existing entries have been substantially revised, e.g. "automatic typification", "description", "epitype", "gathering", "identifier", "isosyntype", "isotype", and "nomenclatural type". The entry for "binary combination" has been changed to "binomial" to accord with the preferred term in the *Code*. The entry for "unispecific" has been deleted because it was used only in Art. 36 Ex. 1, where "unispecific genus" was rephrased as "genus of one species". This reflects the fundamental role of the Glossary, which is strictly to explain terms used in the *Code,* and where possible to do so using the precise wording associated with these terms in the *Code*. The Glossary does not seek to cover all terms useful in the nomenclature of algae, fungi, and plants; for that, users can refer to a work such as Hawksworth, *Terms used in Bionomenclature* (2010; online at https://www.gbif.org/document/80577).

THE APPENDICES

The Appendices of the *Code* have been maintained over the last several years by John Wiersema as an online database currently hosted by the Department of Botany at the Smithsonian National Museum of Natural History in Washington, DC (https://naturalhistory.si.edu/research/botany/codes-proposals). The Appendices will continue to be available online in this form, while the possibility of publication as printed matter or in Portable Document Format (PDF) is not precluded. The online database not only generates any or all of the Appendices of the *Code*, but it also provides an accounting of all proposals to conserve, protect, and reject names, to suppress works, and all requests for binding decisions since the first proposals appeared in 1892. The database can be searched using various criteria, and the matching content is reported in two different ways. A **"Code Appendices"** report reproduces all entries matching any set of search criteria within *Madrid Code* Appendices I–VII. A **"Proposals/**

Requests" report details the history of selected proposals or requests, providing citations of all Nomenclature Committee reports regarding their status, and indicating the final determination, the presence, if any, in earlier *Code* Appendices, and any resulting entries in the *Madrid Code* Appendices.

In support of Art. 14.15, 34.2, 38.5, 53.4, 56.3, F.2.1, and F.7.1, which rule that once proposals or requests have been approved by the General Committee their effects become authorized, any names or works involved in such proposals or requests can be promptly added to the online Appendices, eliminating unnecessary publication delays. Nearly 800 new Appendix entries have thus been generated since the 2018 *Shenzhen Code*, initially preceded by an asterisk (*), but with ratification by the Madrid IBC the asterisks have been deleted.

Lists of protected names of fungi (Art. F.2.1) approved for both the *Shenzhen* and *Madrid Codes,* and incorporated according to their rank in App. IIA, III, and IV, were initially protected only against listed synonyms. Because they are now protected against unlisted synonyms as well, there is no longer a need to list synonyms for protected names, so the Editorial Committee has decided to remove them. No lists of rejected names of fungi (Art. F.7.1) have yet been approved.

FORMATTING AND STANDARDS USED IN THE *CODE*

Recent editions of the *Code* have used three different sizes of type, with the Recommendations and Notes set in smaller type than the Articles, and the Examples and footnotes in smaller type than the Recommendations and Notes. These type sizes, which have been maintained in this edition, reflect the distinctions between mandatory rules (Articles), complementary information (Notes) or non-mandatory advice (Recommendations), and explanatory material (Examples). Notes, unlike Articles, do not introduce any new provision or concept; rather they explain something that may not at first be readily apparent but is covered explicitly or implicitly elsewhere in the *Code*. Consequently, if an apparent conflict is perceived between a Note and an Article, the Article should be followed. Notes are appropriately identified (at least in the print edition of the *Code*) with an "i" for "information", highlighted in the same way as the Article numbers. Examples are distinguished, in addition to the smaller font size, by being indented. An Example prefixed by an asterisk (*) is a voted Example, which is comparable to a rule (see Art. 7 *Ex. 17 footnote). Footnotes provide supplementary material to the Preamble, Articles and Examples in Div. II, and Provisions in Div. III.

As in all recent editions, scientific names under the jurisdiction of the *Code,* irrespective of rank, are consistently given in *italic type.* The *Code* sets no binding standard in this respect, because typography is a matter of editorial style and tradition, not of nomenclature. Nevertheless, in the interest of international uniformity, editors and authors could consider following the practice exemplified by the *Code,* which has been well received in general, is followed in a number of botanical and mycological journals, and was emphatically recommended by Thines & al. (in IMA Fungus 11(25). 2020). To set off scientific names even better, italics are not used for technical terms and other words in Latin, although they are still used for word elements that are part of a scientific name and for non-abbreviated titles of books or journals.

Double quotation marks are used to denote designations, to readily distinguish them from validly published names, e.g. *"Echinocereus sanpedroensis".* Single quotation marks are used to denote incorrect spellings of validly published names, e.g. *Gluta 'benghas'* L. Double quotation marks are also used to indicate text that is quoted verbatim, with single quotation marks enclosing quotes within quotes.

The Editorial Committee has tried hard to achieve uniformity in bibliographic style and formal presentation. Author citations of scientific names appearing in the *Code* are standardized in conformity with Brummitt & Powell, *Authors of plant names* (1992), updated as necessary from the International Plant Names Index (IPNI; https://www.ipni.org/), albeit with additional spacing, as mentioned in Rec. 46A Note 1. The *Code* has placed a space after a full stop (period) in author citations since the 1906 Vienna *Rules* (Briquet, Règles Int. Nomencl. Bot. 1906), whereas *Authors of plant names* recommended against placing spaces after full stops and has been followed by IPNI. The use or non-use of spaces in standard author citations is a matter of editorial style, and neither method is incorrect. The titles of books in bibliographic citations are abbreviated in conformity with *Taxonomic literature,* ed. 2 (*TL-2;* Stafleu & Cowan in Regnum Veg. 94, 98, 105, 110, 112, 115, 116. 1976–1988; Supplements 1–6 by Stafleu &

Mennega in Regnum Veg. 125, 130, 132, 134, 135, 137. 1992–2000; Supplements 7 and 8 by Dorr & Nicolson in Regnum Veg. 149, 150. 2008, 2009; online at https://www.sil.si.edu/DigitalCollections/tl-2/index.cfm) or, when not in *TL-2,* by analogy, but always with capital initial letters. For journal titles, the abbreviations follow BPH Online (Botanico-Periodicum-Huntianum; https://huntbot.org/bph/) or, when not in BPH Online, by analogy. Standard herbarium codes follow Thiers, Index Herbariorum (continuously updated; online at https://sweetgum.nybg.org/science/ih/).

ACKNOWLEDGEMENTS

We first thank our fellow members of the Editorial Committee for their hard work, patience, helpfulness, and friendship, and acknowledge the support for their work on nomenclature by their respective institutions. We also thank our fellow members of the Editorial Committee for Fungi, especially its Secretary, Konstanze Bensch, who is the only member not in common with the Editorial Committee.

We especially wish to express our gratitude to John McNeill, who, having been unable to attend the Nomenclature Section in Madrid, is absent from the Editorial Committee after serving on this Committee for seven *Codes* since the Sydney Congress of 1981, twice as Chair, once as Vice-chair, and twice as Secretary. John has long been, and continues to be, a valuable mentor to both of us and many other Committee members, and we are all indebted to him for the generous sharing of his nomenclatural expertise.

Additionally we give special thanks to Sandra Knapp, President of the Nomenclature Section, for indefatigably leading the five days of deliberations in Madrid with clarity, efficiency, procedural exactitude, and good humour.

We are grateful to the following for their contributions at the Nomenclature Section: José María Martell, Vice-president for Scientific and Technical Research of the CSIC, for his opening speech at the Section; the Recorders, Inés Álvarez and Anna Monro; the Vice-presidents, Werner Greuter, David Mabberley, Gideon Smith, Carmen Ulloa Ulloa, and Karen Wilson; the members of the Nominating Committee, Jana Leong-Škorničková (Secretary), Martin Callmander, Rocío Deanna, Ana Paula Fortuna, Roy Gereau,

Lynn Gillespie, Leslie Landrum, Blanca León, Jinshuang Ma, Rosa Rankin Rodríguez, and Tiina Särkinen; the Tellers, Edeline Gagnon, Briggitthe Melchor-Castro, Peter Moonlight, Gustavo Shimizu, and Jacek Wajer; and of course the Organizing Committee of the IBC, especially Gonzalo Nieto Feliner, Jefferson Prado, and Inés Álvarez, and all the local staff and volunteers in Madrid who helped the Section run smoothly.

We thank Eva Kráľovičová and Matúš Kempa, at the IAPT Secretariat in Bratislava, for their help with the preliminary guiding vote and notifying institutions of their allocation of institutional votes.

We are also grateful to the following for their contributions at the Nomenclature Session at IMC12 in Maastricht: Amy Rossman (Chair), Tom May (Secretary), Konstanze Bensch (Deputy Secretary), Ewald Groenewald and Jos Houbraken (Recorders), Catherine Aime, Lei Cai, Pedro Crous, Irina Druzhinina, and David Hawksworth (Deputy Chairs), Tatiana Gibertoni, Lei Cai, David L. Hawksworth, Meike Piepenbring, and Irina Druzhinina (Nominating Committee), the Organizing Committee of IMC12, and the local staff and volunteers in Maastricht.

We thank the Council and officers of the IAPT, including its President, Lúcia Lohmann, and Secretary-General, Mauricio Bonifacino, for maintaining the IAPT's traditional commitment to nomenclature by funding the Editorial Committee meeting in Berlin as well as supporting our travel to the IBC in Madrid. Travel by the Rapporteur-général to both the IBC and IMC was also supported by the Botanischer Garten und Botanisches Museum Berlin (BGBM), Freie Universität Berlin. We also thank Sarah Eichhorn, at the IAPT US office in Washington, DC, for coordinating expenses.

We gratefully acknowledge Thomas Borsch, Director of the BGBM, for hosting the Editorial Committee meeting, providing a conference room, library facilities, and internet access. We are also grateful to Christine Schröter, in the Directorate at BGBM, for much valuable help with the logistics before and during the meeting. The staff of the BGBM Library are also thanked for providing publications that were not available online.

We thank Paul van Rijckevorsel for editorial corrections, suggestions, and especially for creating such a helpful resource in his "Overview of editions of the *Code*", hosted online by the IAPT (https://www. iapt-taxon.org/historic/index.htm). This has been invaluable in tracing provisions of the *Code* back in time to clarify meanings that were not immediately obvious.

We also gratefully acknowledge others who have provided editorial suggestions or Examples: Dirk Albach, Subir Bandyopadhyay, Avishek Bhattacharjee, Vincent Demoulin, Laurence Dorr, Paul Kirk, Gopal Krishna, Anand Kumar, David Mabberley, John McNeill, Luis Alberto Parra, Shaun Pennycook, Melanie Schori, Julian Shaw, Ian Turner, Karen Wilson, and Peter Wilson.

We thank Franz Stadler, the Production Editor of *Regnum Vegetabile*, for his excellent formatting and page layout of the *Madrid Code*. We also thank Joseph Calamia, Executive Editor, Science and Technology, The University of Chicago Press, for his friendly and helpful collaboration.

The Rapporteur-général thanks his wife, Christine Turland, for remaining so tolerant and understanding during the time he devoted to this edition of the *Code*. The Vice-rapporteur is similarly grateful to his spouse, Zari Noshad, for the considerable patience and understanding she has displayed over the course of this work.

The implementation of the *Code* between Congresses depends continuously on the effort invested by members of the Permanent Nomenclature Committees, currently over 140 individuals, 116 of whom work principally on proposals for conservation, protection, or rejection of names, suppression of works, and requests for binding decisions. There are also the members of the Special-purpose Committees established by an IBC or IMC with a mandate to investigate particular aspects of the nomenclature of algae, fungi, and plants and to provide reports, perhaps with proposals to amend the *Code*, to the next IBC or IMC. Augmenting these efforts with considerable input of time and expertise are the relevant column editors of the journals *Taxon* and *IMA Fungus*, where the proposals, requests for decisions, and committee reports are published. The nomenclature of algae, fungi, and plants is remarkable for being supported by a vast amount of meticulous and effective work undertaken voluntarily by so many taxonomists. All users of this *Code* benefit from these efforts, and we are sincerely grateful to all who participate in this work.

The online version of the main text of the *Code* and the online database of its Appendices are dependent on the continued support of the association and institution that host the websites: the IAPT for the main text, and the Department of Botany at the Smithsonian National Museum of Natural History (Washington, DC) for the Appendices.

The *International Code of Nomenclature for algae, fungi, and plants* is published under the authority of the International Botanical Congress (IBC), while its Chapter F, on names of organisms treated as fungi, is published under the authority of the International Mycological Congress (IMC). Provisions for the amendment of the *Code* are detailed in its Division III. The next IMC, IMC13, is scheduled to take place in Incheon, South Korea from 15–19 August 2027. The next IBC, the XXI IBC, is scheduled to take place in Cape Town, South Africa from 21–28 July 2029, with its Nomenclature Section in the preceding week (16–20 July). Proposals to amend the *Madrid Code* (excluding Chapter F) may be published in *Taxon* starting in 2026 and ending in 2028. In early 2026 a notice will appear in *Taxon* announcing the opening of the Proposals column and providing detailed instructions on the procedure and required format. Proposals to amend the *Maastricht Chapter F,* to be considered at IMC13, may be published in *IMA Fungus,* in which a similar notice will appear.

This *Code,* as with previous editions, is the culmination of a multi-year process of international cooperation and collaboration. Its scientific standing is dependent on the voluntary acceptance of its rules by authors, editors, publishers, and other users of the names of algae, fungi, and plants. We trust that you, as one of these users, will be happy to accept this *Madrid Code.*

Berlin and Washington, 31 December 2024

Nicholas J. Turland   John H. Wiersema

KEY TO RENUMBERING BETWEEN
THE *SHENZHEN CODE* AND THE *MADRID CODE*

This key includes all changes to the renumbering of Articles, voted Examples, Notes, Recommendations, and footnotes between the *Shenzhen Code* and the *Madrid Code*. For Chapter F, only changes since the San Juan Chapter F are listed. Items are listed in the order in which they appear in the respective edition of the *Code* (or Chapter F). Regular Examples are omitted because these can be traced by the scientific names mentioned in the Indices.

1. *SHENZHEN CODE* TO *MADRID CODE*

Art. 7 Note 1	Art. 7 Note 2
Art. 7 Note 2	Art. 7 Note 3
Art. 7 *Ex. 16	Art. 7 *Ex. 17
Art. 7 *Ex. 16 footnote	Art. 7 *Ex. 17 footnote
Rec. 8A.3	Rec. 8A.4
Rec. 8A.4	Rec. 8A.3
Art. 9 Note 1	Art. 9.2
Art. 9.2	Art. 9.24
Art. 9.4(a)	Art. 9.4(e)
Art. 9.4(b)	Art. 9.4(d)
Art. 9.4(c, d)	Art. 9.4(a–c)
Art. 9 Note 2	Art. 9 Note 5
Art. 9 Note 3	Art. 9 Note 4
Art. 9 Note 4	Art. 9 Note 6
Art. 9 Note 5	Art. 9 Note 8
Art. 9 Note 6	Art. 9 Note 9
Art. 9 Note 7	Art. 9 Note 10
Art. 9 Note 8	Art. 9 Note 11
Art. 10.2 last sentence	Art. 10.2(c)
Art. 10 Note 2	Art. 10 Note 4
Art. 10 Note 3	Art. 10 Note 5
Art. 10.9	Art. 10.10
Art. 10.10	Art. 10.11
Art. 11.7	Art. 11.8 (considerably amended)
Art. 11.8	Art. 11.7
Art. 11 Note 5	Art. 53 Note 3
Art. 16.3	Art. 16.3(a–e)
Art. 18 *Ex. 5	Art. 18 *Ex. 6
Art. 20 Note 1	Art. 20 Note 2
Art. 20.4(b)	Art. 20.4(c)
Art. 20 Note 2	Art. 20 Note 3

Rec. 20A.1(b)..deleted
Rec. 20A.1(c–j) ....Rec. 20A.1(b–i)
Art. 23.6(a–d)...Art. 23.7(b–e)
Art. 23.7...Art. 23.8
Art. 23.8...Art. 23.9
Art. 23 *Ex. 23...Art. 23 *Ex. 29
Rec. 31B.2 ....Rec. 31B.1 in part
Art. 37.5...Art. 37.6
Art. 37.6...Art. 37.7
Art. 37.7...Art. 37.8
Art. 37.8...Art. 37.9
Art. 38 Note 2..deleted
Art. 38.4 ....Art. 38.5
Art. 38.5(a–c) ....Art. 38.6(b–d)
Art. 38.5 last sentence................................Art. 38.6(a)
Art. 38.6...Art. 38.7
Art. 38.7...Art. 38.8
Art. 38.8 ....Art. 38.9
Art. 38.9...Art. 38.10
Art. 38.10 ....Art. 38.11
Art. 38.11 ....Art. 38.12
Art. 38.12 ....Art. 38.13
Art. 38.13 ....Art. 38.14
Art. 38.14 ....Art. 38.15
Art. 40.2 ....Art. 40.3 in part
Art. 40 Note 1..Art. 40 Note 3
Art. 40.3 first sentence................................Art. 40.2
Art. 40.3 second sentence.............................Art. 40.3 in part
Art. 40 Note 3..Art. 40 Note 4
Art. 40.4 ....Art. 40.6 in part
Art. 40.5 ....Art. 40.6 in part
Art. 40.6 ....Art. 40.4
Art. 40.7...Art. 40.5
Art. 40 Note 4..Art. 40 Note 5
Art. 40.8 ....Art. 40.7
Rec. 40A.1..deleted
Rec. 40A.2..Rec. 40A.1
Rec. 40A.3..Rec. 40A.2
Rec. 40A.4..Rec. 40A.3
Rec. 40A.5..Rec. 40A.4
Rec. 40A.6..Rec. 40A.5
Art. 41 Note 2..Art. 41 Note 3
Art. 41 Note 3 ....Art. 41 Note 4
Art. 42.3 ....Art. 42.4
Art. 42 Note 1..Art. 42 Note 2
Art. 48.2(a) ....Art. 48.2(a, b)
Art. 48.2(b, c) ....Art. 48.2(c, d)
Art. 52.2(a) ....Art. 52.2(a, b)

2. *MADRID CODE* TO *SHENZHEN CODE*

Renumbering

IMPORTANT DATES IN THE *CODE*

DATES UPON WHICH PARTICULAR PROVISIONS OF THE *CODE* BECOME OR CEASE TO BE EFFECTIVE

1 May 1753	Art. 7.9, 13.1(a, c, e), 13 Note 1, F.1.1
4 August 1789	Art. 13.1(a, c)
1 January 1801	Art. 13.1(b)
31 December 1820	Art. 13.1(f)
1 January 1848	Art. 13.1(e)
1 January 1886	Art. 13.1(e)
1 January 1887	Art. 37.2
1 January 1890	Art. 37.4
1 January 1892	Art. 13.1(e)
1 January 1900	Art. 13.1(e)
17 June 1905	Art. 14 Note 4(a)
1 January 1908	Art. 38.8, 38.9
18 May 1910	Art. 14 Note 4(b)
1 January 1912	Art. 20.2, 43.2
1 January 1921	Art. 10.7(c–f)
1 January 1935	Art. 10.7(a–f), 39.1
1 June 1940	Art. 14 Note 4(c)(1)
20 July 1950	Art. 14 Note 4(c)(2, 3)
1 January 1953	Art. 30.5, 30.7, 30.8, 30.9, 36.3, 37.1, 37.3, 38.14, 41.3, 41.4, 41.5, 41.6, 41.8
1 January 1954	Art. 14.15
1 January 1958	Art. 9 Note 1, 40.1, 40.3, 44.1, 44.2
1 January 1973	Art. 30.7, 33.1
1 January 1990	Art. 9 Note 1, 9.22, 40.4, 40.5
1 January 1996	Art. 43.1
1 January 2001	Art. 7.11, Art. 9 Note 9, 9.15, 9.23, 43.3
1 January 2007	Art. 40.6, 41.5
31 December 2011	Art. 39.1, 44.1
1 January 2012	Art. 29.1, 29 Note 1, 39.2
1 January 2013	Art. 43 Note 2, F.5.2, F.8.1
1 January 2019	Art. 40.7, F.5.4, F.6.1
1 January 2026	Art. 23.2, 40.8, 51.2

PROVISIONS INVOLVING DATES APPLICABLE TO PARTICULAR GROUPS

All groups   Art. 7.11, 9 Note 1, 9 Note 9, 9.22, 9.23, 10.7,
                  14.15, 14 Note 4(a, b), 20.2, 23.2, 29.1, 29 Note 1,
                  30.5, 30.7, 30.8, 30.9, 33.1, 36.3, 37.1, 37.2, 37.3,
                  37.4, 38.8, 38.9, 38.14, 39.2, 40.1, 40.3, 40.4, 40.5,
                  41.3, 41.4, 41.5, 41.6, 41.8, 51.2

Algae   Art. 7.9, 13.1(e), 13 Note 1, 40.6, 40.7, 44.1, 44.2

Bryophytes   Art. 7.9, 13.1(b, c), 13 Note 1, 39.1, 40.6

Fossils   Art. 7.9, 9.15, 13.1(f), 14 Note 4(c)(3), 40.8, 43.1,
                  43.2, 43.3, 43 Note 2

Fungi   Art. 13 Note 1, 14 Note 4(c)(2), 39.1, 40.6, 40.7, 43
                  Note 2, F.1.1, F.5.2, F.5.4, F.6.1, F.8.1

Vascular plants   Art. 13.1(a), 13 Note 1, 14 Note 4(c)(1), 39.1, 40.6

PROVISIONS DEFINING THE DATES OF CERTAIN WORKS

Art. 13.1(a–c, e, f), 13 Note 1, F.1.1

LATIN TERMS AND THEIR ABBREVIATIONS
USED IN THE *CODE*

This list includes Latin terms and their abbreviations, either or both of which appear in this *Code,* except for Latin words abbreviated as part of book or journal titles in bibliographic citations. Terms in bold have entries in the Glossary.

Term	Abbreviation	Meaning
ad interim	ad int.	for the time being
Anonymus	Anon.	Anonymous
auctorum	auct.	of the authors
clarus / clarissimus	cl.	famous / most famous
combinatio nova	comb. nov.	**new combination**
confer	cf.	compare
descriptio generico-specifica	–	description of genus and species, i.e. one description simultaneously validating the name of a genus and the name of its single species
emendavit	emend.	(he / she) has emended
epitypus	–	**epitype**
et alii	et al. / & al.	and others
et cetera	etc.	and the others (remainder, rest)
ex / e	–	from
ex holotypo	–	**ex-holotype**
ex isotypo	–	**ex-isotype**
ex typo	–	**ex-type**
exclusa specie / exclusis speciebus	excl. sp.	excluding the species (singular) / excluding the species (plural)
exclusa varietate / exclusis varietatibus	excl. var.	excluding the variety / excluding the varieties
excluso genere / exclusis generibus	excl. gen.	excluding the genus / excluding the genera

Latin terms

Term	Abbreviation	Meaning
exempli gratia	e.g.	for example
filius	f.	son (in author citations)
forma specialis / formae speciales	f. sp.	**special form** / special forms
genus novum	gen. nov.	new genus
hic designatus	–	designated here
holotypus	–	**holotype**
Hortulanorum	Hort.	of gardeners (in citations of scientific names; could also refer to hortus, garden)
icon	ic.	image, i.e. an **illustration**
id est	i.e.	that is
in litteris	in litt.	in writing
incertae sedis	–	of an uncertain seat, i.e. of uncertain taxonomic position
ineditus	ined.	not published
lectotypus	–	**lectotype**
loco citato	l.c. / loc. cit.	in the place cited
manuscriptum / manuscripta	ms. / mss.	manuscript / manuscripts
mihi	m.	for me, used after a scientific name to indicate that the author has introduced it
mutatis characteribus	mut. char.	with changed features (characters)
nec	–	and not, nor
neotypus	–	**neotype**
nobis	nob.	for us, used in the same way as mihi
nomen alternativum	nom. alt.	alternative name, for eight specified family names and one subfamily name
nomen conservandum	nom. cons.	name to be conserved, i.e. **conserved name**

Term	Abbreviation	Meaning
nomen novum	nom. nov.	new name, i.e. a **replacement name**
nomen nudum	nom. nud.	naked name, i.e. without a validating description or diagnosis or reference to one
nomen rejiciendum	nom. rej.	name to be rejected, i.e. **rejected name**
nomen sanctionatum	nom. sanct.	**sanctioned name**
nomen specificum legitimum	–	legitimate specific name, i.e. Linnaean phrase name
nomen triviale / nomina trivialia	–	trivial name, i.e. Linnaean specific epithet / trivial names
nomen utique rejiciendum	nom. utique rej.	name to be rejected outright (suppressed name)
non	–	not
opera utique oppressa	–	**suppressed works**
opere citato	op. cit.	in the work cited
orthographia conservanda	orth. cons.	orthography to be conserved (see **orthographical variants**)
pro hybrida	pro hybr.	as a hybrid
pro parte	p. p.	in part
pro specie	pro sp.	as a species
pro synonymo	pro syn.	as a synonym
sensu amplo	s. ampl.	in a large (ample) sense
sensu lato	s. l.	in a broad (wide) sense
sensu stricto	s. str.	in a narrow (strict) sense
species nova	sp. nov.	new species
status novus	stat. nov.	new status, i.e. **name at new rank**
tabula	t.	tablet, i.e. plate (illustration)
typus	–	type, i.e. **nomenclatural type**

INTERNATIONAL CODE OF NOMENCLATURE FOR ALGAE, FUNGI, AND PLANTS

PREAMBLE

1. Biology requires a precise and simple system of nomenclature that is used in all countries, dealing on the one hand with the terms that denote the ranks of taxonomic groups or units, and on the other hand with the scientific names that are applied to the individual taxonomic groups. The purpose of giving a name to a taxonomic group is not to indicate its characters or history, but to supply a means of referring to it and to indicate its taxonomic rank. This *Code* aims at the provision of a stable method of naming taxonomic groups, avoiding and rejecting the use of names that may cause error or ambiguity or throw science into confusion. Next in importance is the avoidance of the useless creation of names. Other considerations, such as absolute grammatical correctness, regularity or euphony of names, more or less prevailing custom, regard for persons, etc., despite their undeniable importance, are relatively accessory.

2. Algae, fungi, and plants are the organisms[1] covered by this *Code*.

3. The Principles form the basis of the system of nomenclature governed by this *Code*.

4. The detailed provisions are divided into rules, which are set out in the Articles (Art.) (sometimes with clarification in Notes), and Recommendations (Rec.). Examples (Ex.)[2] are added to the rules and recommendations to illustrate them. A Glossary defining terms used in this *Code* is included.

5. The object of the rules is to put the nomenclature of the past into order and to provide for that of the future; names contrary to a rule cannot be maintained.

---

1   In this *Code*, unless otherwise indicated, the word "organism" applies only to the organisms covered by this *Code*, i.e. those traditionally studied by botanists, mycologists, and phycologists (see Pre. 8).
2   See also Art. 7 *Ex. 17 footnote.

6. The Recommendations deal with subsidiary points; their object is to achieve greater uniformity and clarity, especially in future nomenclature; names contrary to a Recommendation cannot, on that account, be rejected, but they are not examples to be followed.

7. The Provisions regulating the governance of this *Code* form its last Division (Div. III). These Provisions are not retroactive.

8. The provisions of this *Code* apply to all organisms traditionally treated as algae, fungi, or plants, whether fossil or non-fossil, including blue-green algae *(Cyanobacteria)*[1], chytrids, oomycetes, slime moulds, and photosynthetic protists with their taxonomically related non-photosynthetic groups (but excluding *Microsporidia*). Provisions for the names of hybrids appear in Chapter H.

9. Names that have been conserved, protected, or rejected, suppressed works, and binding decisions are given in Appendices I–VII.

10. The Appendices form an integral part of this *Code,* whether published together with, or separately from, the main text.

11. The *International Code of Nomenclature for Cultivated Plants* is prepared under the authority of the International Commission for the Nomenclature of Cultivated Plants and deals with the use and formation of names applied to special categories of organisms in agriculture, forestry, and horticulture.

12. The only proper reasons for changing a name are either a more profound knowledge of the facts resulting from adequate taxonomic study or the necessity of giving up a nomenclature that is contrary to the rules.

13. In the absence of a relevant rule or where the consequences of rules are doubtful, established custom is followed.

14. This edition of the *Code* supersedes all previous editions.

1　For the nomenclature of other prokaryotic groups, see the *International Code of Nomenclature of Prokaryotes. Prokaryotic Code (2008 Revision);* https://doi.org/10.1099/ijsem.0.000778; formerly the *International Code of Nomenclature of Bacteria (Bacteriological Code).*

DIVISION I
PRINCIPLES

PRINCIPLE I
INDEPENDENCE

The nomenclature of algae, fungi, and plants is independent of zoological and prokaryotic nomenclature. This *Code* applies equally to names of taxonomic groups treated as algae, fungi, or plants, whether or not these groups were originally treated as such (see Pre. 8).

PRINCIPLE II
TYPES

The application of names of taxonomic groups is determined by means of nomenclatural types.

PRINCIPLE III
PRIORITY

The nomenclature of a taxonomic group is based upon priority of publication.

PRINCIPLE IV
SINGLE CORRECT NAME

Each taxonomic group with a particular circumscription, position, and rank can have only one correct name, the earliest that is in accordance with the rules, except in specified cases.

PRINCIPLE V
LATIN NAMES

Scientific names of taxonomic groups are treated as Latin regardless of their derivation.

PRINCIPLE VI
RETROACTIVITY

The rules of nomenclature are retroactive unless expressly limited.

DIVISION II
RULES AND RECOMMENDATIONS

CHAPTER I
TAXA AND THEIR RANKS

ARTICLE 1
TAXA

1.1. Taxonomic groups at any rank will, in this *Code,* be referred to as taxa (singular: taxon).

1.2. A taxon (diatom taxa excepted) the name of which is based on a fossil type is a fossil-taxon. A fossil-taxon comprises the remains of one or more parts of the parent organism, or one or more of their life-history stages, in one or more preservational states, as indicated in the original or any subsequent description or diagnosis of the taxon (see also Art. 11.1 and 13.3).

> *Ex. 1.* *Alcicornopteris hallei* J. Walton (in Ann. Bot. (Oxford), ser. 2, 13: 450. 1949) is a fossil-species for which the original description included rachides, sporangia, and spores of a pteridosperm, preserved in part as compressions and in part as petrifactions.

> *Ex. 2.* *Protofagacea allonensis* Herend. & al. (in Int. J. Pl. Sci. 156: 94. 1995) is a fossil-species for which the original description included dichasia of staminate flowers (with anthers containing pollen grains), fruits, and cupules. The fossil-species therefore comprises more than one part and more than one life-history stage.

> *Ex. 3.* *Stamnostoma* A. G. Long (in Trans. Roy. Soc. Edinburgh 64: 212. 1960) is a fossil-genus that was originally described with a single species, *S. huttonense* A. G. Long, comprising anatomically preserved ovules with completely fused integuments forming an open collar around the lagenostome. Rothwell & Scott (in Rev. Palaeobot. Palynol. 72: 281. 1992) subsequently modified the description of the fossil-genus, expanding its circumscription to include also the cupules in which the ovules were borne. The name *Stamnostoma* may be applied to a fossil-genus with either circumscription or to any other fossil-genus that involves other parts, life-history stages, or preservational states, as long as it includes *S. huttonense* but not the type of any earlier legitimate generic name.

ARTICLE 2
RANKS

2.1. Every individual organism is treated as belonging to an indefinite number of taxa at consecutively subordinate ranks, among which the rank of species is basic.

ARTICLE 3
PRINCIPAL RANKS

3.1. The principal ranks of taxa in descending sequence are: kingdom (regnum), division or phylum (divisio or phylum), class (classis), order (ordo), family (familia), genus (genus), and species (species). Thus, each species is assignable to a genus, each genus to a family, etc.

ⓘ **Note 1.** Species and subdivisions of genera must be assigned to genera, and infraspecific taxa must be assigned to species, because their names are combinations (Art. 21.1, 23.1, and 24.1), but this provision does not preclude the placement of taxa as incertae sedis with regard to ranks higher than genus.

> **Ex. 1.** The genus *Haptanthus* Goldberg & C. Nelson (in Syst. Bot. 14: 16. 1989) was originally described without being assigned to a family.

> **Ex. 2.** The fossil-genus *Paradinandra* Schönenberger & E. M. Friis (in Amer. J. Bot. 88: 478. 2001) was assigned to "*Ericales* s.l." but its family placement was given as "incertae sedis".

3.2. The principal ranks of hybrid taxa (nothotaxa) are nothogenus and nothospecies. These ranks are the same as genus and species. The prefix "notho-" indicates the hybrid character (see Art. H.1.1).

ARTICLE 4
SECONDARY AND FURTHER RANKS

4.1. The secondary ranks of taxa in descending sequence are tribe (tribus) between family and genus, section (sectio) and series (series) between genus and species, and variety (varietas) and form (forma) below species.

4.2. If a greater number of ranks of taxa is desired, the terms for these are made by adding the prefix "sub-" to the terms denoting the principal or secondary ranks. An organism may thus be assigned to taxa at the following ranks (in descending sequence): kingdom (regnum), subkingdom (subregnum), division or phylum (divisio or phylum), subdivision or subphylum (subdivisio or subphylum), class (classis), subclass (subclassis), order (ordo), suborder (subordo), family (familia), subfamily (subfamilia), tribe (tribus), subtribe (subtribus), genus (genus), subgenus (subgenus), section (sectio), subsection (subsectio), series (series), subseries (subseries), species (species), subspecies (subspecies), variety (varietas), subvariety (subvarietas), form (forma), and subform (subforma).

ⓘ **Note 1.** Ranks formed by adding "sub-" to the principal ranks (Art. 3.1) may be formed and used whether or not any secondary ranks (Art. 4.1) are adopted.

4.3. Further ranks may also be intercalated or added, provided that confusion or error is not thereby introduced.

4.4. The subordinate ranks of nothotaxa are the same as the subordinate ranks of non-hybrid taxa, except that nothogenus is the highest rank permitted (see Chapter H).

ⓘ **Note 2.** Throughout this *Code* the phrase "subdivision of a family" refers only to taxa at a rank between and not including family and genus, and "subdivision of a genus" refers only to taxa at a rank between and not including genus and species.

ⓘ **Note 3.** For the designation of special categories of organisms used in agriculture, forestry, and horticulture, see Pre. 11 and Art. 28 Notes 2, 4, and 5.

ⓘ **Note 4.** In classifying parasites, especially fungi, authors who do not give specific, subspecific, or varietal value to taxa characterized from a physiological standpoint but scarcely or not at all from a morphological standpoint may distinguish within the species special forms (formae speciales) characterized by their adaptation to different hosts, but the nomenclature of special forms is not governed by the provisions of this *Code*.

ARTICLE 5
ORDER OF RANKS

5.1. The relative order of the ranks specified in Art. 3 and 4 must not be altered (see Art. 37.7 and F.4.1).

Recommendation 5A

5A.1. For purposes of standardization, the following abbreviations are recommended: cl. (class), ord. (order), fam. (family), tr. (tribe), gen. (genus), sect. (section), ser. (series), sp. (species), var. (variety), f. (form). The abbreviations for additional ranks created by the addition of the prefix sub-, or for nothotaxa with the prefix notho-, should be formed by adding the prefixes, e.g. subsp. (subspecies), nothosp. (nothospecies), but subg. (subgenus) not "subgen."

CHAPTER II
STATUS, TYPIFICATION, AND PRIORITY OF NAMES

SECTION 1
STATUS DEFINITIONS

ARTICLE 6

6.1. Effective publication is publication in accordance with Art. 29–31. Except in specified cases (Art. 8.1, 9.4(e), 9.22, Rec. 9A.3, and Art. 40.5), text and illustrations[1] must be effectively published to be taken into account for the purposes of this *Code*.

Ex. 1. The name *Kalanchoe arborescens* Humbert (in Bull. Mus. Natl. Hist. Nat., ser. 2, 5: 163. 1933) was published with two gatherings cited in the protologue: "Delta de la Linta (côte Sud-Ouest), sables (H. Humbert et C. F. Swingle, 5.415, 23 août 1928); Kotoala au S.-W. d'Ambovombe, anciennes dunes (Decary, 9092, 5 août 1931; hauteur 1^m,50)." Two syntype specimens with "H. Humbert et C. F. Swingle, 5.415" and two with "Decary, 9092" exist at P. The handwritten annotation "HOLOTYPE" on one of the syntypes (barcode P00438070) cannot be accepted as a holotype designation because it was not effectively published (Art. 29–31).

6.2. Valid publication of names is publication in accordance with the relevant provisions of Art. 32–45, F.4, F.5.2, F.5.3, and H.9 (see also Art. 61).

---

1   Here and elsewhere in this *Code,* the term "illustration" designates a work of art or a photograph depicting a feature or features of a species or infraspecific taxon (see also Art. 10.4 and 43.2), e.g. a drawing, a picture of a herbarium specimen, or a scanning or transmission electron micrograph, but not a photograph of habitat.

ⓘ **Note 1.** For nomenclatural purposes, valid publication creates a name, and sometimes also an autonym (Art. 22.1 and 26.1), but does not itself imply any taxonomic circumscription beyond inclusion of the type of the name (Art. 7.1).

6.3. In this *Code,* unless otherwise indicated, the word "name" means a name that has been validly published, whether it is legitimate or illegitimate (see Art. 12; but see Art. 14.9 and 14.14).

ⓘ **Note 2.** When the same name, based on the same type, has been published independently at different times, by the same or different authors, then only the earliest of these "isonyms" has nomenclatural status. The name is always to be cited from its original place of valid publication (but see Art. 14.14).

> **Ex. 2.** Baker (Summary New Ferns: 9. 1892) and Christensen (Index Filic.: 44. 1905) independently published the name *Alsophila kalbreyeri* as a replacement for *A. podophylla* Baker (in J. Bot. 19: 202. 1881) non Hook. (in Hooker's J. Bot. Kew Gard. Misc. 9: 334. 1857). As published by Christensen, *A. kalbreyeri* is a later isonym of *A. kalbreyeri* Baker without nomenclatural status (see also Art. 41 Ex. 25).

> **Ex. 3.** In publishing "*Canarium pimela* Leenh. nom. nov.", Leenhouts (in Blumea 9: 406. 1959) reused the illegitimate *C. pimela* K. D. Koenig (in Ann. Bot. (König & Sims) 1: 361. 1805), attributing it to himself and basing it on the same type. He thereby created a later isonym without nomenclatural status.

6.4. An illegitimate name is one that is designated as such in Art. 18.3, 19.6, 52–54, F.3.3, or F.6.1 (see also Art. 21 Note 1 and Art. 24 Note 2). A name that according to this *Code* was illegitimate when published can become legitimate later only if:

(a) it is conserved (Art. 14), protected (Art. F.2), or sanctioned (Art. F.3); or

(b) the name is nomenclaturally superfluous under Art. 52 and its intended basionym is conserved or protected; or

(c) Art. 18.3 or 19.6 so provide.

> **Ex. 4.** The name *Hydrodictyon* Roth (Bemerk. Crypt. Wassergew.: 48. 1797) was nomenclaturally superfluous when published and therefore illegitimate under Art. 52 because the genus included the original type of *Reticula* Adans. (Fam. Pl. 2: 3, 598. 1763). The name *Hydrodictyon* had to be conserved to become available for use (see App. III).

6.5. A legitimate name is one that is in accordance with the rules, i.e. one that is not illegitimate as defined in Art. 6.4.

6.6. At the rank of family or below, the correct name of a taxon with a particular circumscription, position, and rank is the legitimate name that must be adopted for it under the rules (see Art. 11).

> **Ex. 5.** The generic name *Vexillifera* Ducke (in Arch. Jard. Bot. Rio de Janeiro 3: 139. 1922), based on the single species *V. micranthera* Ducke, is legitimate. The same is true of the generic name *Dussia* Krug & Urb. ex Taub. (in Engler & Prantl, Nat. Pflanzenfam. 3(3): 193. 1892), based on the single species *D. martini-censis* Krug & Urb. ex Taub. Both generic names are correct when the genera are thought to be separate. Harms (in Repert. Spec. Nov. Regni Veg. 19: 291. 1924), however, united *Vexillifera* and *Dussia* in a single genus; the latter is the correct name for the genus with that particular circumscription. The legitimate name *Vexillifera* may therefore be correct or incorrect according to different taxonomic concepts.

6.7. The name of a taxon below the rank of genus, consisting of the name of a genus combined with one or two epithets, is termed a combination (see Art. 21, 23, and 24).

> **Ex. 6.** The following names are combinations: *Mouriri* subg. *Pericrene* Morley (in Univ. Calif. Publ. Bot. 26: 280. 1953); *Arytera* sect. *Mischarytera* Radlk. (in Engler, Pflanzenr. IV. 165 (Heft 98f): 1271. 1933); *Dendrophyllanthus* S. Moore sect. *Dendrophyllanthus* (established by Bouman & al. in Phytotaxa 540: 53. 2022); *Theobroma* subsect. *Subcymbicalyx* (R. E. Schult.) Colli-Silva (in Brittonia 76: 59. 2024); *Breynia androgyna* (L.) Chakrab. & N. P. Balakr. (in Bangladesh J. Pl. Taxon. 19: 120. 2012); *Gentiana lutea* L. (Sp. Pl.: 227. 1753); *Thysanothecium casuarinarum* subsp. *nipponicum* (Asahina) Asahina (in J. Jap. Bot. 32: 35. 1957); *Gentiana tenella* var. *occidentalis* J. Rousseau & Raymond (in Naturaliste Canad. 79: 77. 1952); *Equisetum palustre* var. *americanum* Vict. (in Contr. Lab. Bot. Univ. Montréal 9: 51. 1927); *Equisetum palustre* f. *fluitans* Vict. (l.c.: 60. 1927); *Dodonaea viscosa* Jacq. subf. *viscosa* (established by Radlkofer in Engler, Pflanzenr. IV. 165 (Heft 98g): 1369. 1933).

6.8. Autonyms are names that are established automatically under Art. 22.3 and 26.3, whether or not they actually appear in the publication in which they are created (see Art. 32.3, Rec. 22B.1 and 26B.1).

6.9. The name of a new taxon (e.g. genus novum, gen. nov., species nova, sp. nov.) is a name validly published in its own right, i.e. one not based on a previously validly published name; it is not a new combination, a name at new rank, or a replacement name.

> **Ex. 7.** *Cannaceae* Juss. (Gen. Pl.: 62. 1789); *Canna* L. (Sp. Pl.: 1. 1753); *Canna indica* L. (l.c. 1753); *Heterotrichum pulchellum* Fisch. (in Mém. Soc. Imp. Naturalistes Moscou 3: 71. 1812); *Poa sibirica* Roshev. (in Izv. Imp. S.-Peterburgsk. Bot. Sada 12: 121. 1912); *Solanum umtuma* Voronts. & S. Knapp (in PhytoKeys 8: 4. 2012).

6.10. A new combination (combinatio nova, comb. nov.) or name at new rank (status novus, stat. nov.) is a new name based on a legitimate, previously published name, which is the basionym of the new name. The basionym does not itself have a basionym; it provides the final epithet[1], name, or stem of the new combination or name at new rank (see also Art. 41.2).

> *Ex. 8.* The basionym of *Centaurea benedicta* (L.) L. (Sp. Pl., ed. 2: 1296. 1763) is *Cnicus benedictus* L. (Sp. Pl.: 826. 1753), the name that provides the epithet.

> *Ex. 9.* The basionym of *Crupina* (Pers.) DC. (in Ann. Mus. Hist. Nat. 16: 157. 1810) is *Centaurea* subg. *Crupina* Pers. (Syn. Pl. 2: 488. 1807), the epithet of which name provides the generic name; it is not *Centaurea crupina* L. (Sp. Pl.: 909. 1753) (see Art. 41.2(b)).

> *Ex. 10.* The basionym of *Anthemis* subg. *Ammanthus* (Boiss. & Heldr.) R. Fern. (in Bot. J. Linn. Soc. 70: 16. 1975) is *Ammanthus* Boiss. & Heldr. (in Boissier, Diagn. Pl. Orient., ser. 1, 11: 18. 1849), the name that provides the epithet.

> *Ex. 11.* The basionym of *Ricinocarpaceae* Hurus. (in J. Fac. Sci. Univ. Tokyo, Sect. 3, Bot. 6: 224. 1954) is *Ricinocarpeae* Müll. Arg. (in Bot. Zeitung (Berlin) 22: 324. 1864), but not *Ricinocarpos* Desf. (in Mém. Mus. Hist. Nat. 3: 459. 1817) (see Art. 41.2(a); see also Art. 49.2), from which the names of both family and tribe are formed.

ⓘ **Note 3.** A descriptive name (Art. 16.1(b)) used at a rank different from that at which it was first validly published is not a name at new rank because descriptive names may be used unchanged at different ranks.

ⓘ **Note 4.** The phrase "nomenclatural novelty", as used in this *Code,* refers to any or all of the following categories: name of a new taxon, new combination, name at new rank, and replacement name.

ⓘ **Note 5.** A new combination can at the same time be a name at new rank (comb. & stat. nov.); a nomenclatural novelty with a basionym need not be either of these.

> *Ex. 12.* *Aloe vera* (L.) Burm. f. (Fl. Indica: 83. 1768), based on *A. perfoliata* var. *vera* L. (Sp. Pl.: 320. 1753), is both a new combination and a name at new rank.

> *Ex. 13.* *Centaurea jacea* subsp. *weldeniana* (Rchb.) Greuter, "comb. in stat. nov." (in Willdenowia 33: 55. 2003), based on *C. weldeniana* Rchb. (Fl. Germ. Excurs.: 213. 1831), was not a new combination because *C. jacea* var. *weldeniana* (Rchb.) Briq. (Monogr. Centaurées Alpes Marit.: 69. 1902) had been published previously; nor was it a name at new rank, due to the existence of *C. amara* subsp. *weldeniana* (Rchb.) Kušan (in Prir. Istraž. Kral. Jugoslavije 20: 29. 1936); it was nevertheless a nomenclatural novelty.

---

1   Here and elsewhere in this *Code,* the phrase "final epithet" refers to the last epithet in sequence in any particular name, whether of a subdivision of a genus, a species, or an infraspecific taxon.

6.11. A replacement name (nomen novum, nom. nov.) is a new name published as an explicit substitute (avowed substitute; but see Art. 6.12 and 6.13) for a legitimate or illegitimate, previously published name, which is the replaced synonym of the new name. The replaced synonym, when legitimate, does not provide the final epithet, name, or stem of the replacement name (see also Art. 41.2 and 58.1).

> *Ex. 14.* Gussone (Fl. Sicul. Syn. 2: 468. 1844) described plants from the Eolie Islands near Sicily under the name *Helichrysum litoreum* Guss., citing in synonymy *Gnaphalium angustifolium* Lam. (Encycl. 2: 746. 1788), but without indication that the existing *H. angustifolium* (Lam.) DC. (in Candolle & Lamarck, Fl. Franç., ed. 3, 6: 467. 1815) was an illegitimate later homonym of *H. angustifolium* Pers. (in Syn. Pl. 2: 415. 1807) that needed replacement. At the end of the protologue, Gussone wrote: "nomen mutavi confusionis vitendi gratia [I changed the name to avoid confusion]". This makes explicit Gussone's intent to propose *H. litoreum* as a replacement name based on the type of *G. angustifolium* (from Posillipo near Naples), not on the material he described and cited in the protologue.

> *Ex. 15.* *Mycena coccineoides* Grgur. (in Fungal Diversity Res. Ser. 9: 287. 2003) was published as an explicit substitute ("nom. nov.") for *Omphalina coccinea* Murrill (in Britton, N. Amer. Fl. 9: 350. 1916) because *M. coccinea* (Murrill) Singer (in Sydowia 15: 65. 1962) is an illegitimate later homonym of *M. coccinea* (Sowerby) Quél. (in Bull. Soc. Amis Sci. Nat. Rouen, ser. 2, 15: 155. 1880).

> *Ex. 16.* *Centaurea chartolepis* Greuter (in Willdenowia 33: 54. 2003) was published as an explicit substitute ("nom. nov.") for the legitimate name *Chartolepis intermedia* Boiss. (Diagn. Pl. Orient., ser. 2, 3: 64. 1856) because the epithet *intermedia* was unavailable in *Centaurea* due to the previously published *Centaurea intermedia* Mutel (in Rev. Bot. Recueil Mens. 1: 400. 1846).

6.12. A name not explicitly proposed as a substitute for an earlier name is nevertheless a replacement name if it is either:

(a) validated solely by reference to that earlier name; or
(b) treated as a replacement name under the provisions of Art. 7.5.

6.13. A name not explicitly proposed as a substitute for an earlier name and not covered by Art. 6.12 may be treated either as a replacement name or as the name of a new taxon if in the protologue[1] both:

---

1 Protologue (from Greek πρῶτος, protos, first; λόγος, logos, discourse): everything associated with a name at its valid publication, e.g. description, diagnosis, illustrations (see Art. 6.1 footnote), habitat photographs, references, synonymy, geographical data, citation of specimens, discussion, comments.

(a) a potential replaced synonym is cited; and

(b) all requirements for valid publication of the name of a new taxon are independently met.

Decision on the status of such a name is to be based on predominant usage and is to be achieved by means of appropriate type designation (Art. 9 and 10).

> *Ex. 17.* When describing *Astragalus penduliflorus* Lam. (Fl. Franç. 2: 636. 1779) using material from the French Alps, Lamarck also cited in synonymy *Phaca alpina* L. (Sp. Pl.: 755. 1753, non *Astragalus alpinus* L., l.c.: 760. 1753), described from Siberia. It is questionable whether Linnaeus's and Lamarck's plants belong to the same species. Greuter (in Candollea 23: 265. 1969) designated different types for the two names, so that, in conformity with predominant usage, *A. penduliflorus* is to be treated as the name of a new, European species.

6.14. A factually incorrect statement of a name's status, as defined in Art. 6.9–6.11, does not preclude valid publication of that name with a different status; it is treated as a correctable error (see also Art. 41.4 and 41.8).

> *Ex. 18.* *Racosperma nelsonii* was published by Pedley (in Bot. J. Linn. Soc. 92: 249. 1986) as a new combination ("comb. nova") citing *Acacia nelsonii* Maslin (in J. Adelaide Bot. Gard. 2: 314. 1980) as "basionym". However, *A. nelsonii* Maslin is illegitimate under Art. 53.1 because it is a later homonym of *A. nelsonii* Saff. (in J. Wash. Acad. Sci. 4: 363. 1914). *Racosperma nelsonii* Pedley is therefore validly published as a replacement name (Art. 6.11), with *A. nelsonii* Maslin its replaced synonym, and Pedley's statement is treated as a correctable error.

SECTION 2
TYPIFICATION

ARTICLE 7
TYPIFICATION IN GENERAL

7.1. The application of names of taxa at the rank of family or below is determined by means of nomenclatural types (types of names of taxa). The application of names of taxa at the higher ranks is also determined by means of types when the names are formed from a generic name (see Art. 10.11).

7.2. A nomenclatural type (typus) is that element to which the name of a taxon is permanently attached, whether as the correct name or as a synonym. The nomenclatural type is not necessarily the most typical or representative element of a taxon.

ⓘ **Note 1.** A name of a taxon may have a type (see Art. 7.1) but has no circumscription. The taxon itself has a circumscription but no type.

7.3. A new combination or a name at new rank (Art. 6.10) is typified automatically by the type of the basionym even though it may have been applied erroneously to a taxon now considered not to include that type (but see Art. 48.1).

> **Ex. 1.** *Pinus mertensiana* Bong. (in Mém. Acad. Imp. Sci. St.-Pétersbourg, Sér. 6, Sci. Math. 2: 163. 1832) was transferred to the genus *Tsuga* by Carrière (Traité Gén. Conif., ed. 2: 250. 1867), who, as is evident from his description, erroneously applied the new combination *T. mertensiana* to another species of *Tsuga,* namely *T. heterophylla* (Raf.) Sarg. (Silva 12: 73. 1899). The combination *T. mertensiana* (Bong.) Carrière must not be applied to *T. heterophylla* but must be retained for *P. mertensiana* when that species is placed in *Tsuga;* the citation in parentheses (see Art. 49.1) of the name of the original author, Bongard, indicates the basionym, and hence the type, of the name.

> **Ex. 2.** *Delesseria gmelinii* J. V. Lamour. (in Ann. Mus. Hist. Nat. 20: 124. 1813) is a legitimate replacement name (Art. 6.11) for *Fucus palmetta* S. G. Gmel. (Hist. Fuc.: 183. 1768). The change of epithet was necessitated by the simultaneous publication of *D. palmetta* (Stackh.) J. V. Lamour. (see Art. 11 Note 2). All combinations based on *D. gmelinii* (and not excluding the type of *F. palmetta;* see Art. 48.1) have the same type as *F. palmetta* even though material thought to have been used by Lamouroux may be taxonomically assigned to a different species by subsequent authors.

> **Ex. 3.** The new combination *Cystocoleus ebeneus* (Dillwyn) Thwaites (in Ann. Mag. Nat. Hist., ser. 2, 3: 241. 1849) is typified by the type of its basionym *Conferva ebenea* Dillwyn (Brit. Conferv.: t. 101. 1809) even though the material illustrated by Thwaites was of *Racodium rupestre* Pers. (in Neues Mag. Bot. 1: 123. 1794).

7.4. A replacement name (Art. 6.11) is typified automatically by the type of the replaced synonym even though it may have been applied erroneously to a taxon now considered not to include that type (but see Art. 41 Note 4 and 48.1).

> **Ex. 4.** *Myrcia lucida* McVaugh (in Mem. New York Bot. Gard. 18(2): 100. 1969) was published as a replacement name for *M. laevis* O. Berg (in Linnaea 31: 252. 1862), an illegitimate homonym of *M. laevis* G. Don (Gen. Hist. 2: 845. 1832). The type of *M. lucida* is therefore the type of *M. laevis* O. Berg (non G. Don).

7.5. A name that is illegitimate under Art. 52 is a replacement name, typified automatically (Art. 7.4; but see Art. 7.6) by the type of the name (the replaced synonym) that itself or the epithet of which ought to have been adopted under Art. 11, unless a different type was designated or definitely indicated in the protologue, in which case it is either:

(a) a replacement name with a different replaced synonym; or

(b) treated as the name of a new taxon.

Automatic typification does not apply to names sanctioned under Art. F.3.

Ex. 5. Bauhinia semla Wunderlin (in Taxon 25: 362. 1976) is illegitimate under Art. 52 (see Art. 52 Ex. 7), but its publication as a replacement name for *B. retusa* Roxb. ex DC. (Prodr. 2: 515. 1825) non Poir. (in Lamarck, Encycl. Suppl. 1: 599. 1811) is definite indication of a different type (that of *B. retusa* Roxb. ex DC.) from that of the name (*B. roxburghiana* Voigt, Hort. Suburb. Calcutt.: 254. 1845) that ought to have been adopted.

Ex. 6. Hewittia bicolor Wight & Arn. (in Madras J. Lit. Sci. 5: 22. 1837), which provides the type of *Hewittia* Wight & Arn., is illegitimate under Art. 52 because, in addition to the illegitimate intended basionym *Convolvulus bicolor* Vahl (Symb. Bot. 3: 25. 1794) non Desr. (in Lamarck, Encycl. 3: 564. 1792), the legitimate *C. bracteatus* Vahl (Symb. Bot. 3: 25. 1794) was cited as a synonym and the epithet *bracteatus* ought to have been adopted. Wight & Arnott's adoption of the epithet *bicolor* is definite indication that the type of *H. bicolor,* and therefore the type of *Hewittia,* is the type of *C. bicolor* and not that of *C. bracteatus.*

7.6. If the type of the name causing illegitimacy (Art. 52.2) is included in a subordinate taxon that does not include the intended type of the illegitimate name, then typification is not automatic (see Art. 7.5).

Ex. 7. Mason & Grant (in Madroño 9: 212. 1948) validly published the names *Gilia splendens* and *G. splendens* subsp. *grinnellii,* the former without indicating a type (because they believed the name to be already validly published) and the latter for "a long-tubed form of the species". Under Art. 52, *G. splendens* was illegitimate because of the inclusion of the type of *G. grinnellii* Brand (in Engler, Pflanzenr. IV. 250 (Heft 27): 101. 1907), the basionym of *G. splendens* subsp. *grinnellii.* But, because *G. splendens* subsp. *grinnellii* was applied to a subordinate taxon that did not include the intended type of the illegitimate name, the type of *G. splendens* is not automatically that of *G. grinnellii.* The names *G. splendens* and *G. grinnellii* have since been conserved and rejected, respectively (see App. IV and V).

7.7. The type of an autonym (Art. 22.1 and 26.1) is the same as that of the name from which it is derived.

Ex. 8. The type of *Caulerpa racemosa* (Forssk.) J. Agardh var. *racemosa* is that of *C. racemosa;* the type of *C. racemosa* is that of its basionym, *Fucus racemosus* Forssk. (Fl. Aegypt.-Arab.: 191. 1775), i.e. Herb. Forsskål No. 845 (C).

7.8. A name of a new taxon validly published solely by reference to a previously and effectively published description or diagnosis (Art. 38.1(a)) (and not by a reproduction of such a description or diagnosis) is to be typified by an element selected from the entire context of the validating description or diagnosis, unless the validating author has definitely designated a different type, but not by an element explicitly excluded by the validating author (see also Art. 7.9).

Ex. 9. *Adenanthera bicolor* Moon (Cat. Pl. Ceylon: 34. 1824) was validly published solely by reference to the description associated with an illustration devoid of analysis, "Rumph. amb. 3: t. 112", cited by Moon. The specimen collected by Moon (in K, labelled *"Adenanthera bicolor"*) is not available as the type because it was not definitely designated by him as the type. In the absence of the material on which the validating description was based, the lectotype can only be the associated illustration (Rumphius, Herb. Amboin. 3: t. 112. 1743).

Ex. 10. *Echium lycopsis* L. (Fl. Angl.: 12. 1754) was published without a description or diagnosis but with reference to Ray (Syn. Meth. Stirp. Brit., ed. 3: 227. 1724), in which a *"Lycopsis"* species was discussed with no description or diagnosis but with citation of earlier references, including Bauhin (Pinax: 255. 1623). The accepted validating description of *E. lycopsis* is that of Bauhin, and the type must be chosen from the context of his work. Consequently, the Sherard specimen in the Morison herbarium (OXF), selected by Klotz (in Wiss. Z. Martin-Luther-Univ. Halle-Wittenberg, Math.-Naturwiss. Reihe 9: 375–376. 1960), although probably consulted by Ray, is not eligible as type. The first acceptable choice of lectotype is that of the illustration, cited by both Ray and Bauhin, of *"Echii altera species"* in Dodonaeus (Stirp. Hist. Pempt.: 620. 1583), suggested by Gibbs (in Lagascalia 1: 60–61. 1971) and formally made by Stearn (in Ray Soc. Publ. 148, Introd.: 65. 1973).

Ex. 11. *Hieracium oribates* Brenner (in Meddeland. Soc. Fauna Fl. Fenn. 30: 142. 1904) was validly published without accompanying descriptive matter but with reference to the validating description of *H. saxifragum* subsp. *oreinum* Dahlst. ex Brenner (in Meddeland. Soc. Fauna Fl. Fenn. 18: 89. 1892). Because Brenner definitely excluded his earlier infraspecific name and part of its original material, *H. oribates* is the name of a new taxon, not a replacement name, and may not be typified by an excluded element.

7.9. A name of a taxon assigned to a group with a nomenclatural starting-point later than 1 May 1753 (see Art. 13.1) is to be typified by an element selected from the context of its valid publication (Art. 32–45).

ⓘ **Note 2.** The typification of names of fossil-taxa (Art. 1.2) and of any other analogous taxa at or below the rank of genus does not differ from that indicated above.

7.10. For purposes of priority (Art. 9.19, 9.20, and 10.5), designation of a type is achieved only by effective publication (Art. 29–31).

7.11. For purposes of priority (Art. 9.19, 9.20, and 10.5), designation of a type is achieved only if the type is definitely accepted as such by the typifying author, if the type element is clearly indicated by direct citation including the term "type" (typus) or an equivalent, and, on or after 1 January 2001, if the typification statement includes the phrase "designated here" (hic designatus) or a similar expression demonstrating the author's intent to designate a type there (see also Art. 9.21–9.23 and F.5.4).

ⓘ **Note 3.** Art. 7.10 and 7.11 apply only to the designation of lectotypes and neotypes (and their equivalents under Art. 10.2) and epitypes; for holotypes see Art. 9.1 (and for their equivalents see Art. 10.2).

Ex. 12. The original material for the name *Quercus acutifolia* Née includes nine specimens in MA. In 1985, Breedlove labelled one of these (barcode MA 25953) as "Lectotype", but, because this was not effectively published, Breedlove did not achieve a designation of type (see Art. 7.10). Valencia-A. & al. (in Phytotaxa 218: 289–294. 2015) effectively published an intended designation of the same specimen as "lectotype", but did not include the words "designated here" or a similar expression, as required by Art. 7.11. Nixon & Barrie (in Novon 25: 449. 2017) achieved a designation of lectotype when they effectively published the statement "TYPE: Mexico. Guerrero, *Née s.n.* (lectotype, designated here, MA [bc] MA25953 as image!)" fulfilling all requirements of Art. 7.10 and 7.11.

Ex. 13. The protologue of *Dryopteris hirsutosetosa* Hieron. (in Hedwigia 46: 343–344, t. 6. 1907) cited only a locality ("Aequatoria: crescit in altiplanicie supra Allpayacu inter Baños et Jivaría de Píntuc [Ecuador: it grows on the high plain above Allpayacu between Baños and Jivaría de Píntuc]") and Stübel collecting number ("n. 903"), but did not specify a herbarium, thus indicating all specimens of that gathering as syntypes (Art. 40 Note 3). In citing "Type from Ecuador: Baños-Pintuc, Stübel nr. 903 (B!)" Christensen (in Kongel. Danske Vidensk. Selsk. Skr., Naturvidensk. Math. Afd., ser. 8, 6: 112. 1920) designated the specimen in B as the lectotype of *D. hirsutosetosa* satisfying the requirements of Art. 7.11. A duplicate specimen in BM is an isolectotype.

Ex. 14. The absence of any original material (Art. 9.13) for *Ocimum gratissimum* L. (Sp. Pl.: 1197. 1753) means that Cramer's (in Dassanayake & Fosberg, Revis. Handb. Fl. Ceylon 3: 112. 1981) citation of "Type: Hortu[s] Upsal[i]ensi[s], *749.2* (LINN)" is to be accepted as designation (Art. 7.11) of a neotype, antedating the superfluous neotypification by Paton (in Kew Bull. 47: 411. 1992).

Ex. 15. *Chlorosarcina* Gerneck (in Beih. Bot. Centralbl., Abt. 2, 21: 224. 1907) originally comprised two species, *C. minor* Gerneck and *C. elegans* Gerneck. Vischer (in Beih. Bot. Centralbl., Abt. 1, 51: 12. 1933) transferred *C. minor* to *Chlorosphaera* G. A. Klebs and retained *C. elegans* in *Chlorosarcina*. He did not, however, use the term "type" or an equivalent, so that his action does not constitute typification of *Chlorosarcina*. The first to designate a type, as "LT.", was Starr (in ING Card No. 16528, Nov 1962), who selected *Chlorosarcina elegans*.

Ex. 16. The protologue of *Spermacoce tenuior* L. (Sp. Pl.: 102. 1753) cites the illustration *"Spermacoce verticillis tenuioribus"* (Dillenius, Hort. Eltham: t. 277. 1732). Rendle (in J. Bot. 72: 333. 1934), in his attempt to typify the Linnaean name, wrote "the type of *S. tenuior* L. must be regarded as the figure and specimen of Dillenius...". Rendle's type designation is not effective because it did not clearly indicate a single element (Art. 9.17 does not apply).

*Ex. 17.[1] The phrase "standard species" as used by Hitchcock & Green (in Sprague, Nom. Prop. Brit. Bot.: 110–199. 1929) is now treated as equivalent to "type", and hence type designations in that work are acceptable.

Ex. 18. Pfeiffer (Nomencl. Bot. 1: [Praefatio, p. 2]. 1871) explained that he cited species names only when he intended to indicate the type of names of genera and sections: "Species plantarum in libro meo omnino negliguntur, excepta indicatione illarum, quae typum generis novi aut novo modo circumscripti vel sectionis offerunt. [Species of plants are entirely disregarded in my book, except for the indication of those that are presented as the type of a new or re-circumscribed genus or of a section.]" This explanation includes the term type, and the citation of a species name has therefore been accepted as designation of a type.

🛈 **Note 4.** Unless conservation determines otherwise, the effective typification of a name automatically establishes the same typification for all names sharing the same basionym (Art. 7.3) or replaced synonym (Art. 7.4) and for that basionym or replaced synonym. The effective typification of an autonym automatically establishes the same typification for the name from which it is derived (Art. 7.7).

Ex. 19. Traub & Moldenke (Amaryllid.: Tribe Amaryll.: 111. 1949) designated *Amaryllis striata* Lam. as the type of *A.* subg. *Lais* (Salisb.) Traub & Moldenke (in Herbertia 5: 119. 1938), thereby automatically establishing the same typification for *Lais* Salisb. (Gen. Pl.: 134. 1866), *Hippeastrum* sect. *Lais* (Salisb.) Baker (in J. Bot. 16: 81. 1878), and *H.* subg. *Lais* (Salisb.) Baker (Handb. Amaryll.: 41. 1888).

Ex. 20. Gillett (in Kew Bull. 17: 136. 1963) designated *Beshir 135* as the holotype of *Sesbania sudanica* J. B. Gillett subsp. *sudanica*, thereby automatically

---

1 Here and elsewhere in this *Code,* a prefixed asterisk denotes a "voted Example", accepted by an International Botanical Congress in order to govern nomenclatural practice when the corresponding Article of the *Code* is open to divergent interpretation or does not adequately cover the matter. A voted Example is therefore comparable to a rule, as contrasted with other Examples provided by the Editorial Committee solely for illustrative purposes.

establishing the same typification for *S. sudanica* J. B. Gillett. The herbarium in which the holotype was conserved was not specified but this was not required for valid publication of names of new taxa before 1 January 1990 (Art. 40.5).

Recommendation 7A

7A.1. It is strongly recommended that the material on which the name of a taxon is based, especially the holotype, be deposited in a public herbarium or other public collection with a policy of giving researchers access to deposited material, and that it be scrupulously conserved.

7A.2. It is strongly recommended that duplicates (Art. 8.3 footnote) of the material on which the name of a taxon is based (especially of the holotype but also of a neotype or epitype) be conserved in different herbaria, collections, or institutions, preferably in different areas of the world, as far as possible.

7A.3. Authors publishing the name of a new species or infraspecific taxon are encouraged to deposit type material in one or more herbaria, collections, or other specialized institutions in the country or countries of origin of the newly described taxon.

ARTICLE 8
SPECIMENS AND GATHERINGS

8.1. The nomenclatural type (see Art. 7.2) of a name of a species or infraspecific taxon is a holotype (Art. 9.1), lectotype (Art. 9.3), neotype (Art. 9.8), or conserved type (Art. 14.9), any of which may be supported by an epitype (Art. 9.9). Such a type is either a single specimen conserved in one herbarium or other collection or institution, or is a published or unpublished illustration (but see Art. 8.5; see also Art. 40.6 and Art. 40 Ex. 10).

8.2. For the purpose of typification, a specimen is a gathering[1], or part of a gathering, of a single species or infraspecific taxon, disregarding

---

1 Here and elsewhere in this *Code,* the term "gathering" is used for material collected by the same collector(s) at the same time from a single locality and presumed to be of a single taxon. If specimens lack information on collector, date, or locality, this does not necessarily preclude their being part of the same gathering, but the possibility of a mixed gathering is always to be considered, especially when designating a type. For most fossils, a sample of sediment or rock is not presumed to be of a single fossil-taxon, but is a set of gatherings, each gathering consisting of an individual of a fossil-taxon.

admixtures (see Art. 9.14). It may consist of a single part, multiple parts, or the whole of one or more individual organisms. A specimen is usually mounted on a single herbarium sheet or in an equivalent preparation, such as a box, packet, jar, or microscope slide (for fossil-taxa see Art. 8.6).

Ex. 1. The holotype of *Asparagus kansuensis* F. T. Wang & Tang ex S. C. Chen (in Acta Phytotax. Sin. 16(1): 94. 1978), *Hao 416* (PE [barcode 00034519]) belongs to a gathering of a dioecious species made at one time at a single locality. It consists of a staminate branch and a pistillate branch, i.e. parts of two individuals, mounted on a single herbarium sheet.

Ex. 2. The diatom species *Tursiocola denysii* Frankovich & M. J. Sullivan (in Phytotaxa 234: 228. 2015) was described from material collected from neck skin of four loggerhead turtles and the type designated as "Type:—UNITED STATES. Florida: Florida Bay, samples removed from the skin in the dorsal neck area of loggerhead sea turtles *Caretta caretta*, 24° 55' 01" N, 80° 48' 28" W, *B.A. Stacy, 24 June 2015* (holotype CAS! 223049, illustrated as Figs 1–4, 6, 12, 15–30, paratypes ANSP! GC59142, BM! 101 808, illustrated as Figs 7–10, 14, BRM! ZU10/31, Figs 5, 11, 13)." Because the specimens were collected at the same time, at the same place, by the same collector they comprise a single gathering, admixtures excepted, and the authors' citation of "paratypes" is correctable to isotypes under Art. 9.10.

Ex. 3. *"Echinocereus sanpedroensis"* (Raudonat & Rischer in Echinocereenfreund 8(4): 91–92. 1995) was based on a "holotype" consisting of a complete plant with roots, a detached branch, an entire flower, a flower cut in halves, and two fruits that, according to the label, were taken from the same cultivated individual at different times and preserved, in alcohol, in a single jar. Because this material was collected at more than one time, it belongs to more than one gathering and cannot be accepted as a type. Raudonat & Rischer's name is not validly published under Art. 40.3.

🛈 *Note 1.* Field numbers, collecting numbers, accession numbers, or specimen identifiers alone do not necessarily denote the same or different gatherings.

Ex. 4. Among the specimens of *Goodyera hemsleyana* King & Pantl. (in J. Asiat. Soc. Bengal, Pt. 2, Nat. Hist. 64: 342. 1896) collected by Robert Pantling, the notation "No. 215" was used for more than one gathering, one made in Senchal, India in July 1892 (with specimens in BM, BR, CAL, K, L, P, and W) and another in the same place but in July 1898 (with specimens in AMES and L).

Ex. 5. *Solidago* ×*snarskisii* Gudžinskas & Žalneravičius (in Phytotaxa 253: 148. 2016) was validly published (Art. 40.3) with a single gathering in BILAS indicated as type, the parts of which were numbered separately in the field, mounted on separate sheets and designated as follows: "Holotype:—LITHUANIA. Trakai district, Aukštadvaris Regional Park, environs of Zabarauskai village, in an abandoned meadow on the edge of forest (54.555191° N; 24.512987° E), 13 September 2014, *Z. Gudžinskas & E. Žalneravičius 76801* (generative shoot) and

76802 (vegetative shoot) (BILAS, on two cross-referenced sheets). Isotypes:— *Z. Gudžinskas & E. Žalneravičius 76803, 76804* (BILAS)."

8.3. A specimen may be mounted as more than one preparation, as long as the parts are clearly labelled as being part of that same specimen, or have a single, original label in common. Multiple preparations from a single gathering that are not clearly labelled as being part of a single specimen are duplicates[1], regardless of whether the source was one individual or more than one.

Ex. 6. The holotype specimen of *Delissea eleeleensis* H. St. John, *Christensen 261* (BISH), is mounted as two preparations, a herbarium sheet (BISH No. 519675 [barcode BISH1006410]) with the annotation "fl. bottled" and an inflorescence preserved in alcohol in a jar labelled *"Cyanea, Christensen 261"* (*Cyanea eleeleensis* (H. St. John) Lammers is a homotypic synonym). The annotation indicates that the inflorescence is part of the holotype specimen and not a duplicate, nor is it part of the isotype specimen (BISH No. 519676 [barcode BISH1006411]), which is not labelled as including additional material preserved in a separate preparation.

Ex. 7. The holotype of *Cephaelis acanthacea* Standl. ex Steyerm., *Cuatrecasas 16572* (F), consists of a single specimen mounted on two herbarium sheets, labelled "sheet 1" and "sheet 2". Although the two sheets have separate herbarium accession numbers, F No. 1153741 and F No. 1153742, respectively, the cross-labelling indicates that they constitute a single specimen. A third sheet of *Cuatrecasas 16572*, F No. 1153740, is not cross-labelled and is therefore a duplicate. (The valid publication of this name was discussed by Taylor in Novon 25: 331–332. 2017.)

Ex. 8. The holotype specimen of *Eugenia ceibensis* Standl., *Yuncker & al. 8309*, is mounted on a single herbarium sheet in F. A fragment was removed from the specimen after its designation as holotype and is now conserved in LL. The fragment is mounted on a herbarium sheet along with a photograph of the holotype and is labelled "fragment of type!". The fragment is no longer part of the holotype specimen because it is not permanently conserved in the same herbarium as the holotype. It is a duplicate, i.e. an isotype.

Ex. 9. In the Geneva herbaria, a single specimen is often prepared on two or more sheets, which are not therefore duplicates. Although the individual sheets are usually not labelled as being part of the same specimen, they are physically kept together in their own specimen folder and have a single, original label in common.

Ex. 10. Three specimens collected by Martius (Brazil, Maranhão, "in sylvis ad fl. Itapicurú [in woods near the River Itapicurú]", May 1819, *Martius s.n.*, M)

---

1 Here and elsewhere in this *Code*, the word "duplicate" is given its usual meaning in curatorial practice. A duplicate is part of a single gathering of a single species or infraspecific taxon. For most fossils, a single gathering consists of an individual of a fossil-taxon, and hence there are no duplicates.

are syntypes of *Erythrina falcata* Benth. (in Martius, Fl. Bras. 15(1): 172. 1859). Only one of the herbarium sheets (barcode M-0213337) has Martius's original blue label, whereas the other two (barcodes M-0213336 and M-0213338) have been labelled with the locality to identify them as the same gathering. Because the three specimens do not have a single, original label in common, and are not cross-labelled, they are treated as duplicates.

8.4. Type specimens of names of taxa must be preserved permanently and may not be living organisms or cultures. Nevertheless, cultures of algae and fungi, if preserved in a metabolically inactive state (e.g. by lyophilization or deep-freezing to remain alive in that inactive state), are acceptable as types (see also Art. 40.7 and Rec. F.11A.2).

Ex. 11. *"Dendrobium sibuyanense"* (Naranja & al. in Philipp. Agric. Sci. 88: 484–488. 2005) was described with the statement "Type specimen is a living specimen being maintained at the Orchid Nursery, Department of Horticulture, University of the Philippines Los Baños (UPLB). Collectors: Orville C. Baldos & Ramil R. Marasigan, April 5, 2004". Because a living plant cannot be a type, no type was indicated and the name was not validly published (Art. 40.1; see also Art. 40 Ex. 6).

Ex. 12. *Dipodascus australiensis* Arx & J. S. F. Barker (in Arx, Antonie van Leeuwenhoek 43: 335. 1977) was published with a culture cited as "Typus CBS 625.74". This is acceptable as a holotype because, from 1958, cultures were permanently preserved at the Centraalbureau voor Schimmelcultures (CBS) in a metabolically inactive state by lyophilization (Hoog, Centraalbureau voor Schimmelcultures 75 Years Culture Collection: 20. 1979).

8.5. The type, epitypes (Art. 9.9) excepted, of the name of a fossil-taxon at the rank of species or below is always a specimen (see Art. 9.15). One whole specimen is to be considered as the nomenclatural type (see Rec. 8A.4).

8.6. For the purpose of typification of names of fossil-taxa, a specimen is an individual of a fossil-species or infraspecific fossil-taxon selected from a sample of sediment or rock or subsample or preparation thereof. A specimen is usually contained on or in a slab of rock, box, vial, or micropalaeontology slide, or mounted on a scanning electron microscope stub or microscope slide. Each specimen is treated as a separate gathering. A sample (or subsample or preparation thereof) may contain multiple individuals (i.e. specimens) of the same fossil-taxon as well as other fossil-taxa (see also Art. 40.8).

Ex. 13. The specimen designated as the holotype of the fossil spore *Striatella jurassica* Mädler (in Fortschr. Geol. Rheinl. Westfalen 12: 192. 1964) was mounted on a microscope slide as a strew mount together with other individuals of the same and other taxa. The material on the slide represents a subsample of the residue, which in turn is a subsample of a single rock sample from the Thurau 1 core from the Lower Jurassic of Germany. The holotype is indicated by microscope co-ordinates (21:117.7) and a collection inventory number (TK 3154) on the slide as well as in the protologue. Four surrounding dots were added later to indicate the position of this specimen. This indication was translated into a more widely used England Finder reference (N19/4). Besides this holotype, the slide also contains the holotype of *Ephedripites tortuosus* Mädler (l.c.: 194. 1964), TK3159. Other palynomorphs on the same slide conforming to the circumscription of *S. jurassica* and *E. tortuosus,* but not explicitly cited in the protologues, comprise other parts of the original material, i.e. they are independent uncited specimens (see photograph of this example in Gravendyck & al. in Palynology 45: 727, fig. 5Ab. 2021).

ⓘ **Note 2.** Macrofossils are often discovered by splitting rocks to reveal a fossil organ on both parts of the rock (or the entire fossil itself and a mould). These are usually referred to as "part" and "counterpart"; they are parts of the same specimen, not separate specimens, and often complementary to each other in their structural details. For the purpose of typification, both part and counterpart, where available, comprise the type specimen.

Ex. 14. The fossil-species *Diplotropis claibornensis* Herend. & Dilcher (in Syst. Bot. 15: 527, fig. 1 and 2. 1990) was described based on a fossil that consists of part and counterpart. Both part and counterpart comprise the specimen that was designated as the holotype.

Recommendation 8A

8A.1. When a holotype, a lectotype, or a neotype is an illustration, the specimen or specimens upon which that illustration is based should be used to help determine the application of the name (see also Art. 9.15).

8A.2. When an illustration is the type of a name under Art. 40.6, the collection data of the illustrated material should be given (see also Rec. 38D.2).

8A.3. When a single specimen designated as holotype, lectotype, neotype, or epitype is mounted as multiple preparations, this should be stated in the publication containing the type designation, and the preparations appropriately labelled.

8A.4. If the type specimen of a name of a fossil-taxon is cut into pieces (sections of fossil wood, pieces of coalball plants, etc.), all parts originally used in establishing the diagnosis should be clearly marked.

8A.5. If a type specimen is prepared on a microscope slide, it is strongly recommended that the position of the specimen be indicated by an England Finder reference (Graticules Ltd. in J. Sci. Instrum. 39: 250. 1962) or equivalent unambiguous reference (e.g. single-grain mounts or permanent ink circling; see Art. 8 Ex. 13) to facilitate finding it again.

8A.6. For palaeopalynological samples, it is recommended that at least a subsample of rock or sediment or residue from which the type was selected be deposited in the public collection along with the type, thereby permitting future preparations that could interpret or replace degraded type material.

Recommendation 8B

8B.1. Whenever practicable a living culture should be prepared from the holotype material of the name of a newly described taxon of algae or fungi and deposited in at least two institutional culture or genetic resource collections. (Such action does not obviate the requirement for a holotype specimen under Art. 8.4; see also Rec. F.11A.1.)

8B.2. In cases where the type of a name is a culture permanently preserved in a metabolically inactive state (see Art. 8.4), any living isolates obtained from it should be referred to as "ex-type" (ex typo), "ex-holotype" (ex holotypo), "ex-isotype" (ex isotypo), etc., in order to make it clear they are derived from the type but are not themselves the nomenclatural type.

ARTICLE 9
CATEGORIES AND DESIGNATION OF TYPES

9.1. A holotype of a name of a species or infraspecific taxon is the one specimen or illustration (but see Art. 40.6) either:

(a)   designated in the protologue as the nomenclatural type; or
(b)   used by the author(s) in preparing the protologue when no type was designated; or
(c)   as described in Art. 9 Note 1.

As long as the holotype exists, it fixes the application of the name concerned (but see Art. 9.15).

9.2. Any designation of the type made by the original author(s), if definitely expressed in the protologue, is final (but see Art. 9.11, 9.15, and 9.16). Mention of a single specimen or gathering or illustration does not by itself constitute designation of the holotype (but see Art. 9 Note 1). A cited or uncited specimen or illustration must be accepted as the holotype if there is evidence in the protologue or elsewhere to establish that the specimen or illustration was the only one used (Art.

9.1(b)) and no additional, uncited specimens or illustrations (which may have been lost or destroyed) could have been used. If a name of a new taxon is validly published solely by reference to a previously published description or diagnosis, the same considerations apply to specimens or illustrations used by the author(s) of that description or diagnosis (see Art. 7.8; but see Art. 7.9).

ⓘ **Note 1.** A single specimen or illustration (illustrations of fossil-taxa excepted: see Art. 8.5) cited in the protologue of the name of a new species or infraspecific taxon published on or after 1 January 1958 and before 1 January 1990 is the holotype (see Art. 40.3 and 40.4).

> *Ex. 1.* When Tuckerman published *Opegrapha oulocheila* Tuck. (Lich. Calif.: 32. 1866) he referred to "the single specimen, from Schweinitz's herbarium (Herb. Acad. Sci. Philad.) before me". Even though the term "type" or its equivalent was not used in the protologue, Tuckerman's statement is evidence to establish that he used only that specimen (in PH barcode 00007529), which is therefore the holotype.

> *Ex. 2.* In the protologue of *Coronilla argentea* L. (Sp. Pl.: 743. 1753), Linnaeus cited an illustration by Alpini (Pl. Exot.: 16. 1627) and did not designate a type. Mention of the illustration does not by itself constitute designation of the holotype. Although no uncited specimens or illustrations are known to exist, making Alpini's illustration the only known element of original material, it is not the holotype because it cannot be established that Linnaeus used only this one element when preparing the protologue. Linnaeus rarely cited specimens and could have used a specimen that was subsequently lost or destroyed (he is known to have discarded specimens). Alpini's illustration was designated as the lectotype of *C. argentea* by Greuter (in Ann. Mus. Goulandris 1: 44. 1973).

> *Ex. 3.* In the protologue of *Calycanthus praecox* L. (Sp. Pl., ed. 2: 718. 1762), Linnaeus did not designate a type and cited only one element, an illustration by Kaempfer (Amoen. Exot. Fasc.: 879. 1712); he also stated that the plant was unknown to him ("Ignota mihi"). This is evidence establishing that Linnaeus, when preparing the protologue, used only Kaempfer's illustration, which must therefore be accepted as the holotype.

9.3. A lectotype is one specimen or illustration designated from the original material (Art. 9.4) as the nomenclatural type, in conformity with Art. 9.11 and 9.12:

(a) if the name was published without a holotype; or
(b) if the holotype is lost or destroyed; or
(c) if a type is found to belong to more than one taxon (see also Art. 9.14).

For sanctioned names (Art. F.3), a lectotype may be selected from among elements associated with either or both the protologue and the sanctioning treatment (Art. F.3.10).

Ex. 4. *Adansonia grandidieri* Baill. (in Grandidier, Hist. Phys. Madagascar 34: t. 79B bis, fig. 2 & t. 79E, fig. 1. 1893) was validly published when accompanied solely by two illustrations with analysis (see Art. 38.9). Baum (in Ann. Missouri Bot. Gard. 82: 447. 1995) designated one of the sheets of *Grevé 275* (flowering specimen in P [barcode P00037169]), which he presumed to be the specimen from which most or all of the components of t. 79E, fig. 1 were drawn, as the lectotype of this name.

9.4. Original material of a name comprises the following elements:

(a) the holotype (Art. 9.1) and any isotypes (Art. 9.5); and
(b) any syntypes (Art. 9.6) and isosyntypes[1]; and
(c) any paratypes (Art. 9.7); and
(d) any illustrations published as part of the protologue (fossil-taxa excepted: see Art. 8.5); and
(e) those specimens and illustrations (both published and unpublished; illustrations of fossil-taxa excepted: see Art. 8.5) that were associated with the taxon by, and that were available to:
  (1) the publishing author(s) prior to, or at the time of, publication of the protologue; or
  (2) other author(s) to whom the description or diagnosis may have been ascribed (or unequivocally associated) prior to, or at the time of, preparation of the description, diagnosis, or illustration with analysis (Art. 38.8 and 38.9) validating the name (but see Art. 7.8, 7.9, and F.3.10).

🛈 *Note 2.* Original material under Art. 9.4(a)–(c) has not necessarily been seen by either the author of the validating description or diagnosis or the author of the name.

🛈 *Note 3.* Habitat photographs cannot be original material or types because they are not illustrations as defined by this *Code* (Art. 6.1 footnote).

🛈 *Note 4.* For names falling under Art. 7.8, only elements from the context of the validating description are considered as original material. However, if the validating author has designated a different element as the type in the protologue (indication or usage by the validating author is insufficient to

---

1   Duplicate specimens of a syntype, lectotype, neotype, and epitype are isosyntypes, isolectotypes, isoneotypes, and isoepitypes, respectively.

establish a type), the original material is determined in accordance with Art. 9.4 without regard to the previous sentence.

ⓘ **Note 5.** For names falling under Art. 7.9, only elements from the context of the valid publication of those names are considered as original material.

9.5. An isotype is any duplicate of the holotype; it is always a specimen.

ⓘ **Note 6.** The term isotype is also used for a duplicate of the type of the conserved name of a species or infraspecific taxon because, under Art. 14.8, such a type, like a holotype, may only be changed by the procedure of conservation.

9.6. A syntype is any specimen cited in the protologue when there is no holotype, or any one of two or more specimens simultaneously designated in the protologue as types (see also Art. 40 Note 3). Reference to an entire gathering, or a part thereof, is considered to be citation of the included specimens (see also Art. 40 Note 2).

ⓘ **Note 7.** Specimens not cited in the protologue that are original material according to Art. 9.4(e) are not syntypes.

Ex. 5. In the protologue of *Laurentia frontidentata* E. Wimm. (see Art. 40 Ex. 3) a single gathering in two herbaria was designated as the type. Therefore, there must exist at least two specimens and these are syntypes.

Ex. 6. In the protologue of *Campanula pulla* L. (Sp. Pl.: 163. 1753), Linnaeus cited "Burs. IV. 21", referring to a specimen in the Burser Herbarium (UPS), in addition to an illustration in Bauhin (Prodr.: 35. 1620). This single specimen is a syntype because it was cited in the protologue and there is no holotype. Similarly, in the protologue of *Anemone alpina* L. (Sp. Pl.: 539. 1753), two specimens are cited under the (unnamed) varieties β and γ, as "Burs. IX. 80" and "Burs. IX: 81". These specimens, held in the Burser Herbarium (UPS), are syntypes of *A. alpina.*

Ex. 7. Lavalle (in Darwiniana 41: 68. 2003) cited "SINTIPOS [syntypes]" of the name *Marattia cicutifolia* Kaulf. (Enum. Filic.: 32. 1824). However, they cannot be syntypes because they were not cited in the protologue of that name, where Kaulfuss cited only the provenance "Habitat in Brasilia". Instead, they are original material because they satisfy the definition of that term as given in Art. 9.4(e).

9.7. A paratype is any specimen cited in the protologue that is neither the holotype nor an isotype, nor one of the syntypes if in the protologue two or more specimens were simultaneously designated as types.

Ex. 8. The holotype of the name *Rheedia kappleri* Eyma (in Meded. Bot. Mus. Herb. Rijks Univ. Utrecht 4: 26. 1932), which applies to a polygamous species, is a staminate specimen, *Kappler 593a* (U). The author designated a hermaphroditic specimen, *Forestry Service of Surinam B. W. 1618* (U), as a paratype.

ⓘ **Note 8.** In most cases in which no holotype was designated there are also no paratypes because all the cited specimens are syntypes. However, when an author designated two or more specimens as types (Art. 9.6), any remaining cited specimens are paratypes and not syntypes.

> *Ex. 9.* In the protologue of *Eurya hebeclados* Y. Ling (in Acta Phytotax. Sin. 1(2): 208. 1951) the author simultaneously designated two specimens as types, *Y. Ling 5014* as "typus, ♂" and *Y. Y. Tung 315* as "typus, ♀", which are therefore syntypes. Ling also cited the specimen *Y. Ling 5366* but without designating it as a type; it is therefore a paratype.

9.8. A neotype is a specimen or illustration selected to serve as nomenclatural type if no original material exists, or as long as it is missing (see also Art. 9.16, 9.19(c), and Rec. F.11A.2).

9.9. An epitype is a specimen or illustration selected to serve as an interpretative type when the holotype, lectotype, or previously designated neotype, or all original material associated with a validly published name, is demonstrably ambiguous and cannot be critically identified for the purpose of the precise application of the name to a taxon. Designation of an epitype is not achieved unless the holotype, lectotype, or neotype that the epitype supports is explicitly cited (see also Art. 9.20).

> *Ex. 10.* Podlech (in Taxon 46: 465. 1997) designated Herb. Linnaeus No. 926.43 (LINN) as the lectotype of *Astragalus trimestris* L. (Sp. Pl.: 761. 1753). He simultaneously designated an epitype ("Egypt, Dünen oberhalb Rosetta am linken Nilufer bei Schech Mantur [dunes above Rosetta on the left bank of the Nile at Sheikh Mandur], 9 May 1902, *Anonymous* (BM)") because the lectotype lacks fruits, "which show important diagnostic features for this species".

> *Ex. 11.* The lectotype of *Salicornia europaea* L. (Herb. Linnaeus No. 10.1, LINN, designated by Jafri & Rateeb in Jafri & El-Gadi, Fl. Libya 58: 57. 1978) does not show the relevant characters by which it could be identified for the precise application of this name in a critical group of taxa that are best characterized molecularly. In view of this, Kadereit & al. (in Taxon 61: 1234. 2012) designated as the epitype a molecularly tested specimen from the type locality (Sweden, Gotland, W shore of Burgsviken Bay, Näsudden Cape, *Piirainen & Piirainen 4222,* only the plant numbered G38-1, MJG).

> *Ex. 12.* Martínez-Laborde & al. (in Phytotaxa 220: 96. 2015) designated a specimen with pistillate flowers (*Balansa 2342*, P barcode P00080325) as the lectotype of *Hennecartia omphalandra* J. Poiss. (in Bull. Soc. Bot. France 32: 41. 1885). Although fruits and staminate flowers are important diagnostic characters

for this species, the vegetative organs and pistillate flowers in the lectotype are also clearly diagnostic. Therefore, according to Art. 9.9, an epitype is not necessary because the lectotype can be critically identified to precisely apply the name to the taxon.

9.10. The use of a term defined in the *Code* (Art. 9.1, 9.3, and 9.5–9.9) as denoting a type, in a sense other than that in which it is so defined, is treated as an error to be corrected (for example, the use of the term lectotype to denote what is in fact a neotype; see also Art. 40.4).

Ex. 13. Borssum Waalkes (in Blumea 14: 198. 1966) cited Herb. Linnaeus No. 866.7 (LINN) as the "holotype" of *Sida retusa* L. (Sp. Pl., ed. 2: 961. 1763). However, illustrations in Plukenet (Phytographia: t. 9, fig. 2. 1691) and Rumphius (Herb. Amboin. 6: t. 19. 1750) were cited by Linnaeus in the protologue. Therefore, the original material of *S. retusa* comprises three elements (Art. 9.4(e)), and Borssum Waalkes's use of holotype is an error to be corrected to lectotype.

ⓘ **Note 9.** A misused term may be corrected to lectotype, neotype, or epitype only if the requirements of Art. 7.11 are met, in particular inclusion of the phrase "designated here" or a similar expression for typifications on or after 1 January 2001 (for names of organisms treated as fungi see also Art. F.5.4).

Ex. 14. Bohley & al. (in Syst. Bot. 42: 138. 2017) cited the specimen *Balansa 2263* (G) as the "type" and "holotype" of *Cypselea meziana* K. Müll. (in Bot. Jahrb. Syst. 42(Beibl. 97): 72. 1908). However, this use of the term holotype cannot be corrected to lectotype because the requirement of Art. 7.11 to include, on or after 1 January 2001, the phrase "designated here" or a similar expression was not met. Consequently, designation of a lectotype was not achieved until Jocou & Minué (in Phytotaxa 461: 69. 2020) wrote "Lectotype (designated here)" selecting a specimen, in P, from the same *Balansa* gathering.

9.11. A lectotype (Art. 9.3) or, if permissible, a neotype (Art. 9.8) may be designated (see also Art. 9.16):

(a) if the name of a species or infraspecific taxon was published without a holotype (Art. 9.1); or

(b) when the holotype or previously designated lectotype has been lost or destroyed; or

(c) when the material designated as type is found to belong to more than one taxon.

9.12. In lectotype designation, a part of the holotype (if it is taxonomically mixed) that is not in conflict with the validating description or diagnosis must be chosen if such exists, or otherwise an isotype if such exists, or otherwise a syntype or isosyntype if such exists, or otherwise a paratype if such exists. If none of the above specimens exists, the lectotype must be chosen from among the illustrations and uncited specimens that comprise the remaining original material, if such exist.

> *Ex. 15.* Baumann & al. (in J. Eur. Orch. 34: 176. 2002) designated an illustration cited in the protologue of *Gymnadenia rubra* Wettst. (in Ber. Deutsch. Bot. Ges. 7: 312. 1889) as "lectotype". Because Wettstein also cited syntypes, which always have precedence over illustrations in lectotype designation, Baumann's choice was not in conformity with Art. 9.12 and must not be followed. Later, Baumann & Lorenz (in Taxon 60: 1775. 2011) correctly designated one of the syntypes as the lectotype.

9.13. If no original material exists or as long as it is missing, a neotype may be selected. A lectotype always takes precedence over a neotype, except as provided by Art. 9.16 and 9.19(c).

9.14. When a type (herbarium sheet or equivalent preparation) contains parts belonging to more than one taxon (see Art. 9.11), the name must remain attached to the part (specimen as defined in Art. 8.2) that corresponds most nearly with the original description or diagnosis.

> *Ex. 16.* The holotype of the name *Tetrapterys alternifolia* Cuatrec. (in Webbia 13: 435. 1958) is *A. Dugand & R. Jaramillo 2850* (US); the material on this sheet, however, proved to be mixed. Anderson (in Contr. Univ. Michigan Herb. 25: 91. 2007) acted in accordance with Art. 9.14 in designating one part of the sheet ("lectotype, designated here: US!, the stem with flowers") in US (barcode 00108506) as the lectotype.

9.15. The holotype (or lectotype) of a name of a fossil-species or infraspecific fossil-taxon (Art. 8.5) is the specimen (or one of the specimens) on which the validating illustrations (Art. 43.2) are based. If, before 1 January 2001 (see Art. 43.3), a type specimen is indicated (Art. 40.1) but not identified among the validating illustrations in the protologue of a name of a new fossil-taxon at the rank of species or below, a lectotype must be designated from among the specimens illustrated in the protologue. This choice is superseded if it can be demonstrated that the original type specimen corresponds to another validating illustration.

9.16. When a holotype or a previously designated lectotype has been lost or destroyed and it can be shown that all the other original material differs taxonomically from the lost or destroyed type, a neotype

may be selected to preserve the usage established by the previous typification (see also Art. 9.18).

9.17. A designation of a lectotype, neotype, or epitype that is later
found to refer to a single gathering but to more than one specimen
must nevertheless be accepted (subject to Art. 9.19 and 9.20) but may
be further narrowed to a single one of these specimens by way of a
subsequent lectotypification, neotypification, or epitypification (see
also Art. 9.14).

Ex. 17. *Erigeron plantagineus* Greene (in Pittonia 3: 292. 1898) was described
from material collected by R. M. Austin in California. Cronquist (in Brittonia 6:
173. 1947) wrote "Type: *Austin s.n.,* Modoc County, California (ND)", thereby
designating the Austin material in ND as the lectotype (first-step). Strother &
Ferlatte (in Madroño 35: 85. 1988), noting that there were two specimens of this
gathering in ND, designated one of them, "ND-G, 057228 [barcode NDG57228]",
as the (second-step) lectotype. In subsequent references, both lectotypification
steps may be cited in sequence.

Ex. 18. In the protologue of *Eucalyptus oreades* R. T. Baker (in Proc. Linn. Soc.
New South Wales 24: 596. 1900) Baker cited: "*Hab.*—Lawson (*H. G. Smith* and
R.T.B.); Mount Victoria and road to Jenolan Caves (*R. H. Cambage*)." Specimens
of these gatherings are syntypes (see Art. 9.6). Brooker & al. (in Boland & al.,
Forest Trees Australia, ed. 4: 314. 1984) designated a (first-step) lectotype of the
name, fulfilling the requirements of Art. 7.11 by citing "Type: Near Lawson, New
South Wales, Apr. 1899, R. T. Baker and H. G. Smith." No herbarium was specified, but this was not a requirement in 1984 (see Art. 9.22). Bean (in Telopea 12:
316. 2009), noting that the gathering by Baker and Smith was represented by five
specimens, one in K and four in NSW, designated NSW barcode NSW325376 as
the (second-step) lectotype.

9.18. A neotype selected under Art. 9.16 may be superseded if it can
be shown to differ taxonomically from the holotype or lectotype that
it replaced.

9.19. The author who first designates (Art. 7.10, 7.11, and F.5.4) a
lectotype or a neotype in conformity with Art. 9.11–9.13 must be followed, but that choice is superseded if:

(a) the holotype or, in the case of a neotype, any of the original material is found to exist.

The choice may also be superseded if it can be shown that:

(b) it is contrary to Art. 9.14; or

(c) it is in serious conflict with the protologue, in which case an element that is not in conflict with the protologue is to be chosen.

A lectotype may only be superseded by a non-conflicting element of the original material, if such exists; if none exists it may be superseded by a neotype.

> **Ex. 19.** (c) Fischer (in Feddes Repert. 108: 115. 1997) designated Herb. Linnaeus No. 26.58 (LINN) as lectotype of *Veronica agrestis* L. (Sp. Pl.: 13. 1753). However, Martínez-Ortega & al. (in Taxon 51: 763. 2002) established that the designated lectotype was in serious conflict with Linnaeus's diagnosis and that three specimens of original material not conflicting with the protologue were available in the Celsius herbarium. One of them was designated as the new lectotype of *V. agrestis,* superseding the choice of Fischer.

ⓘ **Note 10.** Only a choice of uncited material as lectotype may be superseded under Art. 9.19(c); cited specimens and illustrations are part of the protologue and cannot therefore be in serious conflict with it.

9.20. The author who first designates (Art. 7.10, 7.11, and F.5.4) an epitype must be followed; a different epitype may be designated only if the original epitype is lost or destroyed (see also Art. 9.17). A lectotype or neotype supported by an epitype may be superseded in accordance with Art. 9.19 or, in the case of a neotype, in accordance with Art. 9.18. If it can be shown that an epitype and the type it supports differ taxonomically and that neither Art. 9.18 nor 9.19 applies, the name may be proposed for conservation with a conserved type (Art. 14.9; see also Art. 57.1).

ⓘ **Note 11.** An epitype supports only the type to which it is linked by the typifying author. If the supported type is replaced because it was lost, destroyed, or superseded, the epitype has no standing with respect to the replacement type.

9.21. Designation of an epitype is not achieved unless the herbarium, collection, or institution in which the epitype is conserved is specified or, if the epitype is a published illustration, a full and direct bibliographic reference (Art. 41.5) to it is provided.

9.22. On or after 1 January 1990, lectotypification or neotypification of a name of a species or infraspecific taxon by a specimen or unpublished illustration is not achieved unless the herbarium, collection, or institution in which the type is conserved is specified.

9.23. On or after 1 January 2001, lectotypification, neotypification, or epitypification of a name of a species or infraspecific taxon is not achieved unless indicated by use of the term "lectotypus", "neotypus", or "epitypus", its equivalent in a modern language, or abbreviations of these (see also Art. 7.11 and 9.10).

> ***Ex. 20.*** *Clavaria fumosa* Pers. (in Ann. Bot. (Usteri) 15: 31. 1795) was described without mention of any specimens or a specific locality, and there is no holotype according to Art. 9.1 and 9.2. Kautmanová & al. (in Persoonia 29: 141. 2012) cited a specimen (in L [barcode L 0115746]) that lacks a collecting date and locality details as "Typus (designated here)". This act does not constitute an effective lectotypification or neotypification because it lacks the term "lectotypus" or "neotypus" or a term correctable to one of these terms (Art. 9.10). Later, a different specimen was effectively designated as the neotype of the name *C. fumosa* when Franchi & Marchetti (I Funghi Clavarioidi in Italia: 196. 2022) wrote "Neotypus (hic designatus): BRA CR15656, Slovakia, Západné Tatry Mts, Zuberec village, Mačie diery Nature Reserve, alt. 800 m, in dry mowed meadow, on limestone, leg. I. Kautmanova, 16 June 2007. [typification identifier] IF 557637".

9.24. If a designation of holotype, lectotype, neotype, or epitype made in the publication containing the type designation of the name of a taxon is later found to contain errors (e.g. in locality, date, collector, collecting number, herbarium or collection or institution or its abbreviation, specimen identifier, or citation of an illustration), these errors are to be corrected provided that the intent of the original author(s) is not changed (see also Art. F.5.8). However, omissions of required information under Art. 7.11, 9.21–9.23, 40.4, 40.5, 40.7, 40.8, and F.5.4 are not correctable.

> ***Ex. 21.*** The name *Phoebe calcarea* S. Lee & F. N. Wei (in Guihaia 3: 7. 1983) was validly published with the holotype designated as *Du'an Expedition "4-10-004"* in IBK, but no specimen with this collecting number exists in IBK. However, a specimen in IBK annotated with "*Phoebe calcarea* sp. nov.", "Typus", and matching all other details of the protologue has the collecting number *Du'an Expedition 4-10-0243*. Therefore, the original type citation is obviously erroneous and is to be corrected.

> ***Ex. 22.*** The name *Capparis trichocarpa* B. S. Sun (in Acta Phytotax. Sin. 9: 113. 1964) was validly published with a gathering *C. W. Wang 73796* in PE designated as the type, but in the herbarium PE there are two duplicates of this gathering. Li & al. (in Bull. Bot. Res., Harbin 28: 265. 2008) designated one of these specimens as the lectotype: "China. Yunnan: Fo-hai (= Menghai), alt. 1520 m, March 1936, *C. W. Wang 73796* (lectotype, PE Herb. Bar Code No. 00029137, designated here, PE!; isolectotype, PE!)." However, the collecting date on the label of the lectotype specimen is May 1936. Therefore, the erroneous date in the designation of lectotype is to be corrected.

Recommendation 9A

9A.1. Typification of names published without a holotype should only be carried out with an understanding of the author's method of working; in particular it should be realized that some of the material used by the author in describing the taxon may not be in the author's herbarium or may not even have survived, and conversely, that not all the material surviving in the author's herbarium was necessarily used in describing the taxon.

9A.2. Designation of a lectotype should be undertaken only in the light of an understanding of the group concerned. In choosing a lectotype, all aspects of the protologue should be considered as a basic guide. Mechanical methods, such as the automatic selection of the first element cited or of a specimen collected by the person after whom a species is named, should be avoided as unscientific and leading to possible confusion and further change.

9A.3. In choosing a lectotype, any indication of intent by the author of a name should be given preference unless such indication is contrary to the protologue. Such indications are manuscript notes, annotations on herbarium sheets, recognizable figures, and epithets such as *typicus, genuinus,* etc.

9A.4. When two or more heterogeneous elements were included in or cited with the original description or diagnosis, the lectotype should be selected so as to preserve current usage. If another author has already segregated one or more elements as other taxa, one of the remaining elements should be designated as the lectotype provided that this element is not in conflict with the original description or diagnosis (see Art. 9.19(c)).

Recommendation 9B

9B.1. In selecting a neotype, particular care and critical knowledge should be exercised because there is usually no guide except personal judgement as to what best fits the protologue; if this selection proves to be faulty it may result in further change.

9B.2. Authors designating an epitype should state in what way the holotype, lectotype, neotype, or all original material is ambiguous such that epitypification is necessary.

Recommendation 9C

9C.1. Specification of the herbarium, collection, or institution of deposition should be followed by any available number permanently and unambiguously identifying the lectotype, neotype, or epitype specimen (see also Rec. 40A.5).

9C.2. Author(s) designating a type should notify the curator of the herbarium, collection, or institution to update the labelling of the specimen. This is intended to minimize the chances of superfluous nomenclatural acts.

ARTICLE 10
TYPIFICATION OF NAMES ABOVE THE RANK OF SPECIES

10.1. The type of a name of a genus or of any subdivision of a genus is the type of a name of a species (except as provided by Art. 10.4). For purposes of designation, indication, or citation of a type, the species name alone suffices, i.e. it is considered as the full equivalent of its type (see also Rec. 40A.2).

ⓘ **Note 1.** Terms such as "holotype", "syntype", "lectotype", and "neotype", as defined in Art. 9, apply only to the types of names at the rank of species or below. Although not applicable to the types of names at higher ranks, such terms have sometimes been so used by analogy (e.g. by citation of a "lectotype" for a generic name).

ⓘ **Note 2.** Because the type of a name of a genus or subdivision of a genus is the type of a name of a species, any change in the type of that species name (e.g. by lectotypification or conservation) also affects the application of the generic or subdivisional name.

10.2. If in the protologue of a name of a genus or of any subdivision of a genus the type(s) of one or more previously or simultaneously published species name(s) is definitely included (see Art. 10.3), the type must be chosen from among these types, unless:

(a) the type was indicated (Art. 10.8, 40.1–40.3) or designated by the author of the name (see also Art. 10.9); or
(b) the name was sanctioned (Art. F.3), in which case the type may also be chosen from among the types of species names included in the sanctioning treatment.

If no type of a previously or simultaneously published species name was definitely included:

(c) a type must be otherwise chosen, but the choice is to be superseded if it can be demonstrated that the selected type is not conspecific with any of the material associated with either the protologue or the sanctioning treatment.

Ex. 1. The genus *Anacyclus,* as originally circumscribed by Linnaeus (Sp. Pl.: 892. 1753), comprised three validly named species. Cassini (in Cuvier, Dict. Sci. Nat. 34: 104. 1825) designated *Anthemis valentina* L. (l.c.: 895. 1753) as type of *Anacyclus,* but this was not an original element of the genus. Green (in Sprague, Nom. Prop. Brit. Bot.: 182. 1929) designated *Anacyclus valentinus* L. (l.c.: 892. 1753), "the only one of the three original species still retained in the genus", as the "standard species" (see Art. 7 *Ex. 17), and her choice must be followed (Art. 10.5). Humphries (in Bull. Brit. Mus. (Nat. Hist.), Bot. 7: 109. 1979) designated a specimen in the Clifford Herbarium (BM) as lectotype of *Anacyclus valentinus,* and that specimen thereby became the type of *Anacyclus.*

Ex. 2. Warburg (in Engler, Pflanzenw. Ost-Afrikas Lief. 2/3(C): 179–180. 1895) provided separate descriptions for his new genus *Brochoneura* Warb. and new species *B. usambarensis* Warb. based on specimens from Tanzania, mentioning that three other species from Madagascar also belong to this genus. Subsequently, Warburg (in Nova Acta Acad. Caes. Leop.-Carol. German. Nat. Cur. 68: 128, 234. 1897) transferred three Malagasy species to *Brochoneura* (*B. acuminata* (Lam.) Warb., *B. madagascariensis* (Lam.) Warb., and *B. vouri* (Baill.) Warb.). Although Warburg (l.c. 1895) indicated that four species belonged to his new genus, the original type of *Brochoneura* is the type of *B. usambarensis* because this is the only validly published species name definitely included by Warburg in the protologue.

Ex. 3. (c) *Castanella* Spruce ex Benth. & Hook. f. (Gen. Pl. 1: 394. Aug 1862) was described based on a single specimen collected by Spruce and without mention of a species name. Swart (in ING Card No. 2143. 1957) was the first to designate a type (as "T."): *C. granatensis* Planch. & Linden (in Ann. Sci. Nat., Bot., ser. 4, 18: 365. Dec 1862), based on *Linden 1360.* As long as the Spruce specimen is considered to be conspecific with Linden's material, Swart's type designation cannot be superseded, even though the Spruce specimen became the type of *Paullinia paullinioides* Radlk. (Monogr. Paullinia: 173. 1896), because the latter is not a "previously or simultaneously published species name".

10.3. For the purpose of Art. 10.2, definite inclusion of the type of a name of a species is achieved by citation of, or reference (direct or indirect) to, a validly published species name, whether accepted or synonymized by the author, or by citation of the type of a previously or simultaneously published species name.

Ex. 4. The protologue of *Elodes* Adans. (Fam. Pl. 2: 444, 553. 1763) includes references to *"Elodes"* of Clusius (Alt. App. Rar. Pl. Hist., App. Alt. Auct.: [7]. 1611, i.e. *"Ascyrum supinum ἐλῶδης"*), *"Hypericum"* of Tournefort (Inst. Rei Herb. 1: 255. 1700, i.e. *"Hypericum palustre, supinum, tomentosum"*), and *Hypericum aegypticum* L. (Sp. Pl.: 784. 1753). The last is the only reference to a validly published species name, and neither of the other elements is the type of a species name. The type of *H. aegypticum* is therefore the type of the generic name *Elodes* even though subsequent authors designated *H. elodes* L. (Amoen. Acad.

4: 105. 1759) as the type (see Robson in Bull. Brit. Mus. (Nat. Hist.), Bot. 5: 305, 337. 1977).

ⓘ **Note 3.** For purposes of designating or conserving a type of a name of a genus or subdivision of a genus (Art. 10.2, 10.5–10.7, and 14.9) or for purposes of inclusion or exclusion of the type of a name of a genus or subdivision of a genus (Art. 22, 48, and 52), a species name validly published without a nomenclatural type (Art. 8.1) is treated as having a type (but see Art. 40.1).

> **Ex. 5.** In the protologue of *Decarinium* Raf. (Neogenyton: 1. 1825), *Croton glandulosus* L. was designated as the type. Until a specimen was selected as the lectotype of *C. glandulosus* (Fawcett & Rendle, Fl. Jamaica 4: 285. 1920), neither *Decarinium* nor *C. glandulosus* had a type. Klotzsch (in Arch. Naturgesch. 7: 254. 1841) included *C. glandulosus* in his circumscription of *Geiseleria* Klotzsch and, even though he did not mention *Decarinium*, this is to be treated for the purposes of Art. 52.1 as inclusion of the type of *Decarinium*. As a result, *Geiseleria* is illegitimate (see Art. 58 Ex. 4).

10.4. By and only by conservation (Art. 14.9), the type of a name of a genus may be a specimen or illustration, preferably used by the author in the preparation of the protologue, other than the type of a name of an included species.

ⓘ **Note 4.** If the element designated under Art. 10.4 is the type of a species name, that name may be cited as the type of the generic name. If the element is not the type of a species name, a parenthetical reference to the correct name of the type element may be added.

> **Ex. 6.** *Physconia* Poelt (in Nova Hedwigia 9: 30. 1965) was conserved with the specimen "'*Lichen pulverulentus*', Germania, Lipsia in *Tilia*, 1767, *Schreber* (M)" as the conserved type. That specimen is the type of *P. pulverulacea* Moberg (in Mycotaxon 8: 310. 1979), the name now cited in the type entry in App. III.

> **Ex. 7.** *Pseudolarix* Gordon (Pinetum: 292. 1858) was conserved with a specimen from the Gordon herbarium (K No. 3455 [barcode K000287582]) as its conserved type. Because this specimen is not the type of any species name, its accepted identity "[= *P. amabilis* (J. Nelson) Rehder...]" has been added to the corresponding entry in App. III.

10.5. The author who first designates (Art. 7.10, 7.11, and F.5.4) a type of a name of a genus or subdivision of a genus must be followed, but the choice may be superseded if the author used a largely mechanical method of selection (Art. 10.6). A type chosen using a largely mechanical method of selection is superseded by any later choice of a different type not made using such a method, unless, in the interval, the supersedable choice has been affirmed by its adoption in a publication that did not use a mechanical method of selection.

ⓘ **Note 5.** The effective date of a typification (cf. Art. 22.2, 48.2(c), and 52.2(c)) that could be superseded under Art. 10.5 remains that of the original selection, unless the type has been superseded.

10.6. For the purpose of Art. 10.5, "a largely mechanical method of selection" is defined as one in which the type is selected following a set of objective criteria such as those set out in "Canon 15" of the so-called "Philadelphia Code" (Arthur & al. in Bull. Torrey Bot. Club 31: 255–257. 1904) or in "Canon 15" of the *American Code of Botanical Nomenclature* (Arthur & al. in Bull. Torrey Bot. Club 34: 172–174. 1907).

> **Ex. 8.** The first type designation for *Delphinium* L. was by Britton (in Britton & Brown, Ill. Fl. N. U.S., ed. 2, 2: 93. 1913), who followed the *American Code* and whose selection of *D. consolida* L. is therefore considered to have been largely mechanical. His choice has been superseded under Art. 10.5 by the designation of *D. peregrinum* L. by Green (in Sprague, Nom. Prop. Brit. Bot.: 162. 1929).

10.7. Unless the author(s) specifically state that they are not using a mechanical method of type selection, the following criteria determine whether a particular publication, appearing before 1 January 1935, has adopted a largely mechanical method of type selection:

(a) Any statement to that effect, including that the *American Code* or the "Philadelphia Code" was being followed or that types were determined in a particular mechanical way (e.g. the first species in order).

(b) Adoption of any provision of the "Philadelphia Code" or the *American Code* that was contrary to the provisions of the *International Rules of Botanical Nomenclature* in force at that time, e.g. the inclusion of one or more tautonyms as species names.

Additionally, for publications appearing before 1 January 1921:

(c) An author of the publication was a signatory of the "Philadelphia Code"[1] (and was therefore also a signatory of the *American Code*).

(d) An author of the publication stated publicly (e.g. in another publication) that in the typification of generic names the "Philadelphia Code" or the *American Code* was followed.

---

1 A list of the 23 signatories of the "Philadelphia Code" was published in Taxon 65: 1448. 2016, as well as in Bull. Torrey Bot. Club 31: 250. 1904.

(e) An author of the publication was an employee or a recognized associate of the New York Botanical Garden.

(f) An author of the publication was an employee of the United States government.

Ex. 9. (a) Fink (in Contr. U. S. Natl. Herb. 14: 2. 1910) specified that he was "stating the types of the genera according to the 'first species' rule". His type designations may therefore be superseded under Art. 10.5. For example, Fink had designated *Biatorina griffithii* (Ach.) A. Massal. as the type of *Biatorina* A. Massal.; but his choice was superseded when the next subsequent designation, by Santesson (in Symb. Bot. Upsal. 12: 428. 1952), stated a different type, *B. atropurpurea* (Schaer.) A. Massal.

Ex. 10. (a) Underwood (in Mem. Torrey Bot. Club 6: 247–283. 1899) wrote (p. 251): "For each genus established the first named species will be regarded as the type". Therefore, his designation (p. 276) of *Caenopteris furcata* Bergius as type of *Caenopteris* Bergius (in Acta Acad. Sci. Imp. Petrop. 1782(2): 249. 1786) is supersedable; this was done by Copeland (Gen. Filicum: 166. 1947), who designated *C. rutifolia* Bergius as type.

Ex. 11. (a) Murrill (in J. Mycol. 9: 87. 1903), referring to generic types, wrote: "The principles by which I have been chiefly guided are also quite well known, having been stated and explained by Underwood" (see Art. 10 Ex. 10). Accordingly, Murrill (l.c.: 95, 98. 1903) listed the first-named species treated by Quélet (Enchir. Fung.: 175. 1886), *Coriolus lutescens* (Pers.) Quél., as type of *Coriolus* Quél. (l.c. 1886) and later (in Bull. Torrey Bot. Club 32: 640. 1906) listed *Polyporus zonatus* Nees as type because it was "the first species accompanied by a correct citation of a figure". Both lectotypifications are considered to be mechanical and were superseded by the choice of *Polyporus versicolor* (L.) Fr. by Donk (Revis. Niederl. Homobasidiomyc.: 180. 1933).

Ex. 12. (b) Britton & Wilson (Bot. Porto Rico 6: 262. 1925) designated *Cucurbita lagenaria* L. as type of *Cucurbita* L. (Sp. Pl.: 1010. 1753). However, because they were evidently following the *American Code* (they included many tautonyms in their publication, e.g. "*Abrus Abrus* (L.) W. Wight", "*Acisanthera Acisanthera* (L.) Britton", and "*Ananas Ananas* (L.) Voss"), their type selections used a largely mechanical method. Their selection of *C. lagenaria* (currently treated as *Lagenaria siceraria* (Molina) Standl.) has been superseded by the selection of *C. pepo* L. by Green (in Sprague, Nom. Prop. Brit. Bot.: 190. 1929).

Ex. 13. (d) In considering the typification of *Achyranthes* L. in a preliminary to his account of *Amaranthaceae* in the *North American Flora,* Paul C. Standley (in J. Wash. Acad. Sci. 5: 72. 1915) selected *A. repens* L. as type stating that "there seems, moreover, no doubt as to the type of the genus *Achyranthes* under the *American Code* of nomenclature", noting that, as a result, "the name *Achyranthes* must be used in a sense other than that in which it has generally been employed in recent years". As a result of this published statement of acceptance of the *American Code,* not only is Standley's selection of *A. repens* superseded by that of *A. aspera* L. by Hitchcock (in Sprague, Nom. Prop. Brit. Bot.: 135. 1929), but types cited in Standley's other publications (e.g. in Britton,

N. Amer. Fl. 21: 1–254. 1916–1918) are supersedable under Art. 10.5. Therefore, Standley's statement (l.c.: 134. 1917) that *A. repens* was the type of *Achyranthes* does not constitute affirmation of his earlier selection; similarly, his publication of type designations previously made by Britton & Brown, such as *Chenopodium rubrum* L. (l.c.: 9. 1916) and *Amaranthus caudatus* L. (l.c.: 102. 1917), does not constitute affirmation of their selection; the typification of *Chenopodium* L. has been superseded by the selection of *C. album* L. by Hitchcock (in Sprague, l.c.: 137. 1929) and that of *Amaranthus* L. was first affirmed by Green (in Sprague, l.c.: 188. 1929).

10.8. When the epithet in the name of a subdivision of a genus is identical with or derived from the epithet in one of the originally included species names, the type of the higher-ranking name is the same as that of the species name, unless the original author of the higher-ranking name designated another type.

Ex. 14. The type of *Euphorbia* subg. *Esula* Pers. (Syn. Pl. 2: 14. 1806) is the type of *E. esula* L., one of the species names originally included by Persoon; the designation of *E. peplus* L. (also included by Persoon) as type by Croizat (in Revista Sudamer. Bot. 6: 13. 1939) has no standing.

Ex. 15. The type of *Cassia* [unranked] *Chamaecrista* L. (Sp. Pl.: 379. 1753) is the type of *C. chamaecrista* L., nom. rej. (App. V), one of the five species names included by Linnaeus.

10.9. When the name of a new species is validly published solely by reference to a description or diagnosis of a genus under Art. 38.13, making that genus monotypic (see Art. 38.7), the type of the generic name is automatically the same as that of the species name.

Ex. 16. The generic name *Antirhea* Comm. ex Juss. (Gen. Pl.: 204. 1789) was published without any included species names. Because *A. borbonica* J. F. Gmel. (Syst. Nat. 2: 244. 1791) was validly published solely by reference to the description of *Antirhea,* making the genus monotypic, the type of *Antirhea* is automatically that of *A. borbonica.*

10.10. The type of a name of a family or of any subdivision of a family is automatically the same as that of the generic name from which it is formed (see Art. 18.1). For purposes of designation or citation of a type, the generic name alone suffices, i.e. it is considered as the full equivalent of its type. The type of a name of a family or subfamily not formed from a generic name is the same as that of the corresponding alternative name (Art. 18.5 and 19.8).

10.11. The principle of typification does not apply to names of taxa above the rank of family, except for names that are automatically typified by being formed from generic names (see Art. 16.1(a)), the type of which is the same as that of the generic name.

Recommendation 10A

10A.1. When a combination at the rank of a subdivision of a genus has been published under a generic name that has not yet been typified, the type of the generic name should be selected from the subdivision of the genus that was designated as nomenclaturally typical, if that is apparent.

10A.2. In citing a type chosen using a largely mechanical method of selection that has since been affirmed by an author not using such a method, both the place of original selection and that of affirmation should be cited, e.g. "*Quercus* L. ... Type: *Q. robur* L. designated by Britton & Brown (Ill. Fl. N. U.S., ed. 2, 1: 616. 1913); affirmed by Green (in Sprague, Nom. Prop. Brit. Bot.: 189. 1929)".

SECTION 3
PRIORITY AND STATUS OF NAMES

ARTICLE 11
PRIORITY OF NAMES

11.1. Each family or lower-ranked taxon with a particular circumscription, position, and rank can have only one correct name. Special exceptions are made for nine families and one subfamily for which alternative names are permitted (see Art. 18.5 and 19.8). The use of separate names is allowed for fossil-taxa that represent different parts, life-history stages, or preservational states of what may have been a single organismal taxon or even a single individual (Art. 1.2).

Ex. 1. The generic name *Sigillaria* Brongn. (in Bull. Sci. Soc. Philom. Paris 1822: 26. 1822) was established for fossils of "bark" fragments, but Brongniart (in Arch. Mus. Hist. Nat. 1: 405. 1839) subsequently included stems with preserved anatomy within his concept of *Sigillaria*. Cones with preserved anatomy that may in part represent the same biological taxon are referred to as *Mazocarpon* M. J. Benson (in Ann. Bot. (Oxford) 32: 569. 1918), whereas such cones preserved as adpressions are known as *Sigillariostrobus* Schimp. (Traité Paléont. Vég. 2: 105. 1870). All these generic names can be used concurrently even though they could, at least in part, apply to the same organism.

11.2. A name has no priority outside the rank at which it is published (but see Art. 53.3).

> *Ex. 2.* When *Campanula* sect. *Campanopsis* R. Br. (Prodr.: 561. 1810) is treated as a genus, it is called *Wahlenbergia* Roth (Nov. Pl. Sp.: 399. 1821), a name conserved against the heterotypic (taxonomic) synonym *Cervicina* Delile (Descr. Egypte, Hist. Nat.: 150. 1813), and not *Campanopsis* (R. Br.) Kuntze (Revis. Gen. Pl. 2: 378. 1891).

> *Ex. 3.* *Solanum* subg. *Leptostemonum* Bitter (in Bot. Jahrb. Syst. 55: 69. 1919) is the correct name of the subgenus that includes its type, *S. mammosum* L., because it is the earliest available name at that rank. The homotypic *S.* sect. *Acanthophora* Dunal (Hist. Nat. Solanum: 131, 218. 1813), the inclusion of which caused the illegitimacy of *S.* sect. *Leptostemonum* Dunal (in Candolle, Prodr. 13(1): 29, 183. 1852), has no priority outside its own rank.

> *Ex. 4.* *Helichrysum stoechas* subsp. *barrelieri* (Ten.) Nyman (Consp. Fl. Eur.: 381. 1879) when treated at specific rank is called *H. conglobatum* (Viv.) Steud. (Nomencl. Bot., ed. 2, 1: 738. 1840), based on *Gnaphalium conglobatum* Viv. (Fl. Libyc. Spec.: 55. 1824), and not *H. barrelieri* (Ten.) Greuter (in Boissiera 13: 138. 1967), based on *G. barrelieri* Ten. (Fl. Napol. 5: 220. 1835–1838).

> *Ex. 5.* *Magnolia virginiana* var. *foetida* L. (Sp. Pl.: 536. 1753) when treated at specific rank is called *M. grandiflora* L. (Syst. Nat., ed. 10: 1082. 1759), not *M. foetida* (L.) Sarg. (in Gard. & Forest 2: 615. 1889).

ⓘ **Note 1.** The provisions of Art. 11 determine priority between different names applicable to the same taxon; they do not concern homonymy.

11.3. For any taxon from family to genus, inclusive, the correct name is the earliest legitimate one with the same rank, except in cases of limitation of priority by conservation or protection (see Art. 14 and F.2) or where Art. 11.7, 11.8, 19.4, 19.5, 56, 57, F.3, or F.7 apply.

> *Ex. 6.* When *Aesculus* L. (Sp. Pl.: 344. 1753), *Pavia* Mill. (Gard. Dict. Abr., ed. 4: *Pavia.* 1754), *Macrothyrsus* Spach (in Ann. Sci. Nat., Bot., ser. 2, 2: 61. 1834), and *Calothyrsus* Spach (l.c.: 62. 1834) are referred to a single genus, its correct name is *Aesculus*.

11.4. For any taxon below the rank of genus, the correct name is the combination of the final epithet of the earliest legitimate name of the taxon at the same rank, with the correct name of the genus or species to which it is assigned, except:

(a) in cases of limitation of priority under Art. 14, 56, 57, F.2, F.3, or F.7; or

(b) if Art. 11.7, 11.8, 22.1, or 26.1 rules that a different combination be used; or *(c)* if the resulting combination could not be validly published under Art. 32.1(c) or would be illegitimate under Art. 53.

If (c) applies, the final epithet of the next earliest legitimate name at the same rank is to be used instead or, if there is no final epithet available, a replacement name or the name of a new taxon may be published.

Ex. 7. When *Aeginetia acaulis* (Roxb.) Walp. (in Repert. Bot. Syst. 3: 481. 1844) and *A. pedunculata* Wall. (Pl. Asiat. Rar. 3: 13. 1831) are considered to apply to the same species, *A. acaulis* is the correct name because it is the combination of the final epithet of *Orobanche acaulis* Roxb. (Pl. Coromandel 3: 89. 1820), the earliest legitimate name of the taxon at specific rank, with *Aeginetia* L., the correct name of the genus to which the species is assigned.

Ex. 8. *Primula* sect. *Dionysiopsis* Pax (in Jahresber. Schles. Ges. Vaterl. Cult. 87(IIb): 20. 1909) when transferred to *Dionysia* Fenzl becomes *D.* sect. *Dionysiopsis* (Pax) Melch. (in Mitt. Thüring. Bot. Vereins 50: 164–168. 1943); the replacement name *D.* sect. *Ariadna* Wendelbo (in Bot. Not. 112: 496. 1959) is illegitimate under Art. 52.1.

Ex. 9. *Antirrhinum spurium* L. (Sp. Pl.: 613. 1753) when transferred to *Linaria* Mill. is called *L. spuria* (L.) Mill. (Gard. Dict., ed. 8: *Linaria* No. 15. 1768).

Ex. 10. When transferring *Serratula chamaepeuce* L. (Sp. Pl.: 819. 1753) to *Ptilostemon* Cass., Cassini illegitimately (Art. 52.1) named the species *P. muticus* Cass. (in Cuvier, Dict. Sci. Nat. 44: 59. 1826). In *Ptilostemon,* the correct name is *P. chamaepeuce* (L.) Less. (Gen. Cynaroceph.: 5. 1832).

Ex. 11. The correct name for *Rubus aculeatiflorus* var. *taitoensis* (Hayata) T. S. Liu & T. Y. Yang (in Annual Taiwan Prov. Mus. 12: 12. 1969) is *R. taitoensis* Hayata var. *taitoensis* because *R. taitoensis* Hayata (in J. Coll. Sci. Imp. Univ. Tokyo 30(1): 96. 1911) has priority over *R. aculeatiflorus* Hayata (Icon. Pl. Formosan. 5: 39. 1915).

Ex. 12. When transferring *Spartium biflorum* Desf. (Fl. Atlant. 2: 133. 1798) to *Cytisus* Desf., Ball correctly proposed the replacement name *C. fontanesii* Spach ex Ball (in J. Linn. Soc., Bot. 16: 405. 1878) because of the previously and validly published *C. biflorus* L'Hér. (Stirp. Nov.: 184. 1791); the combination *C. biflorus* based on *S. biflorum* would be illegitimate under Art. 53.1.

Ex. 13. *Spergula stricta* Sw. (in Kongl. Vetensk. Acad. Nya Handl. 20: 235. 1799) when transferred to *Arenaria* L. is called *A. uliginosa* Schleich. ex Lam. & DC. (Fl. Franç., ed. 3, 4: 786. 1805) because of the existence of the name *A. stricta* Michx. (Fl. Bor.-Amer. 1: 274. 1803), based on a different type; but on further transfer to the genus *Minuartia* L. the epithet *stricta* is again available and the species is called *M. stricta* (Sw.) Hiern (in J. Bot. 37: 320. 1899).

Ex. 14. *Arum dracunculus* L. (Sp. Pl.: 964. 1753) when transferred to *Dracunculus* Mill. is named *D. vulgaris* Schott (Melet. Bot. 1: 17. 1832). The use of the

Linnaean epithet in *Dracunculus* would result in a tautonym (Art. 23.4), which would not be validly published (Art. 32.1(c)).

Ex. 15. *Cucubalus behen* L. (Sp. Pl.: 414. 1753) when transferred to *Behen* Moench was legitimately renamed *B. vulgaris* Moench (Methodus: 709. 1794) to avoid the tautonym *"B. behen"*. In *Silene* L., the epithet *behen* is unavailable because of the existence of *S. behen* L. (l.c.: 418. 1753). Therefore, the replacement name *S. cucubalus* Wibel (Prim. Fl. Werth.: 241. 1799) was proposed. This, however, is illegitimate (Art. 52.1) because the specific epithet *vulgaris* was available. In *Silene,* the correct name of the species is *S. vulgaris* (Moench) Garcke (Fl. N. Mitt.-Deutschland, ed. 9: 64. 1869).

Ex. 16. *Helianthemum italicum* var. *micranthum* Gren. & Godr. (Fl. France 1: 171. 1847) when transferred as a variety to *H. penicillatum* Thibaud ex Dunal retains its varietal epithet and is named *H. penicillatum* var. *micranthum* (Gren. & Godr.) Grosser (in Engler, Pflanzenr. IV. 193 (Heft 14): 115. 1903).

Ex. 17. The final epithet in the combination *Thymus praecox* subsp. *arcticus* (Durand) Jalas (in Veröff. Geobot. Inst. ETH Stiftung Rübel Zürich 43: 190. 1970), based on *T. serpyllum* var. *arcticus* Durand (Pl. Kaneanae Groenl.: 196. 1856), was first used at the rank of subspecies in the combination *T. serpyllum* subsp. *arcticus* (Durand) Hyl. (in Uppsala Univ. Årsskr. 1945(7): 276. 1945). But if *T. britannicus* Ronniger (in Repert. Spec. Nov. Regni Veg. 20: 330. 1924) is included in *T. praecox,* the correct name at subspecific rank is *T. praecox* subsp. *britannicus* (Ronniger) Holub (in Preslia 45: 359. 1973), for which the final epithet was first used at this rank in the combination *T. serpyllum* subsp. *britannicus* (Ronniger) Litard. (in Arch. Bot. Mém. 2: 6. 1928).

Ex. 18. Transfer of *Polypodium tenerum* Roxb. (in Calcutta J. Nat. Hist. 4: 490. 1844) to *Cyclosorus* Link (Hort. Berol. 2: 128. 1833) would result in a later homonym due to the existence of *C. tener* (Fée) Christenh. (in Bot. J. Linn. Soc. 161: 250. 2009), based on *Goniopteris tenera* Fée (Mém. Foug. 11: 60. 1866). The correct name is a heterotypic synonym, *C. ciliatus* (Wall. ex Benth.) Panigrahi (in Res. J. Pl. Environm. 9: 66. 1993), based on the next earliest legitimate name of the taxon at the same rank, *Aspidium ciliatum* Wall. ex Benth. (Fl. Hongkong.: 455. 1861).

🛈 **Note 2.** The valid publication of a name at a rank lower than genus precludes any simultaneous homonymous combination (Art. 53), regardless of the priority of other names with the same final epithet that may require transfer to the same genus or species.

Ex. 19. Tausch included two species in his new genus *Alkanna: A. tinctoria* Tausch (in Flora 7: 234. 1824), a new species based on *"Anchusa tinctoria"* in the sense of Linnaeus (Sp. Pl., ed. 2: 192. 1762), and *A. matthioli* Tausch (l.c.: 235. 1824), a replacement name based on *Lithospermum tinctorium* L. (Sp. Pl.: 132. 1753). Both names are legitimate and take priority from 1824.

Ex. 20. Raymond-Hamet transferred to the genus *Sedum* both *Cotyledon sedoides* DC. (in Mém. Agric. Econ. Soc. Agric. Seine 11: 11. 1808) and *Sempervivum sedoides* Decne. (in Jacquemont, Voy. Inde 4(Bot.): 63. 1844). He combined the epithet of the later name, *Sempervivum sedoides,* under *Sedum,* as *S. sedoides* (Decne.) Raym.-Hamet (in Candollea 4: 26. 1929), and published a replacement name, *S. candollei* Raym.-Hamet (l.c. 1929), for the earlier name. Both of Raymond-Hamet's names are legitimate.

11.5. When, for any taxon at the rank of family or below, a choice is possible between legitimate names of equal priority at the same rank, or between available final epithets of names of equal priority at the same rank, the first such choice to be effectively published (Art. 29–31) establishes the priority of the chosen name, and of any legitimate combination with the same type and final epithet at that rank, over the other competing name(s) (but see Art. 11.6; see also Rec. F.5A.2).

ⓘ *Note 3.* A choice as provided for in Art. 11.5 is made by adopting one of the competing names, or its final epithet in the required combination, and simultaneously rejecting or relegating to synonymy the other(s) or their homotypic (nomenclatural) synonyms.

Ex. 21. When *Dentaria* L. (Sp. Pl.: 653. 1753) and *Cardamine* L. (l.c.: 654. 1753) are united, the resulting genus is called *Cardamine* because that name was chosen by Crantz (Cl. Crucif. Emend.: 126. 1769), who first united them.

Ex. 22. When *Claudopus* Gillet (Hyménomycètes: 426. 1876), *Eccilia* (Fr.) P. Kumm. (Führer Pilzk.: 23. 1871), *Entoloma* (Fr. ex Rabenh.) P. Kumm. (l.c.: 23. 1871), *Leptonia* (Fr.) P. Kumm. (l.c.: 24. 1871), and *Nolanea* (Fr.) P. Kumm. (l.c.: 24. 1871) are united, one of the four generic names simultaneously published by Kummer must be used for the combined genus. Donk (in Bull. Jard. Bot. Buitenzorg, ser. 3, 18(1): 157. 1949) selected *Entoloma,* which is therefore treated as having priority over the other names.

Ex. 23. Brown (in Tuckey, Narr. Exped. Zaire: 484. 1818) was the first to unite *Waltheria americana* L. (Sp. Pl.: 673. 1753) and *W. indica* L. (l.c. 1753). He adopted the name *W. indica* for the combined species, and this name is accordingly treated as having priority over *W. americana.*

Ex. 24. Baillon (in Adansonia 3: 162. 1863), when uniting for the first time *Sclerocroton integerrimus* Hochst. (in Flora 28: 85. 1845) and *S. reticulatus* Hochst. (l.c. 1845), adopted the name *Stillingia integerrima* (Hochst.) Baill. for the combined taxon. Consequently, *Sclerocroton integerrimus* is treated as having priority over *S. reticulatus* regardless of the genus (*Sclerocroton, Stillingia,* or any other) to which the species is assigned.

Ex. 25. Linnaeus (Sp. Pl.: 902. 1753) simultaneously published the names *Verbesina alba* and *V. prostrata.* Later (Mant. Pl.: 286. 1771), he published *Eclipta erecta,* an illegitimate name because *V. alba* was cited in synonymy, and *E. prostrata,* based on *V. prostrata.* The first author to unite these taxa was Roxburgh (Fl. Ind., ed. 1832, 3: 438. 1832), who adopted the name *E. prostrata* (L.) L. Therefore, *V. prostrata* is treated as having priority over *V. alba.*

Ex. 26. *Donia speciosa* and *D. formosa,* which were simultaneously published by Don (Gen. Hist. 2: 468. 1832), were illegitimately renamed *Clianthus oxleyi* and *C. dampieri,* respectively, by Lindley (in Trans. Hort. Soc. London, ser. 2, 1: 522. 1835). Brown (in Sturt, Narr. Exped. C. Australia 2: 71. 1849) united both in a single species, adopting the illegitimate name *C. dampieri* and citing *D. speciosa* and *C. oxleyi* as synonyms. His choice is not of the kind provided for by Art. 11.5. *Clianthus speciosus* (G. Don) Asch. & Graebn. (Syn. Mitteleur. Fl. 6(2): 725. 1909), published with *D. speciosa* and *C. dampieri* listed as synonyms, is an illegitimate later homonym of *C. speciosus* (Endl.) Steud. (Nomencl. Bot., ed. 2, 1: 384. 1840). Again, conditions for a choice under Art. 11.5 were not satisfied. Ford & Vickery (in Contr. New South Wales Natl. Herb. 1: 303. 1950) published the legitimate combination *C. formosus* (G. Don) Ford & Vickery and cited *D. formosa* and *D. speciosa* as synonyms, but because the epithet of the latter was unavailable in *Clianthus* Sol. ex Lindl. a choice was not possible and again Art. 11.5 does not apply. The first acceptable choice was by Thompson (in Telopea 4: 4. 1990), who published the combination *Swainsona formosa* (G. Don) Joy Thomps. and indicated that *D. speciosa* was a synonym of it.

11.6. An autonym is treated as having priority over the name(s) of the same date and rank that upon their valid publication established the autonym (see Art. 22.3 and 26.3).

ⓘ *Note 4.* When the final epithet of an autonym is used in a new combination under the requirements of Art. 11.6, the basionym of that combination is the name from which the autonym is derived, or its basionym if it has one.

Ex. 27. The publication of *Synthyris* subg. *Plagiocarpus* Pennell (in Proc. Acad. Nat. Sci. Philadelphia 85: 86. 1933) simultaneously established the autonym *Synthyris* Benth. (in Candolle, Prodr. 10: 454. 1846) subg. *Synthyris.* If *Synthyris,* including *S.* subg. *Plagiocarpus,* is recognized as a subgenus of *Veronica* L. (Sp. Pl.: 9. 1753), the correct name is *V.* subg. *Synthyris* (Benth.) M. M. Mart. Ort. & al. (in Taxon 53: 440. 2004), which has precedence over a combination in *Veronica* based on *S.* subg. *Plagiocarpus.*

Ex. 28. *Heracleum sibiricum* L. (Sp. Pl.: 249. 1753) includes *H. sibiricum* subsp. *lecokii* (Godr. & Gren.) Nyman (Consp. Fl. Eur.: 290. 1879) and *H. sibiricum* L. subsp. *sibiricum* automatically established at the same time. When *H. sibiricum,* so circumscribed, is included in *H. sphondylium* L. (l.c. 1753) as a single subspecies, the correct name of that subspecies is *H. sphondylium* subsp. *sibiricum* (L.) Simonk. (Enum. Fl. Transsilv.: 266. 1887), not *"H. sphondylium* subsp. *lecokii".*

Ex. 29. The publication of *Salix tristis* var. *microphylla* Andersson (Salices Bor.-Amer.: 21. 1858) simultaneously established the autonym *S. tristis* Aiton (Hort. Kew. 3: 393. 1789) var. *tristis.* If *S. tristis,* including *S. tristis* var. *microphylla,* is recognized as a variety of *S. humilis* Marshall (Arbust. Amer.: 140. 1785), the correct name is *S. humilis* var. *tristis* (Aiton) Griggs (in Proc. Ohio Acad. Sci. 4: 301. 1905). However, if both these varieties of *S. tristis* are recognized as varieties of *S. humilis,* then the names *S. humilis* var. *tristis* and *S. humilis* var. *microphylla* (Andersson) Fernald (in Rhodora 48: 46. 1946) are used.

11.7. When the names of a non-fossil taxon and a fossil-taxon (diatoms excepted) of the same rank are treated as synonyms, the correct name of the non-fossil taxon must be accepted, even if it is antedated by that of the fossil-taxon.

Ex. 30. If *Platycarya* Siebold & Zucc. (in Abh. Math.-Phys. Cl. Königl. Bayer. Akad. Wiss. 3: 741. 1843), based on a non-fossil type, and *Petrophiloides* Bowerb. (Hist. Fruits London Clay: 43. 1840), based on a fossil type, are treated as heterotypic synonyms for a non-fossil genus, the name *Platycarya* is correct even though it is antedated by *Petrophiloides*.

Ex. 31. The generic name *Metasequoia* Miki (in Jap. J. Bot. 11: 261. 1941) was based on the fossil type of *M. disticha* (Heer) Miki. After the discovery of the non-fossil species *M. glyptostroboides* Hu & W. C. Cheng, conservation of *Metasequoia* Hu & W. C. Cheng (in Bull. Fan Mem. Inst. Biol., Bot., ser. 2, 1: 154. 1948) as based on the non-fossil type was approved. Without conservation, any new generic name based on *M. glyptostroboides* would have been treated as having priority over *Metasequoia* Miki.

Ex. 32. *Hyalodiscus* Ehrenb. (in Ber. Bekanntm. Verh. Königl. Preuss. Akad. Wiss. Berlin 1845: 71. 1845), based on the fossil type of *H. laevis* Ehrenb. (l.c.: 78. 1845), is the name of a diatom genus that includes non-fossil species. If later synonymous generic names based on a non-fossil type exist, they are not treated as having priority over *Hyalodiscus* because Art. 11.7 excepts diatoms.

Ex. 33. Boalch & Guy-Ohlson (in Taxon 41: 529–531. 1992) synonymized the two non-diatom algal generic names *Pachysphaera* Ostenf. (in Knudsen & Ostenfeld, Iagtt. Overfladevand. Temp. Salth. Plankt. 1898: 52. 1899) and *Tasmanites* E. J. Newton (in Geol. Mag. 12: 341. 1875). *Pachysphaera* is based on a non-fossil type and *Tasmanites* on a fossil type. Under the *Code* in effect in 1992, *Tasmanites* had priority and was therefore adopted. Under the current Art. 11.7, which excepts only diatoms and not algae in general, *Pachysphaera* is the correct name for a non-fossil genus to which both of these names are applied as heterotypic synonyms.

11.8. Dual nomenclature in fossil-taxa (diatoms excepted) accommodates taxonomic equivalence between a fossil-taxon and a morphologically similar or identical part or life-history stage of a non-fossil taxon at the same rank, when the names of these two taxa are not considered to be synonyms.

Ex. 34. The name *Polysphaeridium zoharyi* (M. Rossignol) J. P. Bujak & al. (in Special Pap. Palaeontol. 24: 34. 1980), based on *Hystrichosphaeridium zoharyi* M. Rossignol (in Pollen & Spores 4: 132. 1962), may be retained under dual nomenclature for a fossil-species of dinoflagellate cyst even though apparently morphologically identical resting cysts form part of the life cycle of the non-fossil species *Pyrodinium bahamense* L. Plate (in Arch. Protistenk. 7: 427. 1906).

Ex. 35. The fossil dinoflagellate *Votadinium spinosum* P. C. Reid (in Nova Hedwigia 29: 445. 1977) was considered by Reid to represent the resting cyst of the non-fossil dinoflagellate *Peridinium claudicans* Paulsen (in Meddel. Kommiss. Havundersøgelser, Serie: Plankton 1(5): 16. 1907). *Votadinium spinosum* can be used as the equivalent correct name for the fossil-species given that Reid did not explicitly consider it a synonym of *P. claudicans*.

11.9. For purposes of priority, names given to hybrids are subject to the same rules as are those of non-hybrid taxa at equivalent rank (but see Art. H.8).

Ex. 36. The name ×*Solidaster* H. R. Wehrh. (in Bonstedt, Pareys Blumengärtn. 2: 525. 1932) has priority over ×*Asterago* Everett (in Gard. Chron., ser. 3, 101: 6. 1937) for the hybrids between *Aster* L. and *Solidago* L.

Ex. 37. *Anemone* ×*hybrida* Paxton (in Paxton's Mag. Bot. 15: 239. 1849) has priority over *A.* ×*elegans* Decne. (pro sp.) (in Rev. Hort. (Paris) 1852: 41. 1852). The former is correct when both are considered to apply to the same hybrid, *A. hupehensis* (Lemoine & É. Lemoine) Lemoine & É. Lemoine × *A. vitifolia* Buch.-Ham. ex DC. (Art. H.4.1).

Ex. 38. Aimée Camus (in Bull. Mus. Natl. Hist. Nat. 33: 538. 1927) published the name ×*Agroelymus* E. G. Camus ex A. Camus without a description or diagnosis, mentioning only the names of the parental genera (*Agropyron* Gaertn. and *Elymus* L.). Because this name was not validly published under the *Code* then in force, Rousseau (in Mém. Jard. Bot. Montréal 29: 10–11. 1952) published a Latin diagnosis. However, under the present *Code* (Art. H.9), the date of valid publication of ×*Agroelymus* is 1927, not 1952, and therefore it has priority over the name ×*Elymopyrum* Cugnac (in Bull. Soc. Hist. Nat. Ardennes 33: 14. 1938).

11.10. The principle of priority does not apply above the rank of family (but see Rec. 16A.1).

ARTICLE 12
STATUS OF NAMES

12.1. A name of a taxon has no status under this *Code* unless it is validly published (see Art. 6.3; but see Art. 14.9 and 14.14).

SECTION 4
LIMITATION OF PRIORITY

ARTICLE 13
NOMENCLATURAL STARTING-POINTS

13.1. Valid publication of names for organisms of different groups is treated as beginning at the following dates (for each group a work is mentioned that is treated as having been published on the date given for that group):

Non-fossil organisms:

(a) SPERMATOPHYTA and PTERIDOPHYTA, names at ranks of genus and below, 1 May 1753 (Linnaeus, *Species plantarum,* ed. 1); suprageneric names, 4 August 1789 (Jussieu, *Genera plantarum*).

(b) MUSCI (except *Sphagnaceae*), 1 January 1801 (Hedwig, *Species muscorum frondosorum*).

(c) SPHAGNACEAE and HEPATICAE (including *Anthocerotae*), names at ranks of genus and below, 1 May 1753 (Linnaeus, *Species plantarum,* ed. 1); suprageneric names, 4 August 1789 (Jussieu, *Genera plantarum*).

(d) FUNGI (Pre. 8), see Art. F.1.1.

(e) ALGAE, 1 May 1753 (Linnaeus, *Species plantarum,* ed. 1). Exceptions:

NOSTOCACEAE HOMOCYSTEAE, 1 January 1892 (Gomont, "Monographie des Oscillariées", in Ann. Sci. Nat., Bot., ser. 7, 15: 263–368; 16: 91–264). The two parts of Gomont's "Monographie", which appeared in 1892 and 1893, respectively, are treated as having been published simultaneously on 1 January 1892.

NOSTOCACEAE HETEROCYSTEAE, 1 January 1886 (Bornet & Flahault, "Révision des Nostocacées hétérocystées", in Ann. Sci. Nat., Bot., ser. 7, 3: 323–381; 4: 343–373; 5: 51–129; 7: 177–262). The four parts of the "Révision", which appeared in 1886, 1886, 1887, and 1888, respectively, are treated as having been published simultaneously on 1 January 1886.

DESMIDIACEAE (s. l.), 1 January 1848 (Ralfs, *British* Desmidieae).

OEDOGONIACEAE, 1 January 1900 (Hirn, "Monographie und Iconographie der Oedogoniaceen", in Acta Soc. Sci. Fenn. 27(1)).

Fossil organisms (diatoms excepted):

(f) ALL GROUPS, 31 December 1820 (Sternberg, *Flora der Vorwelt, Versuch* 1: 1–24, t. 1–13).

13.2. The group to which a name is assigned for the purpose of Art. 13.1 and F.1 is determined by the accepted taxonomic position of the type of the name.

> *Ex. 1.* The genus *Porella* L. and the single species that was described under it, *P. pinnata* L., were placed by Linnaeus (Sp. Pl.: 1106. 1753) in the *Musci.* Because the type of the name *P. pinnata* is now accepted as belonging to the *Hepaticae,* the two names were validly published in 1753.

> *Ex. 2.* The type of *Lycopodium* L. (Sp. Pl.: 1100. 1753) is *L. clavatum* L. (l.c.: 1101. 1753). Even though Linnaeus placed *Lycopodium* in the *Musci,* because the type of the name *L. clavatum* is now accepted as a pteridophyte s. l. (lycophytes), the generic name and names of the pteridophyte species treated under it were validly published in 1753.

13.3. For nomenclatural purposes, a name is treated as pertaining to a non-fossil taxon unless its type is fossil in origin (Art. 1.2). Fossil material is distinguished from non-fossil material by stratigraphic relations at the site of original occurrence. In cases of doubtful stratigraphic relations, and for all diatoms, provisions for non-fossil taxa apply.

> *Ex. 3.* The holotype of *Echinidinium granulatum* K. A. F. Zonn. ex M. J. Head & al. (in J. Quatern. Sci. 16: 633. 2001) does not have a stratigraphic context because it was collected from a sediment trap suspended within the water column. It is therefore treated as a non-fossil dinoflagellate. The name was not validly published when Zonneveld (in Rev. Palaeobot. Palynol. 97: 325. 1997) provided only an English diagnosis because at that date Latin was required for non-fossil algae (Art. 44.1). The Latin diagnosis provided by Head & al. (l.c. 2001) validated the name.

> *Ex. 4.* The holotype of *Algidasphaeridium spongium* K. A. F. Zonn. (in Rev. Palaeobot. Palynol. 97: 325. 1997) had stratigraphic context because it was collected from surface (upper centimetre) sediments of the Arabian Sea. It is therefore treated as a fossil dinoflagellate. Because an English diagnosis was provided and this is a fossil-taxon, the name was validly published (Art. 43.1).

13.4. Generic names that appear in Linnaeus's *Species plantarum,* ed. 1 (1753) and ed. 2 (1762–1763), are associated with the first subsequent description given under those names in Linnaeus's *Genera plantarum,* ed. 5 (1754) and ed. 6 (1764). The spelling of the generic names included in *Species plantarum,* ed. 1, is not to be altered because a different spelling has been used in *Genera plantarum,* ed. 5.

ⓘ **Note 1.** The two volumes of Linnaeus's *Species plantarum,* ed. 1 (1753), which appeared in May and August, 1753, respectively, are treated as having been published simultaneously on 1 May 1753 (Art. 13.1).

> ***Ex. 5.*** The generic names *Thea* L. (Sp. Pl.: 515. 24 May 1753; Gen. Pl., ed. 5: 232. 1754) and *Camellia* L. (Sp. Pl.: 698. 16 Aug 1753; Gen. Pl., ed. 5: 311. 1754) are treated as having been published simultaneously on 1 May 1753. Under Art. 11.5, the combined genus has the name *Camellia* because Sweet (Hort. Suburb. Lond.: 157. 1818), who was the first to unite the two genera, chose that name and cited *Thea* as a synonym.

> ***Ex. 6.*** *Sideroxylon* L. (Sp. Pl.: 192. 1753) is not to be altered because Linnaeus spelled it *'Sideroxylum'* in *Genera plantarum,* ed. 5 (p. 89. 1754); usage of *Brunfelsia* L. (Sp. Pl.: 191. 1753, orth. cons., *'Brunsfelsia'*), which Linnaeus adopted in 1754, has been made possible only through conservation (see App. III).

ARTICLE 14
CONSERVED NAMES

14.1. In order to avoid disadvantageous nomenclatural changes entailed by the strict application of the rules, and especially of the principle of priority in starting from the dates given in Art. 13 and F.1, this *Code* provides, in App. II–IV, lists of names of families, genera, and species that are conserved (nomina conservanda; see Rec. 50E.1). Conserved names are legitimate even though initially they may have been illegitimate. The name of a subdivision of a genus or of an infraspecific taxon may be conserved with a conserved type and listed in App. III and IV, respectively, when it is the basionym or replaced synonym of a name of a genus or species that could not continue to be used in its current sense without conservation (see also Art. 19.5 for names of subdivisions of families).

14.2. Conservation aims at retention of those names that best serve stability of nomenclature.

14.3. The application of both conserved and rejected names is determined by nomenclatural types. The type of the species name cited as the type of a conserved generic name may, if desirable, be conserved and listed in App. IV. Application of conserved and rejected names of nothogenera is determined by a statement of parentage (Art. H.9.1).

14.4. A conserved name of a family or genus is conserved against all other names at the same rank with the same type (homotypic, i.e. nomenclatural, synonyms, which are to be rejected) whether or not these are cited in the corresponding list as rejected names, and against those names with different types (heterotypic, i.e. taxonomic, synonyms) that are listed as rejected.[1] A conserved name of a species is conserved against all names listed as rejected, and against all new combinations based on the rejected names.

ⓘ **Note 1.** Except as by Art. 14.14 (see also Art. 14.9(b)), the *Code* does not provide for conservation of a name against itself, i.e. against an isonym (Art. 6 Note 2). Only the earliest known isonyms are listed in App. IIA, III, and IV.

ⓘ **Note 2.** A species name listed as conserved or rejected in App. IV may have been published as the name of a new taxon, or as a combination based on an earlier name. Rejection of a name based on an earlier name does not by itself preclude the use of the earlier name because that name is not "a combination based on a rejected name" (Art. 14.4).

Ex. 1. Rejection of *Lycopersicon lycopersicum* (L.) H. Karst. (Deut. Fl.: 966. 1882) in favour of *L. esculentum* Mill. (Gard. Dict., ed. 8: *Lycopersicon* No. 2. 1768) does not preclude the use of the homotypic *Solanum lycopersicum* L. (Sp. Pl.: 185. 1753).

14.5. When a conserved name competes with one or more names based on different types and against which it is not explicitly conserved, the earliest of the competing names is adopted in accordance with Art. 11, except for the conserved family names listed in App. IIB, which are conserved against unlisted names.

Ex. 2. If *Mahonia* Nutt. (Gen. N. Amer. Pl. 1: 211. 1818) is united with *Berberis* L. (Sp. Pl.: 330. 1753), the earlier name *Berberis* must be adopted for the combined genus, although *Mahonia* is conserved and *Berberis* is not.

Ex. 3. *Nasturtium* W. T. Aiton (Hort. Kew., ed. 2, 4: 109. 1812) was conserved only against the homonym *Nasturtium* Mill. (Gard. Dict. Abr., ed. 4: *Nasturtium*. 1754) and the homotypic (nomenclatural) synonym *Cardaminum* Moench (Methodus:

---

1 The *International Code of Zoological Nomenclature* uses the terms "objective synonym" and "subjective synonym" for homotypic synonym and heterotypic synonym, respectively.

262. 1794). Consequently, if *Nasturtium* is reunited with *Rorippa* Scop. (Fl. Carniol.: 520. 1760), the name *Rorippa* must be adopted for the combined genus.

Ex. 4. *Combretaceae* R. Br. (Prodr.: 351. 1810) is conserved against the unlisted earlier heterotypic name *Terminaliaceae* J. St.-Hil. (Expos. Fam. Nat. 1: 178. 1805).

14.6. When a name of a taxon has been conserved against an earlier heterotypic synonym, the latter is to be restored, subject to Art. 11, if it is considered to be the name of a taxon at the same rank distinct from that of the conserved name.

Ex. 5. The generic name *Luzuriaga* Ruiz & Pav. (Fl. Peruv. 3: 65. 1802) is conserved against the earlier names *Enargea* Banks ex Gaertn. (Fruct. Sem. Pl. 1: 283. 1788) and *Callixene* Comm. ex Juss. (Gen. Pl.: 41. 1789). If, however, *Enargea* is considered to be a separate genus, the name *Enargea* is retained for it.

Ex. 6. To preserve the name *Roystonea regia* (Kunth) O. F. Cook (in Science, n.s., 12: 479. 1900), its basionym *Oreodoxa regia* Kunth (in Humboldt & al., Nov. Gen. Sp. 1, ed. qu.: 305; ed. fol.: 244. 1816) is conserved against *Palma elata* W. Bartram (Travels Carolina: iv, 115–116. 1791). However, the name *R. elata* (W. Bartram) F. Harper (in Proc. Biol. Soc. Wash. 59: 29. 1946) can be used for a species distinct from *R. regia*.

14.7. A rejected name, or a combination based on a rejected name, may not be restored for a taxon that includes the type of the corresponding conserved name.

Ex. 7. *Enallagma* (Miers) Baill. (Hist. Pl. 10: 54. 1888) is conserved against *Dendrosicus* Raf. (Sylva Tellur.: 80. 1838), but not against *Amphitecna* Miers (in Trans. Linn. Soc. London 26: 163. 1868); if *Enallagma, Dendrosicus,* and *Amphitecna* are united, the combined genus must have the name *Amphitecna,* although the latter is not explicitly conserved against *Dendrosicus*.

14.8. The listed type and spelling of a conserved name (evident misspellings excepted) may only be changed by the procedure outlined in Art. 14.12.

Ex. 8. Bullock & Killick (in Taxon 6: 239. 1957) published a proposal that the listed type of *Plectranthus* L'Hér. be changed from *P. punctatus* (L. f.) L'Hér. to *P. fruticosus* L'Hér. This proposal was approved by the appropriate committees and by an International Botanical Congress (see App. III).

14.9. A name may be conserved with a different type from that designated by the author or determined by application of the *Code* (see also Art. 10.4). Such a name may be conserved either:

(a) from its place of valid publication (even though the type may not then have been included in the named taxon); or

(b) from a later publication by an author who did include the type as conserved.

In the second case (b) the name as conserved is treated as validly published in the later publication, whether or not the name as conserved was accompanied by a description or diagnosis of the taxon named; the original name and the name as conserved are treated as homonyms (see Art. 14.10).

> **Ex. 9.** (a) *Bromus sterilis* L. (Sp. Pl.: 77. 1753) has been conserved from its place of valid publication even though its conserved type, a specimen (*Hubbard 9045,* E) collected in 1932, was not originally included in Linnaeus's species.

> **Ex. 10.** (b) *Protea* L. (Sp. Pl.: 94. 1753) did not include the conserved type of the generic name, *P. cynaroides* (L.) L. (Mant. Pl.: 190. 1771), which in 1753 was placed in the genus *Leucadendron*. *Protea* was therefore conserved from the 1771 publication, and *Protea* L. (l.c.: 187. 1771), although not intended to be a new generic name and still including the original type elements, is treated as if it were a validly published homonym of *Protea* L. (1753).

14.10. A conserved name, with any corresponding autonym, is conserved against all earlier homonyms. An earlier homonym of a conserved name is not made illegitimate by that conservation but is unavailable for use; if not otherwise illegitimate, it may serve as basionym of another name or combination based on the same type (see also Art. 55.3).

> **Ex. 11.** The generic name *Smithia* Aiton (Hort. Kew. 3: 496. 1789), conserved against *Damapana* Adans. (Fam. Pl. 2: 323, 548. 1763), is conserved automatically against the earlier, listed homonym *Smithia* Scop. (Intr. Hist. Nat.: 322. 1777). *Blumea* DC. (in Arch. Bot. (Paris) 2: 514. 1833) is conserved automatically against *Blumea* Rchb. (Consp. Regn. Veg.: 209. 1828–1829), although the latter name is not listed alongside the former in App. III.

14.11. A name may be conserved in order to preserve a particular spelling or gender. A name so conserved is to be attributed without change of date to the author who validly published it, not to an author who later introduced the conserved spelling or gender.

> **Ex. 12.** The spelling *Rhodymenia,* used by Montagne (in Ann. Sci. Nat., Bot., ser. 2, 12: 44. 1839), has been conserved against the original spelling '*Rhodomenia*', used by Greville (Alg. Brit.: xlviii, 84. 1830). The name is cited as *Rhodymenia* Grev. (1830).

ⓘ **Note 3.** The date upon which a name was conserved does not affect its priority (Art. 11), which is determined only based on the date of its valid publication (Art. 32–45; see also Art. F.4, F.5.2, F.5.3, and H.9; but see Art. 14.9 and 14.14).

14.12. The lists of conserved names will remain permanently open for additions and changes (but see Art. 14.14). Any proposal of an additional name must be accompanied by a detailed statement of the cases both for and against conservation. Such proposals must be submitted to the General Committee, which will refer them for examination to the specialist committees for the various taxonomic groups (see Rec. 14A.1, Div. III Prov. 2.2, 7.10(b), 7.11, and 8.13(a); see also Art. 34.1 and 56.2).

14.13. Entries of conserved names may not be deleted.

> **Ex. 13.** In the Seattle *Code* of 1972 (p. 254), "*Alternaria* C. G. Nees ex Wallroth, Fl. Crypt. Germ. 148. 1833" was listed as conserved against "*Macrosporium* E. M. Fries, Syst. Mycol. 3: 373. 1832" because *Macrosporium* Fr. antedated *Alternaria* "C. G. Nees ex Wallroth" in relation to the then starting-point work for fungi (Fries, Syst. Mycol. 1. 1821). Conservation became unnecessary following the abolition of later starting-point dates for fungi at the Sydney Congress of 1981 and in the Sydney *Code* of 1983, which resulted in *Alternaria* being recognized as having been validly published by Nees (Syst. Pilze: 72. 1816). In addition, it was realized that *Alternaria* had been adopted by Fries in the introduction to the sanctioning work (Syst. Mycol. 1: xlvi. 1821; Art. F.3.1). Because the entry cannot be deleted, *Alternaria* Nees continues to be listed in App. III, but without a corresponding rejected name.

14.14. The places of publication cited for conserved names of families in App. IIB are treated as correct in all circumstances and consequently are not to be changed, even when otherwise such a name would not be validly published or when it is a later isonym.

14.15. When a proposal for the conservation (Art. 14) or protection (Art. F.2) of a name has been approved by the General Committee after study by the specialist committee for the taxonomic group concerned, retention of that name as approved is authorized subject to the decision of a later International Botanical Congress (see also Art. 34.2, 38.5, 53.4, and 56.3). Before 1 January 1954, conservation takes effect on the date of decision by the relevant International Botanical Congress. On or after that date, conservation or protection takes effect on the date of effective publication (Art. 29–31) of the General Committee's approval.

ⓘ **Note 4.** The effective dates for International Botanical Congress (IBC) decisions on conservation of names made before 1954 are as follows:

(a) Conservation of names in the 1906 Vienna *Rules* became effective on 17 June 1905 at the II IBC in Vienna (see Verh. Int. Bot. Kongr. Wien 1905: 135–137. 1906).

(b) Conservation of names in the 1912 Brussels *Rules* became effective on 18 May 1910 at the III IBC in Brussels (see Actes Congr. Int. Bot. Bruxelles 1910: 67–83. 1912).

(c) Conservation of names for the 1952 Stockholm *Code* includes:

    (1) Those of the Special Committee for Phanerogamae and Pteridophyta, which became effective on 1 June 1940 under the authority of the VI IBC held in Amsterdam in 1935 (see Bull. Misc. Inform. Kew 1940: 81–134. 1940).

    (2) Those of the Special Committee for Fungi, which became effective on 20 July 1950 at the VII IBC in Stockholm (see Regnum Veg. 1: 549–550. 1953).

    (3) Those of the Special Committee for Palaeobotanical Nomenclature, which also became effective on 20 July 1950 at the VII IBC in Stockholm (see Regnum Veg. 1: 548. 1953), but were omitted from both the Stockholm *Code* and the 1961 Paris *Code.*

The date, from 1954 onward, of the General Committee's approval of a particular conservation or protection proposal can be determined by consulting the *International Code of Nomenclature for algae, fungi, and plants* Appendices database (https://naturalhistory.si.edu/research/botany/codes-proposals).

Recommendation 14A

14A.1. When a proposal for the conservation (Art. 14) or protection (Art. F.2) of a name has been referred to the appropriate specialist committee for study, authors should follow existing usage of names as far as possible pending the General Committee's recommendation on the proposal (see also Rec. 34A.1 and 56A.1).

ARTICLE 15
SANCTIONED NAMES
SEE ARTICLE F.3

CHAPTER III
NOMENCLATURE OF TAXA ACCORDING TO THEIR RANK

SECTION 1
NAMES OF TAXA ABOVE THE RANK OF FAMILY

ARTICLE 16
NAMES ABOVE THE RANK OF FAMILY

16.1. The name of a taxon above the rank of family is treated as a noun in the plural and is written with an initial capital letter. Such names may be either:

(a) automatically typified names (Art. 10.11), formed from a generic name in the same way as family names (Art. 18.1; but see Art. 16.4) by adding the appropriate rank-denoting termination (Art. 16.3 and 17.1), preceded by the connecting vowel -*o*- if the termination begins with a consonant; or

(b) descriptive names, not so formed, which may be used unchanged at different ranks (see also Art. 6 Note 3).

Ex. 1. Automatically typified names above the rank of family: *Lycopodiophyta*, formed from *Lycopodium; Magnoliophyta*, from *Magnolia; Gnetophytina*, from *Gnetum; Pinopsida*, from *Pinus; Marattiidae*, from *Marattia; Caryophyllidae* and *Caryophyllales*, from *Caryophyllus; Fucales*, from *Fucus; Bromeliineae*, from *Bromelia*.

Ex. 2. Descriptive names above the rank of family: *Angiospermae, Anthophyta, Ascomycetes, Ascomycota, Ascomycotina, Centrospermae, Chlorophyta, Coniferae, Enantioblastae, Gymnospermae, Lycophyta, Parietales*.

16.2. For automatically typified names, the name of the subdivision or subphylum that includes the type of the adopted name of a division or phylum, the name of the subclass that includes the type of the adopted name of a class, and the name of the suborder that includes the type of the adopted name of an order are to be formed from the same generic name (see also Art. 16.4) as the corresponding higher-ranked name.

Ex. 3. Pteridophyta Schimp. (in Zittel, Handb. Palaeont., Palaeophyt.: 1. 1879) and *Pteridophytina* B. Boivin (in Bull. Soc. Bot. France 103: 493. 1956); *Gnetopsida* Prantl (Lehrb. Bot., ed. 5: 194. 1883) and *Gnetidae* Pax (in Prantl, Lehrb. Bot., ed. 9: 210. 1894); *Liliales* Perleb (Lehrb. Naturgesch. Pflanzenr.: 129. 1826) and *Liliineae* Rchb. (Deut. Bot. Herb.-Buch: xxxvii. 1841).

16.3. Automatically typified names end as follows:

(a) The name of a division or phylum ends in *-phyta,* unless it is referable to the fungi in which case it ends in *-mycota.*

(b) The name of a subdivision or subphylum ends in *-phytina,* unless it is referable to the fungi in which case it ends in *-mycotina.*

(c) The name of a class in the algae ends in *-phyceae,* and of a subclass in *-phycidae.*

(d) The name of a class in the fungi ends in *-mycetes,* and of a subclass in *-mycetidae.*

(e) The name of a class in the plants ends in *-opsida,* and of a subclass in *-idae* (but not *-viridae*).

Automatically typified names with a termination not in accordance with this rule or Art. 17.1 are to be corrected, without change of authorship or date of publication (see Art. 32.2). However, if such names are published with a non-Latin termination they are not validly published.

Ex. 4. '*Cacteae*' Juss. ex Bercht. & J. Presl (Přir. Rostlin: 238. 1820, formed from *Cactus* L.) and '*Coriales*' Lindl. (Nix. Pl.: 11. 1833, formed from *Coriaria* L.), both published for taxa at the rank of order, are to be corrected to *Cactales* Juss. ex Bercht. & J. Presl (1820) and *Coriariales* Lindl. (1833), respectively.

Ex. 5. Ptéridées (Kirschleger, Fl. Alsace 2: 379. 1853–Jul 1857), published for a taxon at the rank of order, is not to be accepted as "*Pteridales* Kirschl." because it has a French rather than a Latin termination. Later, the name *Pteridales* Doweld (Prosyll. Tracheophyt., Tent. Syst. Pl. Vasc.: xi. 2001) was validly published.

ⓘ *Note 1.* The terms "divisio" and "phylum", and their equivalents in modern languages, are treated as referring to one and the same rank (Art. 3.1). When "divisio" and "phylum" are used simultaneously to denote different non-consecutive ranks, this is to be treated as informal usage of rank-denoting terms (see Art. 37.9; see also Art. 37 Note 1).

16.4. At ranks higher than order, the word elements *-clad-, -cocc-, -cyst-, -monad-, -mycet-, -nemat-,* or *-phyt-,* all of which are genitive singular stems of the second part of a name of an included genus, may be omitted before the rank-denoting termination. Such names are automatically typified when their derivation is obvious or is indicated in the protologue.

Ex. 6. The name *Raphidophyceae* Chadef. ex P. C. Silva (in Regnum Veg. 103: 78. 1980) was indicated by its author to be formed from *Raphidomonas* F. Stein (Organismus Infus. 3(1): x, 69, 152, 153. 1878). The name *Saccharomycetes* G. Winter (Rabenh. Krypt.-Fl., ed. 2, 1(1): 32. 1880) is regarded as being formed from *Saccharomyces* Meyen (in Arch. Naturgesch. 4(2): 100. 1838). The name *Trimerophytina* H. P. Banks (in Taxon 24: 409. 1975) was indicated by its author to be formed from *Trimerophyton* Hopping (in Proc. Roy. Soc. Edinburgh, B, Biol. 66: 25. 1956).

ⓘ ***Note 2.*** The principle of priority does not apply above the rank of family (Art. 11.10; but see Rec. 16A.1).

Recommendation 16A

16A.1. In choosing among typified names for a taxon above the rank of family, authors should generally follow the principle of priority.

ARTICLE 17
NAMES OF ORDERS AND SUBORDERS

17.1. Automatically typified names of orders or suborders are to end in *-ales* (but not *-virales*) and *-ineae,* respectively (see Art. 16.3 and 32.2).

17.2. Names intended as names of orders, but published with their rank denoted by a term such as "cohors", "nixus", "alliance", or "Reihe" instead of "order", are treated as having been published as names of orders.

Recommendation 17A

17A.1. A new name should not be published for an order for which a name already exists that is based on the same type as the name of an included family.

SECTION 2
NAMES OF FAMILIES, SUBFAMILIES, TRIBES, AND SUBTRIBES

ARTICLE 18
NAMES OF FAMILIES

18.1. The name of a family is a plural adjective used as a noun; it is formed from the genitive singular of a name of an included genus by replacing the genitive singular inflection (Latin *-ae, -i, -us, -is;* transcribed Greek *-ou, -os, -es, -as,* or *-ous,* and its equivalent *-eos*) with the termination *-aceae* (but see Art. 18.5). For generic names of non-classical origin, when analogy with classical names is insufficient to determine the genitive singular, *-aceae* is added to the full word. Likewise, when formation from the genitive singular of a generic name results in a homonym, *-aceae* may be added to the nominative singular. For generic names with alternative genitives the one implicitly used by the original author must be maintained, except that the genitive of names ending in *-opsis* is always *-opsidis*.

🛈 **Note 1.** The generic name from which the name of a family is formed provides the type of the family name (Art. 10.10) but is not a basionym of that name (Art. 6.10; see Art. 41.2(a)).

> *Ex. 1.* Family names formed from a generic name of classical origin: *Rosaceae* (from *Rosa,* genitive singular: *Rosae*), *Salicaceae* (from *Salix, Salicis*), *Plumbaginaceae* (from *Plumbago, Plumbaginis*), *Rhodophyllaceae* (from *Rhodophyllus, Rhodophylli*), *Rhodophyllidaceae* (from *Rhodophyllis, Rhodophyllidos*), *Sclerodermataceae* (from *Scleroderma, Sclerodermatos*), *Aextoxicaceae* (from *Aextoxicon, Aextoxicou*), *Potamogetonaceae* (from *Potamogeton, Potamogetonos*).

> *Ex. 2.* Family names formed from a generic name of non-classical origin: *Nelumbonaceae* (from *Nelumbo, Nelumbonis,* declined by analogy with *umbo, umbonis*), *Ginkgoaceae* (from *Ginkgo,* indeclinable).

> *Ex. 3.* Family name formed from the nominative singular of a generic name because of an earlier potential homonym: *Trigoniumaceae* Glezer (in Taxon 68: 415. 2019), formed from *Trigonium* Cleve, non *Trigoniaceae* A. Juss. (in Orbigny, Dict. Univ. Hist. Nat. 12: 670. 1849) formed from *Trigonia* Aubl.

🛈 **Note 2.** The name of a family may be formed from any validly published name of an included genus, even one that is unavailable for use, although the provisions of Art. 18.3 apply if the generic name is illegitimate.

> *Ex. 4. Cactaceae* Juss. (Gen. Pl.: 310. 1789) formed from *Cactus* L. (Sp. Pl.: 466. 1753), a generic name now rejected in favour of *Mammillaria* Haw. (Syn. Pl. Succ.: 177. 1812).

18.2. Names intended as names of families, but published with their rank denoted by one of the terms "order" (ordo) or "natural order" (ordo naturalis) instead of "family", are treated as having been published as names of families (see also Art. 19.2), unless this treatment would result in a taxonomic sequence with a misplaced rank-denoting term.

Ex. 5. *Cyperaceae* Juss. (Gen. Pl.: 26. 1789), *Lobeliaceae* Juss. (in Bonpland, Descr. Pl. Malmaison: [19]. 1813), and *Xylomataceae* Fr. (Scleromyceti Sueciae 2: [2]. 1820), nom. sanct., were published as "ordo *Cyperoideae*", "ordo naturalis *Lobeliaceae*", and "ordo *Xylomaceae*", respectively.

i **Note 3.** If the term "family" is simultaneously used to denote a rank different from "order" or "natural order", a name published for a taxon at the latter rank cannot be considered to have been published as the name of a family.

***Ex. 6.** Names published at the rank of order ("řad") by Berchtold & Presl (*O přirozenosti rostlin* ... 1820) are not to be treated as having been published at the rank of family, because the term family ("čeleď") was sometimes used to denote a rank below order.

18.3. A name of a family formed from an illegitimate generic name is illegitimate unless and until it or the generic name from which it is formed is conserved or protected.

Ex. 7. *Caryophyllaceae* Juss. (Gen. Pl.: 299. 1789), nom. cons., formed from *Caryophyllus* Mill. non L.; *Winteraceae* R. Br. ex Lindl. (Intr. Nat. Syst. Bot.: 26. 1830), nom. cons., formed from *Wintera* Murray, an illegitimate replacement name for *Drimys* J. R. Forst. & G. Forst.

Ex. 8. *Nartheciaceae* Fr. ex Bjurzon (Skand. Vaxtfam.: 64. 1846), formed from *Narthecium* Huds., nom. cons. (Fl. Angl.: 127. 1762), became legitimate when the generic name was conserved against its earlier homonym *Narthecium* Gérard (Fl. Gallo-Prov.: 142. 1761) (see App. III).

18.4. When a name of a family has been published with an improper Latin termination, the termination must be changed to conform with Art. 18.1, without change of authorship or date (see Art. 32.2). However, if such a name is published with a non-Latin termination, it is not validly published.

Ex. 9. '*Coscinodisceae*' Kütz. (Kieselschal. Bacill.: 130. 1844), published to designate a family, is to be accepted as *Coscinodiscaceae* Kütz. (1844) and not attributed to De Toni, who first used the correct termination (in Notarisia 5: 915. 1890).

Ex. 10. '*Atherospermeae*' R. Br. (Gen. Rem.: 21. 1814), published to designate a family, is to be accepted as *Atherospermataceae* R. Br. (1814) and not attributed to Airy Shaw (in Willis, Dict. Fl. Pl., ed. 7: 104. 1966), who first used the correct spelling, nor to Lindley (Veg. Kingd.: 300. 1846), who used the spelling *'Atherospermaceae'*.

Ex. 11. Tricholomées (Roze in Bull. Soc. Bot. France 23: 49. 1876), published to designate a family, is not to be accepted as "*Tricholomataceae* Roze" because it has a French rather than a Latin termination. The name *Tricholomataceae* was validly published by Pouzar (in Česká Mykol. 37: 175. 1983; see App. IIA).

18.5. The following names, of long usage, are treated as validly published: *Compositae* (nom. alt.: *Asteraceae;* type: *Aster* L.); *Cruciferae* (nom. alt.: *Brassicaceae;* type: *Brassica* L.); *Gramineae* (nom. alt.: *Poaceae;* type: *Poa* L.); *Guttiferae* (nom. alt.: *Clusiaceae;* type: *Clusia* L.); *Labiatae* (nom. alt.: *Lamiaceae;* type: *Lamium* L.); *Leguminosae* (nom. alt.: *Fabaceae;* type: *Faba* Mill.); *Palmae* (nom. alt.: *Arecaceae;* type: *Areca* L.); *Papilionaceae* (nom. alt.: *Fabaceae;* type: *Faba* Mill.); *Umbelliferae* (nom. alt.: *Apiaceae;* type: *Apium* L.). When the *Papilionaceae* are regarded as a family distinct from the remainder of the *Leguminosae,* the name *Papilionaceae* is conserved against *Leguminosae.*

18.6. The use, as alternatives, of the eight family names indicated as "nom. alt." (nomen alternativum) in Art. 18.5 is authorized.

ARTICLE 19
NAMES OF SUBFAMILIES, TRIBES, AND SUBTRIBES

19.1. The name of a subfamily is a plural adjective used as a noun; it is formed in the same manner as the name of a family (Art. 18.1) but by adding the termination *-oideae* instead of *-aceae.*

19.2. Names intended as names of subfamilies, but published with their rank denoted by the term "suborder" (subordo) instead of subfamily, are treated as having been published as names of subfamilies (see also Art. 18.2), unless this would result in a taxonomic sequence with a misplaced rank-denoting term.

Ex. 1. Cyrilloideae Torr. & A. Gray (Fl. N. Amer. 1: 256. 1838) and *Sphenocleoideae* Lindl. (Intr. Nat. Syst. Bot., ed. 2: 238. 1836) were published as "suborder *Cyrilleae*" and "Sub-Order ? *Sphenocleaceae*", respectively.

ⓘ **Note 1.** If the term "subfamily" is simultaneously used to denote a rank different from "suborder", a name published for a taxon at the latter rank cannot be considered to have been published as the name of a subfamily.

19.3. The name of a tribe or subtribe is formed in the same manner as the name of a subfamily (Art. 19.1), except that the termination is *-eae* for a tribe and *-inae* (but not *-virinae*) for a subtribe.

> **Ex. 2.** The generic name *Mareyopsis* Pax. & K. Hoffm. is the type of the subtribe *Mareyopsidinae* G. L. Webster (in Kubitzki, Fam. Gen. Vasc. Pl. 11: 115. 2014), spelled '*Mareyopsinae*' in the protologue. Names of families and subfamilies are formed in the same manner as names of tribes and subtribes, and under Art. 18.1 the genitive of names ending in *-opsis* is always *-opsidis*. Because the genitive of *Mareyopsis* is *Mareyopsidis* with *Mareyopsidi-* as the stem, '*Mareyopsinae*' is to be corrected to *Mareyopsidinae*.

19.4. The name of any subdivision of a family that includes the type of the adopted, legitimate name of the family to which it is assigned is to be formed from the generic name equivalent to that type (Art. 10.10; but see Art. 19.8).

> **Ex. 3.** The type of the family name *Rosaceae* Juss. is *Rosa* L. and hence the subfamily and tribe assigned to *Rosaceae* that include *Rosa* are to be called *Rosoideae* Endl. and *Roseae* DC., respectively.

> **Ex. 4.** The type of the family name *Gramineae* Juss. (nom. alt.: *Poaceae* Barnhart, see Art. 18.5) is *Poa* L. and hence the subfamily, tribe, and subtribe assigned to *Gramineae* that include *Poa* are to be called *Pooideae* Asch., *Poeae* R. Br., and *Poinae* Dumort., respectively.

ⓘ **Note 2.** Art. 19.4 applies only to the names of those subordinate taxa that include the type of the adopted name of the family (but see Rec. 19A.2).

> **Ex. 5.** The type of the family name *Ericaceae* Juss. is *Erica* L. and hence the subfamily and tribe assigned to *Ericaceae* that include *Erica* are to be called *Ericoideae* Endl. and *Ericeae* D. Don, respectively, despite the priority of any competing names. The subfamily that includes *Rhododendron* L. is called *Rhododendroideae* Endl. However, the correct name of the tribe of *Ericaceae* that includes both *Rhododendron* and *Rhodora* L. is *Rhodoreae* D. Don (in Edinburgh New Philos. J. 17: 152. 1834).

ⓘ **Note 3.** A name of a subdivision of a family that includes the type of the adopted, legitimate name of the family to which it is assigned, but is not formed from the generic name equivalent to that type, is incorrect but may nevertheless be validly published and may become correct in a different context.

> **Ex. 6.** When published, the name *Lippieae* Endl. (Gen. Pl.: 633. 1838) was applied to a tribe of *Verbenaceae* J. St.-Hil. that included *Verbena* L., the type of

the family name, as well as *Lippia* L. Although originally incorrect, *Lippieae* may become correct if used for a tribe of *Verbenaceae* that includes *Lippia* but excludes *Verbena*.

19.5. The name of any subdivision of a family that includes the type of a name listed in App. IIB (i.e. a name of a family conserved against all unlisted names, see Art. 14.5) is to be formed from the generic name equivalent to that type (Art. 10.10), unless this is contrary to Art. 19.4 (see also Art. 19.8). If more than one such type is included, the correct name is determined by precedence in App. IIB of the corresponding family names.

Ex. 7. A subfamily assigned to *Rosaceae* Juss. that includes *Malus* Mill., the type of *Malaceae* Small (Fl. S.E. U.S.: 495, 529. 1903) listed in App. IIB, is to be called *Maloideae* C. Weber (in J. Arnold Arbor. 45: 164. 1964) unless it also includes *Rosa* L., i.e. the type of *Rosaceae*, or the type of another name listed in App. IIB that takes precedence over *Malaceae*. This is so even if the subfamily also includes *Spiraea* L. and/or *Pyrus* L. because, although *Spiraeoideae* Arn. (in Hooker & Arnott, Bot. Beechey Voy.: 107. 1832) and *Pyroideae* Burnett (Outlines Bot.: 695, 1137. 1835) were published earlier than *Maloideae*, neither *Spiraeaceae* nor *Pyraceae* is listed in App. IIB. However, if *Amygdalus* L. is included in the same subfamily as *Malus*, the name *Amygdaloideae* Arn. (Botany: 107. 1832) takes precedence because *Amygdalaceae* Marquis (Esq. Règne Vég.: 49. 1820) is listed in App. IIB with priority over *Malaceae*.

Ex. 8. *Monotropaceae* Nutt. (Gen. N. Amer. Pl. 1: 272. 1818) and *Pyrolaceae* Lindl. (Syn. Brit. Fl.: 175. 1829) are both listed in App. IIB, but *Pyrolaceae* is conserved against *Monotropaceae*. Therefore, a subfamily including both *Monotropa* L. and *Pyrola* L. is called *Pyroloideae* Beilschm. (in Flora 16(Beibl. 1): 72, 109. 1833).

19.6. A name of a subdivision of a family formed from an illegitimate generic name is illegitimate unless and until that generic name or the corresponding family name is conserved or protected.

Ex. 9. The name *Caryophylloideae* Arn. (Botany: 99. 1832), formed from the illegitimate *Caryophyllus* Mill. non L., is legitimate because the corresponding family name, *Caryophyllaceae* Juss., is conserved.

Ex. 10. *Thunbergioideae* T. Anderson (in Thwaites, Enum. Pl. Zeyl.: 223. 1860), formed from *Thunbergia* Retz., nom. cons. (in Physiogr. Sälsk. Handl. 1(3): 163. 1780), became legitimate when the generic name was conserved against its earlier homonym *Thunbergia* Montin (in Kongl. Vetensk. Acad. Handl. 34: 288. 1773) (see App. III).

19.7. When a name of a subdivision of a family has been published with an improper Latin termination, such as *-eae* for a subfamily or *-oideae* for a tribe, the termination must be changed to accord with

Art. 19.1 and 19.3, without change of authorship or date (see Art. 32.2). However, if such a name is published with a non-Latin termination, it is not validly published.

Ex. 11. *'Climacieae'* Grout (Moss Fl. N. Amer. 3: 4. 1928), published to designate a subfamily, is to be accepted as *Climacioideae* Grout (l.c. 1928).

Ex. 12. Melantheen (Kittel in Richard, Nouv. Elém. Bot., Germ. Transl., ed. 3: 727. 1840), published to designate a tribe, is not to be accepted as *"Melanthieae* Kitt." because it has a German rather than a Latin termination. The name *Melanthieae* was validly published by Grisebach (Spic. Fl. Rumel. 2: 377. 1846).

19.8. When the *Papilionaceae* are included in the family *Leguminosae* (nom. alt.: *Fabaceae;* see Art. 18.5) as a subfamily, the name *Papilionoideae* may be used as an alternative to *Faboideae*.

Recommendation 19A

19A.1. When a family is changed to the rank of a subdivision of a family, or the inverse change occurs, and no legitimate name is available at the new rank, the name should be retained, with only the termination *(-aceae, -oideae, -eae, -inae)* altered.

19A.2. When a subdivision of a family is changed to another such rank, and no legitimate name is available at the new rank, its name, Art. 19.5 permitting, should be formed from the same generic name as the name at the former rank.

Ex. 1. The subtribe *Drypetinae* Griseb. (Fl. Brit. W. I.: 31. 1859) when raised to the rank of tribe was named *Drypeteae* Small (Man. S.E. Fl.: 775. 1933); the subtribe *Antidesmatinae* Müll. Arg. (in Linnaea 34: 64. 1865) when raised to the rank of subfamily was named *Antidesmatoideae* Hurus. (in J. Fac. Sci. Univ. Tokyo, Sect. 3, Bot. 6: 322, 340. 1954).

SECTION 3
NAMES OF GENERA AND SUBDIVISIONS OF GENERA

ARTICLE 20
NAMES OF GENERA

20.1. The name of a genus is a noun in the nominative singular, or a word treated as such, and is written with an initial capital letter (see Art. 60.2). It may be taken from any source whatever (but see Rec.

51A) and may even be composed arbitrarily (but see Art. 60.1), but it must not end in *-virus*.

Ex. 1. Bartramia, Convolvulus, Gloriosa, Hedysarum, Ifloga (an anagram of *Filago*), *Impatiens, Liquidambar, Manihot, Rhododendron, Rosa.*

20.2. The name of a genus published before 1 January 1912 may co-incide with a Latin technical term in use in morphology at the time of publication only if it was accompanied by a species name published in accordance with the binary system of Linnaeus.

Ex. 2. "Radicula" (Hill, Brit. Herb.: 264. 1756) coincides with the Latin technical term "radicula" (radicle) and was not accompanied by a species name in accordance with the binary system of Linnaeus. The name *Radicula* is correctly attributed to Moench (Methodus: 262. 1794), who first combined it with specific epithets.

Ex. 3. Tuber F. H. Wigg., nom. sanct., when published in 1780, was accompanied by a binary species name (*Tuber gulosorum* F. H. Wigg., Prim. Fl. Holsat.: 109. 1780) and is therefore validly published even though it coincides with a Latin technical term.

ⓘ **Note 1.** Editions of the *Code* prior to the *Madrid Code* of 2025 included a provision precluding the valid publication after 1911 of the name of a genus that coincided with a technical term in use in morphology at the time of publication. While publication of such names is not recommended (see Rec. 20A.1(j)), in the interest of nomenclatural stability under the current *Code,* binding decisions on the valid publication of each generic "name" or any case with identical spelling where this former provision has been applied are listed in App. VI and take retroactive effect.

20.3. The name of a genus must consist of one word (which may be formed by combining two or more words into one) or of two words joined by a hyphen (but see Art. 60.13 for names of fossil-genera and Art. H.6.5 for nothogeneric names).

Ex. 4. Names validly published that consisted of one word when originally published: *Quisqualis* L. (Sp. Pl., ed. 2: 556. 1762, '*Qvisqvalis*') (formed by combining two words into one); *Asplenium* L. (Sp. Pl.: 1078. 1753); *Leucodon* Schwägr. (Sp. Musc. Frond. Suppl. 1(2): 1. 1816).

Ex. 5. Designation not validly published (Art. 32.1(c)) because it was composed of two separate words not joined by a hyphen: "*Uva ursi*" (Miller, Gard. Dict. Abr., ed. 4: *Uva ursi.* 1754); the corresponding name is correctly attributed to Duhamel (Traité Arbr. Arbust. 2: 371. 1755) as *Uva-ursi* (hyphenated when published).

Ex. 6. Names validly published that consisted of two words joined by a hyphen when originally published: *Neves-armondia* K. Schum. (in Engler & Prantl, Nat.

Pflanzenfam. Nachtr. 1: 302. 1897), *Sebastiano-schaueria* Nees (in Martius, Fl. Bras. 9: 158. 1847), and *Solms-laubachia* Muschl. ex Diels (in Notes Roy. Bot. Gard. Edinburgh 5: 205. 1912).

ⓘ **Note 2.** The names of intergeneric hybrids are formed according to the provisions of Art. H.6.

20.4. The following are not to be regarded as generic names:

(a) Words not intended as names.

> **Ex. 7.** The designation "Anonymos" was applied by Walter (Fl. Carol.: 2, 4, 9, etc. 1788) to 28 different genera to indicate that they were without names (see Sprague in Bull. Misc. Inform. Kew 7: 318–319, 331–334. 1939).

> **Ex. 8.** "Schænoides" and "Scirpoides", as used by Rottbøll (Descr. Pl. Rar.: 14, 27. 1772) to indicate unnamed genera resembling *Schoenus* and *Scirpus* that, as stated on p. 7, he intended to name later, are token words and not generic names. These unnamed genera were subsequently named *Kyllinga* Rottb. (Descr. Icon. Rar. Pl.: 12. 1773) and *Fuirena* Rottb. (l.c.: 70. 1773), respectively.

(b) Words that have been widely used in pharmacopoeia or as descriptive morphological terms: "Balsamum", "Bulbus", "Caulis", "Cortex", "Flos", "Herba", "Lignum", "Oleum", "Radix", "Spina".

(c) Unitary designations of species.

ⓘ **Note 3.** Unitary designations such as *"Leptostachys"* and *"Anthopogon"*, listed in editions of the *Code* prior to the *Tokyo Code* of 1994 were from publications that are now suppressed (see App. I).

Recommendation 20A

20A.1. Authors forming generic names should comply with the following:

(a) Use Latin terminations insofar as possible.
(b) Not make names that are very long.
(c) Not make names by combining words from different languages.
(d) Indicate, if possible, by the formation or ending of the name the affinities or analogies of the genus.
(e) Avoid adjectives used as nouns.
(f) Not use a name similar to or derived from the epithet in the name of one of the species of the genus.
(g) Not dedicate genera to persons quite unconnected with botany, mycology, phycology, or natural science in general.
(h) Give a feminine form to all personal generic names, whether they commemorate a man or a woman (see Rec. 60B; see also Rec. 62A.1).

(i) Not form generic names by combining parts of two existing generic names, because such names are likely to be confused with nothogeneric names (see Art. H.6).

(j) Not publish names for genera that coincide with technical terms currently in use in morphology.

ARTICLE 21
NAMES OF SUBDIVISIONS OF GENERA

21.1. The name of a subdivision of a genus is a combination of a generic name and a subdivisional epithet. A connecting term (subgenus, sectio, series, etc.) is used to denote the rank.

ⓘ **Note 1.** Names of subdivisions of the same genus, even if they differ in rank, are homonyms if they have the same epithet but are based on different types (Art. 53.3), because the rank-denoting term is not part of the name.

21.2. The epithet in the name of a subdivision of a genus is either of the same form as a generic name, or a noun in the genitive plural, or a plural adjective agreeing in gender with the generic name (see Art. 32.2), but not a noun in the genitive singular. It is written with an initial capital letter (see Art. 60.2).

Ex. 1. Epithet of the same form as a generic name: *Euphorbia* sect. *Tithymalus, Ricinocarpos* sect. *Anomodiscus;* epithet a genitive plural noun: *Pleione* subg. *Scopulorum;* epithet a plural adjective: *Arenaria* ser. *Anomalae, Euphorbia* subsect. *Tenellae, Sapium* subsect. *Patentinervia.*

Ex. 2. "*Vaccinium* sect. *Vitis idaea*" (Koch, Syn. Fl. Germ. Helv.: 474. 1837) is not a validly published name because the intended epithet consisted of two separate words not joined by a hyphen (Art. 20.3 and 32.1(c)); "*Vitis idæa*" is a pre-Linnaean, binary generic name. *Vaccinium* sect. *Vitis-idaea* was validly published by Asa Gray (in Mem. Amer. Acad. Arts, n.s., 3: 53. 1846, hyphenated when published).

21.3. The epithet in the name of a subdivision of a genus is not to be formed from the name of the genus to which it belongs by adding the prefix *Eu-* (see also Art. 22.2).

Ex. 3. *Costus* subg. *Metacostus; Valeriana* sect. *Valerianopsis;* but not "*Carex* sect. *Eucarex*".

21.4. A name with a binary combination instead of a subdivisional epithet, but otherwise in accordance with this *Code,* is treated as validly published in the form determined by Art. 21.1 without change of authorship or date.

> *Ex. 4. Sphagnum* "b. *Sph. rigida*" (Lindberg in Öfvers. Förh. Kongl. Svenska Vetensk.-Akad. 19: 135. 1862) and *S.* sect. *"Sphagna rigida"* (Limpricht, Laubm. Deutschl. 1: 116. 1885) are to be cited as *Sphagnum* [unranked] *Rigida* Lindb. and *S.* sect. *Rigida* (Lindb.) Limpr., respectively.

Note 2. Names of hybrids at the rank of a subdivision of a genus are formed according to the provisions of Art. H.7.

Recommendation 21A

21A.1. When it is desired to indicate the name of a subdivision of the genus to which a particular species belongs in connection with the generic name and specific epithet, the subdivisional epithet should be placed in parentheses between the two; when desirable, the subdivisional rank may also be indicated.

> *Ex. 1. Astragalus (Cycloglottis) contortuplicatus; A. (Phaca) umbellatus; Loranthus* (sect. *Ischnanthus*) *gabonensis.*

Recommendation 21B

21B.1. Recommendations made for forming the name of a genus (Rec. 20A) apply equally to an epithet of a subdivision of a genus, unless Rec. 21B.2–4 recommend otherwise.

21B.2. The epithet in the name of a subgenus or section is preferably a noun; that in the name of a subsection or lower-ranked subdivision of a genus is preferably a plural adjective.

21B.3. Authors, when proposing new epithets for names of subdivisions of genera, should avoid those in the form of a noun when other co-ordinate subdivisions of the same genus have them in the form of a plural adjective, and vice versa. They should also avoid, when proposing an epithet for a name of a subdivision of a genus, one already used for a subdivision of a closely related genus, or one that is identical with the name of such a genus.

21B.4. When a section or a subgenus is raised to the rank of genus, or the inverse change occurs, the original name or epithet should be retained unless the resulting name would be contrary to the *Code.*

ARTICLE 22
AUTONYMS OF SUBDIVISIONS OF GENERA

22.1. The name of any subdivision of a genus that includes the type of the adopted, legitimate name of the genus to which it is assigned is to repeat that generic name unaltered as its epithet, not followed by an author citation (see Art. 46). Such names are autonyms (Art. 6.8; see also Art. 7.7).

Ex. 1. The subgenus that includes the type of the name *Rhododendron* L. is to be named *Rhododendron* L. subg. *Rhododendron.*

Ex. 2. The subgenus that includes the type of *Malpighia* L. (*M. glabra* L.) is to be called *M.* subg. *Malpighia,* not *M.* subg. *Homoiostylis* Nied.; and the section that includes the type of *Malpighia* is to be called *M.* sect. *Malpighia,* not *M.* sect. *Apyrae* DC.

ⓘ **Note 1.** Art. 22.1 applies only to the names of those subordinate taxa that include the type of the adopted name of the genus (but see Rec. 22A).

Ex. 3. The correct name of the subgenus of the genus *Solanum* L. that includes *S. pseudocapsicum* L., the type of *S.* sect. *Pseudocapsicum* (Medik.) Roem. & Schult. (Syst. Veg. 4: 569 *('Pseudocapsica'),* 584 *('Pseudo-Capsica').* 1819), if considered as distinct from *S.* subg. *Solanum,* is *S.* subg. *Minon* Raf. (Autikon Bot.: 108. 1840), the earliest legitimate name at that rank, and not *"S.* subg. *Pseudocapsicum".*

22.2. A name of a subdivision of a genus that includes the type (i.e. the original type, all elements eligible as type, or the previously designated or conserved type) of the adopted, legitimate name of the genus is not validly published unless its epithet repeats the generic name unaltered. For the purpose of this provision, explicit indication that the nomenclaturally typical element is included is considered as equivalent to inclusion of the type, whether or not it has been previously designated (see also Art. 21.3).

Ex. 4. *"Dodecatheon* sect. *Etubulosa"* (Knuth in Engler, Pflanzenr. IV. 237 (Heft 22): 234. 1905) was not validly published because it was proposed for a section that included *D. meadia* L., the original type of the generic name *Dodecatheon* L.

Ex. 5. *Cactus* [unranked] *Melocactus* L. (Gen. Pl., ed. 5: 210. 1754) was proposed for one of four unranked (Art. 37.3), named subdivisions of the genus *Cactus,* comprising *C. melocactus* L. (its type under Art. 10.8) and *C. mammillaris* L. It is validly published even though *C. mammillaris* was subsequently designated as the type of *Cactus* L. (by Coulter in Contr. U. S. Natl. Herb. 3: 95. 1894).

22.3. The first instance of valid publication of a name of a subdivision of a genus under a legitimate generic name automatically establishes the corresponding autonym (see also Art. 11.6 and 32.3).

> *Ex. 6.* Publication of *Tibetoseris* sect. *Simulatrices* Sennikov (in Komarovia 5: 91. 2008) automatically established the autonym *Tibetoseris* Sennikov sect. *Tibetoseris*. Publication of *Pseudoyoungia* sect. *Simulatrices* (Sennikov) D. Maity & Maiti (in Compositae Newslett. 48: 31. 2010) automatically established the autonym *Pseudoyoungia* D. Maity & Maiti sect. *Pseudoyoungia*.

> *Ex. 7.* *Umbilicaria* Hoffm. (Descr. Pl. Cl. Crypt. 1: 8. 1789), a later homonym (non *Umbilicaria* Heist. ex Fabr., Enum.: 42. 1759), was conserved with *U. hyperborea* (Ach.) Hoffm. as the conserved type (Nicolson in Taxon 45: 527. 1996). Both *U.* sect. *Gyrophora* (Ach.) Endl. (Gen. Pl.: 13. 1836) and *U.* subg. *Gyrophora* (Ach.) Frey (Rabenh. Krypt.-Fl., ed. 2, 9(4, 1): 209. 1933) included *U. hyperborea*. Both were validly published but neither established an autonym because the name *Umbilicaria* Hoffm. remained illegitimate until 1996. After *Umbilicaria* became legitimate through conservation, the autonym *U.* subg. *Umbilicaria* was established by the valid publication of *U.* subg. *Floccularia* Davydov & al. (in Taxon 66: 1297. 2017).

22.4. The epithet in the name of a subdivision of a genus may not repeat unchanged the correct name of the genus unless the two names have the same type.

22.5. The epithet in the name of a subdivision of a genus may not repeat unchanged the generic name if the latter is illegitimate.

> *Ex. 8.* When Kuntze (in Post & Kuntze, Lex. Gen. Phan.: 106. 1903) published *Caulinia* sect. *Hardenbergia* (Benth.) Kuntze under *Caulinia* Moench (Suppl. Meth.: 47. 1802), a later homonym of *Caulinia* Willd. (in Mém. Acad. Roy. Sci. Hist. (Berlin) 1798: 87. 1801), he did not establish the autonym *"Caulinia* sect. *Caulinia"*.

Recommendation 22A

22A.1. A section including the type of the correct name of a subgenus, but not including the type of the correct name of the genus, should, where there is no obstacle under the rules, be given a name with the same epithet and type as the subgeneric name.

22A.2. A subgenus not including the type of the correct name of the genus should, where there is no obstacle under the rules, be given a name with the same epithet and type as the correct name of one of its subordinate sections.

> *Ex. 1.* When Brizicky raised *Rhamnus* sect. *Pseudofrangula* Grubov to the rank of subgenus, instead of using a new epithet he named the taxon *R.* subg. *Pseudofrangula* (Grubov) Brizicky so that the type of both names is the same.

Recommendation 22B

22B.1. When publishing a name of a subdivision of a genus that will also establish an autonym, the author should mention that autonym in the publication.

SECTION 4
NAMES OF SPECIES

ARTICLE 23

23.1. The name of a species is a binary combination consisting of the name of the genus followed by a single specific epithet in the form of an adjective, a noun in the genitive, an adverb, or a word in apposition (see also Art. 23.7). If an epithet consisted originally of two or more words, these are to be united or hyphenated. An epithet not so joined when originally published is not to be rejected but, when used, is to be united or hyphenated, as specified in Art. 60.12.

23.2. The epithet in the name of a species may be taken from any source whatever (but see Rec. 51A and Art. 61.6), and may even be composed arbitrarily (but see Art. 60.1). The epithet in the name of a species published on or after 1 January 2026 must consist of at least two but not more than 30 characters.

> ***Ex. 1.*** *Adiantum capillus-veneris, Atropa bella-donna, Cornus sanguinea, Dianthus monspessulanus, Embelia sarasiniorum, Fumaria gussonei, Geranium robertianum, Impatiens noli-tangere, Papaver rhoeas, Spondias mombin* (an indeclinable epithet), *Uromyces fabae.*

23.3. Symbols forming part of specific epithets proposed by Linnaeus do not prevent valid publication of the relevant names but must be transcribed.

> ***Ex. 2.*** *Scandix* 'pecten ♀' L. is to be transcribed as *Scandix pecten-veneris; Veronica* 'anagallis ▽' L. is to be transcribed as *Veronica anagallis-aquatica.*

23.4. The specific epithet, with or without the addition of a transcribed symbol, may not exactly repeat the generic name (a designation formed by such repetition is a tautonym).

> ***Ex. 3.*** *"Linaria linaria"* and *"Nasturtium nasturtium-aquaticum"* are tautonyms and cannot be validly published.

Ex. 4. *Linum radiola* L. (Sp. Pl.: 281. 1753) when transferred to *Radiola* Hill may not be named *"Radiola radiola"*, as was done by Karsten (Deut. Fl.: 606. 1882), because that combination is a tautonym and cannot be validly published. The next earliest name, *L. multiflorum* Lam. (Fl. Franç. 3: 70. 1779), is an illegitimate superfluous name for *L. radiola*. In *Radiola*, the species has been given the legitimate name *R. linoides* Roth (Tent. Fl. Germ. 1: 71. 1788).

23.5. The specific epithet, when adjectival in form and not used as a noun, agrees with the gender of the generic name; when the epithet is a noun in apposition or a genitive noun or a noun and its accompanying adjective in the genitive case, it retains its own gender and termination regardless of the gender of the generic name; when the epithet is an adverb, its termination is independent of the gender of the generic name. Epithets not conforming to this rule are to be corrected (see Art. 32.2) to the proper form of the termination (Latin or transcribed Greek) of the original author(s). In particular, the usage of the word element *-cola* (dweller) as an adjective is a correctable error.

Ex. 5. Names with Latin adjectival epithets: *Helleborus niger* L., *Brassica nigra* (L.) W. D. J. Koch, *Verbascum nigrum* L.; *Rumex cantabricus* Rech. f., *Daboecia cantabrica* (Huds.) K. Koch (*Vaccinium cantabricum* Huds.); *Vinca major* L., *Tropaeolum majus* L.; *Bromus mollis* L., *Geranium molle* L.; *Peridermium balsameum* Peck (derived from the epithet of *Abies balsamea* (L.) Mill. treated as an adjective).

Ex. 6. Names with transcribed Greek adjectival epithets: *Brachypodium distachyon* (L.) P. Beauv. (*Bromus distachyos* L.), *Oxycoccus macrocarpos* (Aiton) Pursh (*Vaccinium macrocarpon* Aiton).

Ex. 7. Names with a noun in apposition for an epithet: *Convolvulus cantabrica* L., *Gentiana pneumonanthe* L., *Liriodendron tulipifera* L., *Lythrum salicaria* L., *Schinus molle* L., all with epithets featuring pre-Linnaean generic names.

Ex. 8. Names with a genitive noun for an epithet: *Bromus tectorum* L., *Capsicum caatingae* Barboza & Agra, *Cistus clusii* Dunal, *Gloeosporium balsameae* Davis (derived from the epithet of *Abies balsamea* (L.) Mill. treated as a noun).

Ex. 9. Names with an adverb for an epithet: *Acrostichum deorsum* H. Karst., *Brachyotum seorsum* Wurdack, *Caladenia postea* Hopper & A. P. Br., *Phaca unde* Rydb., *Rubus satis* L. H. Bailey.

Ex. 10. Correctable errors in Latin adjectival epithets: *Zanthoxylum trifoliatum* L. (Sp. Pl.: 270. 1753) upon transfer to *Acanthopanax* (Decne. & Planch.) Miq. (masculine, see Art. 62.2(a)) is correctly *A. trifoliatus* (L.) Voss (Vilm. Blumengärtn., ed. 3: 1: 406. 1894, '*trifoliatum*'); *Mimosa latisiliqua* L. (Sp. Pl.: 519. 1753) upon transfer to *Lysiloma* Benth. (neuter) is correctly *L. latisiliquum* (L.) Benth. (in Trans. Linn. Soc. London 30: 534. 1875, '*latisiliqua*'); *Corydalis chaerophylla* DC. (Prodr. 1: 128. 1824) upon transfer to *Capnoides* Mill. (feminine, see Art. 62.4) is correctly *Capnoides chaerophylla* (DC.) Kuntze (Revis. Gen. Pl. 1: 14.

1891, '*chaerophyllum*'); *Areca monostachya* Mart. (Hist. Nat. Palm. 3: 178. 1838) upon transfer to *Linospadix* H. Wendl. (masculine) is correctly *L. monostachyus* (Mart.) H. Wendl. (in Linnaea 39: 199. 1875, '*monostachyos*').

Ex. 11. Correctable errors in transcribed Greek adjectival epithets: *Andropogon distachyos* L. (Sp. Pl.: 1046. 1753, '*distachyon*'); *Bromus distachyos* L. (Fl. Palaest.: 13. 1756) upon transfer to *Brachypodium* P. Beauv. (neuter) is correctly *B. distachyon* (L.) P. Beauv. (Ess. Agrostogr.: 155. 1812, '*distachyum*') or to *Trachynia* Link (feminine) is correctly *T. distachyos* (L.) Link (Hort. Berol. 1: 43. 1827, '*distachya*'); *Vaccinium macrocarpon* Aiton (Hort. Kew. 2: 13. 1789) upon transfer to *Oxycoccus* Hill (masculine) is correctly *O. macrocarpos* (Aiton) Pursh (Fl. Amer. Sept. 1: 263. 1813, '*macrocarpus*') or to *Schollera* Roth (feminine) is correctly *S. macrocarpos* (Aiton) Steud. (Nomencl. Bot.: 746. 1821, '*macrocarpa*').

Ex. 12. Correctable errors in epithets that are nouns: the epithet of *Polygonum segetum* Kunth (in Humboldt & al., Nov. Gen. Sp. 2, ed. qu.: 177; ed. fol.: 142. 1817) is a genitive plural noun (of the corn fields); when Small (Fl. S.E. U.S.: 378. 1903) proposed the new combination *Persicaria* '*segeta*', it was a correctable error for *Persicaria segetum* (Kunth) Small. In *Masdevallia echidna* Rchb. f. (in Bonplandia 3: 69. 1855), the epithet corresponds to the generic name of an animal; when Garay (in Svensk Bot. Tidskr. 47: 201. 1953) proposed the new combination *Porroglossum* '*echidnum*', it was a correctable error for *P. echidna* (Rchb. f.) Garay.

Ex. 13. Correctable errors in Latin epithets that consist of a noun and its accompanying adjective in the genitive case: *Agrostophyllum montis-jayani* Ormerod (in Orchadian 17: 379. 2013, '*montis-jayanum*'); *Loranthus cycnei-sinus* Blakely (in Proc. Linn. Soc. New South Wales 47: 392. 1922, '*Cycneus-Sinus*'); *Salicornia sinus-persici* Akhani (in Pakistan J. Bot. 40: 1638. 2008, '*sinus-persica*').

Ex. 14. Correctable error in the usage of -*cola* as an adjective: when Blanchard (in Rhodora 8: 170. 1906) proposed *Rubus* '*amnicolus*', it was a correctable error for *R. amnicola* Blanch.

23.6. When the final epithet of a name can be interpreted as belonging to two different grammatical categories (e.g. an adjective and a noun), and both are correct under the rules, a subsequent author may choose (directly or indirectly) one of those categories. The first such choice to be effectively published (Art. 29–31) is to be followed.

Ex. 15. The final epithet in *Ruellia hybrida* Pursh (Fl. Amer. Sept. 2: 420. 1813) may be considered as either a noun in apposition or a feminine adjective, because neither option was indicated in the protologue. When the final epithet was combined as *Dipteracanthus ciliosus* var. *hybridus* (Pursh) Nees (in Candolle, Prodr. 11: 123. 1847), Nees chose to treat it as an adjective, and his choice is to be followed.

Ex. 16. When *Peziza lachnoderma* Berk. (in Hooker, Bot. Antarct. Voy., III, Fl. Tasman. 2: 274. 1859) was published, Berkeley did not indicate whether the final epithet was a noun in apposition or a feminine adjective, and both interpretations are possible. A combination made by Rehm (in Ber. Naturhist. Augsburg 26: 76. 1881), currently accepted in *Dasyscyphus*, did not constitute a choice because, at that time, Rehm combined the final epithet under a feminine orthographical variant of the generic name, *"Dasyscypha"*. The first choice was made by Kuntze (Revis. Gen. Pl. 3: 446. 1898), who effectively published the combination *Atractobolus lachnoderma* (Berk.) Kuntze under a masculine generic name, unambiguously using the final epithet as a noun in apposition; Kuntze's choice is to be followed.

23.7. The following designations are not to be regarded as species names:

(a) Designations consisting of a generic name followed by an epithet in the form of a phrase in the ablative case (but see Art. 23.8).

> *Ex. 17. Solanum "fructu-tecto"* (Cavanilles, Icon. 4: 5. 1797) is a generic name followed by an epithet in the form of a phrase in the ablative case. It is not to be regarded as a species name.

(b) Designations consisting of a generic name followed by a phrase name (Linnaean "nomen specificum legitimum") often composed of one or more nouns and associated adjectives in the ablative case, but also including any single-word phrase names in works in which phrase names of two or more words predominate.

> *Ex. 18. Smilax "caule inermi"* (Aublet, Hist. Pl. Guiane 2, Tabl.: 27. 1775) is an abbreviated descriptive reference to an imperfectly known species, which is not given a binomial in the text but referred to merely by a phrase name cited from Burman.

> *Ex. 19.* In Miller, *The gardeners dictionary ... abridged,* ed. 4 (1754), phrase names of two or more words largely predominate over those that consist of a single word and that are thereby similar to Linnaean nomina trivialia (specific epithets) but are not distinguished typographically or in any other way from other phrase names. Therefore, designations in that work such as *"Alkekengi officinarum", "Leucanthemum vulgare", "Oenanthe aquatica",* and *"Sanguisorba minor"* are not validly published names.

(c) Other designations of species consisting of a generic name followed by one or more words not intended as a specific epithet.

> *Ex. 20. Viola "qualis"* (qualis, of what sort) (Krocker, Fl. Siles. 2: 512, 517. 1790). *Urtica "dubia?"* (dubia, doubtful) (Forsskål, Fl. Aegypt.-Arab.: cxxi. 1775); the word "dubia?" was repeatedly used in Forsskål's work for species that could not be reliably identified.

> *Ex. 21. Atriplex "nova"* (Winterl, Index Hort. Bot. Univ. Hung.: fol. A [8] recto et verso. 1788); the word "nova" (new) was here used in connection with four

different species of *Atriplex.* However, in *Artemisia nova* A. Nelson (in Bull. Torrey Bot. Club 27: 274. 1900), the species was newly distinguished from others and *nova* was intended as a specific epithet.

Ex. 22. *Cornus "gharaf"* (Forsskål, Fl. Aegypt.-Arab.: xci, xcvi. 1775) is a designation not intended as a species name. Such usage in Forsskål's work is an original designation for an accepted taxon with an epithet-like vernacular name that is not used as an epithet in the "Centuriae" part of the work. *Elcaja "roka"* (Forsskål, l.c.: xcv. 1775) is another example of such a designation; in other parts of the work (l.c.: c, cxvi, 127) this species is not named.

Ex. 23. In *Agaricus "octogesimus nonus"* and *Boletus "vicesimus sextus"* (Schaeffer, Fung. Bavar. Palat. Nasc. 1: t. 100. 1762; 2: t. 137. 1763), the generic names are followed by ordinal adjectives used for enumeration that are not intended as epithets. The corresponding species were given validly published names, *A. cinereus* Schaeff., nom. sanct., and *B. ungulatus* Schaeff., in the final volume of the same work (l.c. 4: 100, 88. 1774).

Ex. 24. In *Agrostis,* Honckeny (Verz. Gew. Teutschl. 1782; see Art. 46 Ex. 47) used species designations such as *"A. Reygeri I.", "A. Reyg. II.", "A. Reyg. III."* (all referring to species described but not named in Reyger, Tent. Fl. Gedan.: 36–37. 1763), and also *"A. alpina. II"* for a newly described species listed after *A. alpina* Scop. These enumerations were not intended as epithets; there is no provision for expansion of the binomials into, e.g., *"Agrostis reygeri-prima".*

(d) **Designations of species consisting of a generic name followed by two or more adjectival words in the nominative case.**

Ex. 25. *"Salvia africana caerulea"* (Linnaeus, Sp. Pl.: 26. 1753) and *"Gnaphalium fruticosum flavum"* (Forsskål, Fl. Aegypt.-Arab.: cxix. 1775) are generic names followed by two adjectival words in the nominative case. They are not to be regarded as species names.

Ex. 26. *Rhamnus 'vitis idaea'* Burm. f. (Fl. Ind.: 61. 1768) is to be regarded as a species name because the generic name is followed by a noun and an adjective, both in the nominative case; these words are to be hyphenated *(R. vitis-idaea)* under the provisions of Art. 23.1 and 60.12. In *Anthyllis 'Barba jovis'* L. (Sp. Pl.: 720. 1753) the generic name is followed by a noun in the nominative case and a noun in the genitive case, and they are to be hyphenated *(A. barba-jovis).* Likewise, *Hyacinthus 'non scriptus'* L. (l.c.: 316. 1753), where the generic name is followed by a negative particle and a past participle used as an adjective, is corrected to *H. non-scriptus,* and *Impatiens 'noli tangere'* L. (l.c.: 938. 1753), where the generic name is followed by two verbs, is corrected to *I. noli-tangere.*

Ex. 27. In *Narcissus 'Pseudo Narcissus'* L. (Sp. Pl.: 289. 1753) the generic name is followed by a prefix (a word that cannot stand independently) and a noun in the nominative case, and the name is to be corrected to *N. pseudonarcissus* under the provisions of Art. 23.1 and 60.12.

(e) Formulae designating hybrids (see Art. H.10.2).

23.8. Names in which Linnaeus used phrases in the ablative case as specific epithets ("nomina trivialia") are to be corrected in accordance with later usage by Linnaeus himself.

Ex. 28. *Apocynum* '*fol. [foliis] androsaemi*' L. is cited as *A. androsaemifolium* L. (Sp. Pl.: 213. 1753 [corr. L., Syst. Nat., ed. 10: 946. 1759]); and *Mussaenda* '*fr. [fructu] frondoso*' L., as *M. frondosa* L. (Sp. Pl.: 177. 1753 [corr. L., Syst. Nat., ed. 10: 931. 1759]).

23.9. Where the status of a designation of a species is uncertain under Art. 23.7, established custom is to be followed (Pre. 13).

*Ex. 29. *Polypodium* '*F. mas*', *P.* '*F. femina*', and *P.* '*F. fragile*' (Linnaeus, Sp. Pl.: 1090–1091. 1753) are, in accordance with established custom, to be treated as *P. filix-mas* L., *P. filix-femina* L., and *P. fragile* L., respectively. Likewise, *Cambogia* '*G. gutta*' is to be treated as *C. gummi-gutta* L. (Gen. Pl., ed. 5: [522]. 1754). The intercalations *"Trich." [Trichomanes]* and *"M." [Melilotus]* in the names of Linnaean species of *Asplenium* and *Trifolium*, respectively, are to be deleted, so that names in the form *Asplenium* '*Trich. dentatum*' and *Trifolium* '*M. indica*', for example, are treated as *A. dentatum* L. and *T. indicum* L. (Sp. Pl.: 765, 1080. 1753).

Recommendation 23A

23A.1. Names of persons and also of countries and localities used in specific epithets should take the form of nouns in the genitive *(clusii, porsildiorum, saharae)* or of adjectives *(clusianus, dahuricus)* (see also Art. 60, Rec. 60C, and 60D).

23A.2. The use of the genitive and the adjectival form of the same word to designate two different species of the same genus should be avoided (e.g. *Lysimachia hemsleyana* Oliv. and *L. hemsleyi* Franch.).

23A.3. In forming specific epithets, authors should comply also with the following:

(a) Use Latin terminations insofar as possible.
(b) Avoid epithets that are very long.
(c) Not make epithets by combining words from different languages.
(d) Avoid those formed of two or more hyphenated words.
(e) Avoid those that have the same meaning as the generic name.
(f) Avoid those that express a character common to all or nearly all the species of a genus.
(g) Avoid in the same genus those that are very much alike, especially those that differ only in their last letters or in the arrangement of two letters.
(h) Avoid those that have been used before in any closely allied genus.

(i) Not adopt epithets from unpublished names found in correspondence, travellers' notes, herbarium labels, or similar sources, attributing them to their authors, unless these authors have approved publication (see Rec. 50G).

(j) Avoid using the names of little-known or very restricted localities unless the species is quite local.

SECTION 5
NAMES OF TAXA BELOW THE RANK OF SPECIES (INFRASPECIFIC TAXA)

ARTICLE 24
NAMES OF INFRASPECIFIC TAXA

24.1. The name of an infraspecific taxon is a combination of the name of a species and an infraspecific epithet. A connecting term is used to denote the rank.

> **Ex. 1.** *Saxifraga aizoon* subf. *surculosa* Engl. & Irmsch. This taxon may also be referred to as *Saxifraga aizoon* var. *aizoon* subvar. *brevifolia* f. *multicaulis* subf. *surculosa* Engl. & Irmsch.; in this way a full classification of the subform within the species is given, not only its name.

24.2. Infraspecific epithets are formed like specific epithets (Art. 23) and, when adjectival in form and not used as nouns or adverbs, they agree grammatically with the generic name (see Art. 23.5 and 32.2).

> **Ex. 2.** *Solanum melongena* var. *insanum* (L.) Prain (Bengal Pl.: 746. 1903, '*insana*').

24.3. Infraspecific names with final epithets such as *genuinus, originalis, originarius, typicus, verus,* and *veridicus,* or with the prefix *eu-,* when purporting to indicate the taxon containing the type of the name of the next higher-ranked taxon, are not validly published unless they have the same final epithet as the name of the corresponding higher-ranked taxon (see Art. 26.2, Rec. 26A.1, and 26A.3).

> **Ex. 3.** *"Hieracium piliferum* var. *genuinum"* (Rouy, Fl. France 9: 270. 1905) was based on *"H. armerioides* var. *genuinum"* of Arvet-Touvet (Hieracium Alpes Franç.: 37. 1888), a designation not validly published under Art. 26.2. As circumscribed by Rouy, the taxon does not include the type of *H. piliferum* Hoppe, but it does include the type of the name of the next higher-ranked taxon, *H. piliferum*

subsp. *armerioides* (Arv.-Touv.) Rouy. Therefore, *"H. piliferum* var. *genuinum"* is not a validly published name of a new variety.

Ex. 4. *"Narcissus bulbocodium* var. *eu-praecox"* and *"N. bulbocodium* var. *eu-albidus"* were not validly published by Emberger & Maire (in Jahandiez & Maire, Cat. Pl. Maroc: 961. 1941) because they were placed, respectively, in *N. bulbocodium* subsp. *praecox* Gattef. & Maire (in Bull. Soc. Hist. Nat. Afrique N. 28: 540. 1937) and *N. bulbocodium* subsp. *albidus* (Emb. & Maire) Maire (in Jahandiez & Maire, l.c.: 138. 1931) and their epithet purports inclusion of the type of the higher-ranked name in the subordinate variety.

Ex. 5. *"Lobelia spicata* var. *originalis"* (McVaugh in Rhodora 38: 308. 1936) was not validly published (see Art. 26 Ex. 1), whereas the autonyms *Galium verum* L. subsp. *verum* and *G. verum* var. *verum* are validly published.

Ex. 6. *Aloe perfoliata* var. *vera* L. (Sp. Pl.: 320. 1753) is validly published because it does not purport to contain the type of *A. perfoliata* L. (l.c. 1753).

24.4. A name with a binary combination instead of an infraspecific epithet, but otherwise in accordance with this *Code,* is treated as validly published in the form determined by Art. 24.1 without change of authorship or date.

Ex. 7. *Salvia grandiflora* subsp. *"S. willeana"* (Holmboe in Bergens Mus. Skr., ser. 2, 1(2): 157. 1914) is to be altered to *S. grandiflora* subsp. *willeana* Holmboe.

Ex. 8. *Phyllerpa prolifera* var. *"Ph. firma"* (Kützing, Sp. Alg.: 495. 1849) is to be altered to *P. prolifera* var. *firma* Kütz.

Ex. 9. *Cynoglossum cheirifolium* "β. *Anchusa (lanata)"* (Lehmann, Pl. Asperif. Nucif.: 141. 1818), a new combination based on *Anchusa lanata* L. (Syst. Nat., ed. 10, 2: 914. 1759), is to be altered to *C. cheirifolium* var. *lanatum* (L.) Lehm.

ⓘ **Note 1.** Infraspecific taxa within different species may have names with the same final epithet; those within one species may have names with the same final epithet as the names of other species (but see Rec. 24B.1).

Ex. 10. *Rosa glutinosa* var. *leioclada* H. Christ (in Boissier, Fl. Orient. Suppl.: 222. 1888) and *Rosa jundzillii* f. *leioclada* Borbás (in Math. Term. Közlem. 16: 376, 383. 1880) are both permissible, as is *Viola tricolor* var. *hirta* Ging. (in Candolle, Prodr. 1: 304. 1824), despite the previous existence of *Viola hirta* L. (Sp. Pl.: 934. 1753).

ⓘ **Note 2.** Names of infraspecific taxa within the same species, even if they differ in rank, are homonyms if they have the same final epithet but are based on different types (Art. 53.3), because the rank-denoting term is not part of the name.

Recommendation 24A

24A.1. Recommendations made for forming specific epithets (Rec. 23A) apply equally for infraspecific epithets.

Recommendation 24B

24B.1. Authors proposing new infraspecific names should avoid final epithets previously used as specific epithets in the same genus.

24B.2. When an infraspecific taxon is raised to the rank of species, or the inverse change occurs, the final epithet of its name should be retained unless the resulting combination would be contrary to the *Code*.

ARTICLE 25
SUM OF SUBORDINATE TAXA

25.1. For nomenclatural purposes, a species or any taxon below the rank of species is regarded as the sum of its subordinate taxa, if any.

> **Ex. 1.** When *Montia parvifolia* (DC.) Greene is treated as comprising two subspecies, the name *M. parvifolia* applies to the species in its entirety, i.e. including both *M. parvifolia* subsp. *parvifolia* and *M. parvifolia* subsp. *flagellaris* (Bong.) Ferris, and its use for *M. parvifolia* subsp. *parvifolia* alone may lead to confusion.

ARTICLE 26
AUTONYMS OF INFRASPECIFIC TAXA

26.1. The name of any infraspecific taxon that includes the type of the adopted, legitimate name of the species to which it is assigned is to repeat the specific epithet unaltered as its final epithet, not followed by an author citation (see Art. 46). Such names are autonyms (Art. 6.8; see also Art. 7.7).

> **Ex. 1.** The variety that includes the type of the name *Lobelia spicata* Lam. is to be named *Lobelia spicata* Lam. var. *spicata* (see also Art. 24 Ex. 5).

ⓘ **Note 1.** Art. 26.1 applies only to the names of those subordinate taxa that include the type of the adopted name of the species (but see Rec. 26A).

26.2. A name of an infraspecific taxon that includes the type (i.e. the holotype or all syntypes or the previously designated or conserved type) of the adopted, legitimate name of the species to which it is assigned is not validly published unless its final epithet repeats the specific epithet unaltered. For the purpose of this provision, explicit indication that the nomenclaturally typical element of the species is included is considered as equivalent to inclusion of the type, whether or not it has been previously designated (see also Art. 24.3).

Ex. 2. The intended combination "*Vulpia myuros* subsp. *pseudomyuros* (Soy.-Will.) Maire & Weiller" was not validly published in Maire (Fl. Afrique N. 3: 177. 1955) because it included in synonymy "*F. myuros* L., Sp. 1, p. 74 (1753) sensu stricto", i.e. *Festuca myuros* L., the basionym of *Vulpia myuros* (L.) C. C. Gmel.

Ex. 3. Linnaeus (Sp. Pl.: 3. 1753) recognized two named varieties under *Salicornia europaea*. Because *S. europaea* has neither a holotype nor syntypes, both varietal names are validly published even though the lectotype of *S. europaea* (designated by Jafri & Rateeb in Jafri & El-Gadi, Fl. Libya 58: 57. 1979) can be attributed to *S. europaea* var. *herbacea* L. (l.c. 1753) and the varietal name was subsequently lectotypified (by Piirainen in Ann. Bot. Fenn. 28: 82. 1991) with the same specimen as the species name.

Ex. 4. Linnaeus (Sp. Pl.: 779–781. 1753) recognized 13 named varieties under *Medicago polymorpha*. Because *M. polymorpha* L. has neither a holotype nor syntypes, all varietal names are validly published, and the lectotype subsequently designated for the species name (by Heyn in Bull. Res. Council Israel, Sect. D, Bot., 7: 163. 1959) is not part of the original material for any of the varietal names of 1753.

26.3. The first instance of valid publication of a name of an infraspecific taxon under a legitimate species name automatically establishes the corresponding autonym (see also Art. 11.6 and 32.3).

Ex. 5. The publication of the name *Lycopodium inundatum* var. *bigelovii* Tuck. (in Amer. J. Sci. Arts 45: 47. 1843) automatically established the name of another variety, *L. inundatum* L. var. *inundatum,* the autonym, the type of which is that of the name *L. inundatum* L. (Art. 7.7).

Ex. 6. Pangalo (in Trudy Prikl. Bot. 23: 258. 1930), when describing *Cucurbita mixta* Pangalo, distinguished two varieties, *C. mixta* var. *cyanoperizona* Pangalo and *C. mixta* var. *stenosperma* Pangalo, together encompassing the entire circumscription of the species. Although Pangalo did not mention the autonym (see Rec. 26B.1), *C. mixta* var. *mixta* was automatically established at the same time. Because neither a holotype nor any syntypes were indicated for *C. mixta,* both varietal names were validly published (see Art. 26.2). Merrick & Bates (in Baileya 23: 96, 101. 1989), in the absence of known type material, neotypified *C. mixta* by an element that can be attributed to *C. mixta* var. *stenosperma.* As long as their choice of neotype is followed, under Art. 11.6 the correct name for that variety recognized under *C. mixta* is *C. mixta* var. *mixta,* dating from 1930, not *C. mixta* var. *stenosperma.*

Recommendation 26A

26A.1. A variety including the type of the correct name of a subspecies, but not including the type of the correct name of the species, should, where there is no obstacle under the rules, be given a name with the same final epithet and type as the subspecific name.

> ***Ex. 1.*** Fernald treated *Stachys palustris* subsp. *pilosa* (Nutt.) Epling (in Repert. Spec. Nov. Regni Veg. Beih. 80: 63. 1934) as composed of five varieties, for one of which (that including the type of *S. palustris* subsp. *pilosa*) he made the combination *S. palustris* var. *pilosa* (Nutt.) Fernald (in Rhodora 45: 474. 1943) because there was no legitimate varietal name available.

26A.2. A subspecies not including the type of the correct name of the species should, where there is no obstacle under the rules, be given a name with the same final epithet and type as a name of one of its subordinate varieties.

> ***Ex. 2.*** Because there was no legitimate name available at the rank of subspecies, Bonaparte made the combination *Pteridium aquilinum* subsp. *caudatum* (L.) Bonap. (Notes Ptérid. 1: 62. 1915), using the same final epithet that Sadebeck had used earlier in the combination *P. aquilinum* var. *caudatum* (L.) Sadeb. (in Jahrb. Hamburg. Wiss. Anst. Beih. 14(3): 5. 1897), with both combinations based on *Pteris caudata* L. Each name is legitimate, and both can be used, as was done by Tryon (in Rhodora 43: 52–54. 1941), who treated *P. aquilinum* var. *caudatum* as one of four varieties under subsp. *caudatum* (see also Art. 36.3).

26A.3. A taxon at a rank lower than variety that includes the type of the correct name of a subspecies or variety, but not the type of the correct name of the species, should, where there is no obstacle under the rules, be given a name with the same final epithet and type as the name of the subspecies or variety. On the other hand, a subspecies or variety that does not include the type of the correct name of the species should not be given a name with the same final epithet as a name of one of its subordinate taxa below the rank of variety.

Recommendation 26B

26B.1. When publishing a name of an infraspecific taxon that will also establish an autonym, the author should mention that autonym in the publication.

ARTICLE 27
FINAL EPITHET IN NAMES OF INFRASPECIFIC TAXA

27.1. The final epithet in the name of an infraspecific taxon may not repeat unchanged the epithet of the correct name of the species to which the taxon is assigned unless the two names have the same type.

27.2. The final epithet in the name of an infraspecific taxon may not repeat unchanged the epithet of the species name if that species name is illegitimate.

> **Ex. 1.** When Honda (in Bot. Mag. (Tokyo) 41: 385. 1927) published *Agropyron japonicum* var. *hackelianum* Honda under the illegitimate *A. japonicum* Honda (l.c.: 384. 1927), which is a later homonym of *A. japonicum* (Miq.) P. Candargy (in Arch. Biol. Vég. Pure Appl. 1: 42. 1901), he did not validly publish an autonym "*A. japonicum* var. *japonicum*" (see also Art. 55 Ex. 3).

SECTION 6
NAMES OF ORGANISMS IN CULTIVATION

ARTICLE 28

28.1. Organisms brought from the wild into cultivation retain the names that are applied to them when growing in nature.

ⓘ **Note 1.** Hybrids, including those arising in cultivation, may receive names as provided in Chapter H (see also Art. 11.9, 32.4, and 50.1).

ⓘ **Note 2.** Additional, independent designations for special categories of organisms used in agriculture, forestry, and horticulture (and arising either in nature or cultivation) are dealt with in the *International Code of Nomenclature for Cultivated Plants (ICNCP),* which defines the cultivar as its basic category (see Pre. 11).

ⓘ **Note 3.** Nothing precludes the use, for cultivated organisms, of names published in accordance with the requirements of this *Code*.

ⓘ **Note 4.** Epithets in names published in conformity with this *Code* are retained as cultivar epithets, included in single quotation marks, under the rules of the *ICNCP* when it is considered appropriate to treat the taxon concerned under that *Code*.

> **Ex. 1.** *Mahonia japonica* DC. (Syst. Nat. 2: 22. 1821) may be treated as a cultivar, which is then designated as *Mahonia* 'Japonica'; *Taxus baccata* var. *variegata* Weston (Bot. Univ. 1: 292, 347. 1770), when treated as a cultivar, is designated as *Taxus baccata* 'Variegata'.

ⓘ **Note 5.** The *ICNCP* also provides for the establishment of epithets differing markedly from epithets provided for under this *Code*.

> **Ex. 2.** ×*Disophyllum* 'Frühlingsreigen'; *Eriobotrya japonica* 'Golden Ziad' and *E. japonica* 'Maamora Golden Yellow'; *Phlox drummondii* 'Sternenzauber'; *Quercus frainetto* 'Hungarian Crown'.

Ex. 3. Juniperus ×pfitzeriana 'Wilhelm Pfitzer' (P. A. Schmidt in Folia Dendrol. 10: 292. 1998) was established for a tetraploid cultivar presumed to result from the original cross between *J. chinensis* L. and *J. sabina* L.

CHAPTER IV
EFFECTIVE PUBLICATION

ARTICLE 29
DEFINITION AND CONDITIONS OF EFFECTIVE PUBLICATION

29.1. Publication is effected, under this *Code,* by distribution of printed matter (through sale, exchange, or gift) to the general public or at least to scientific institutions with generally accessible libraries. Publication is also effected by distribution on or after 1 January 2012 of electronic material in Portable Document Format (PDF; see also Art. 29.3 and Rec. 29A.1) in an online publication (Art. 29.2) with an International Standard Serial Number (ISSN) or an International Standard Book Number (ISBN) (see also Art. 30).

Ex. 1. The paper containing the new combination *Anaeromyces polycephalus* (Y. C. Chen & al.) Fliegerová & al. (Kirk in Index Fungorum 1: 1. 2012), based on *Piromyces polycephalus* Y. C. Chen & al. (in Nova Hedwigia 75: 411. 2002), was effectively published when it was issued online in Portable Document Format with an ISSN on 1 January 2012.

Ex. 2. Intended nomenclatural novelties by Ruck & al. (in Molec. Phylogen. Evol. 103: 155–171. 22 Jul 2016) appeared only in supplementary material published online in Microsoft Word document format and were not therefore effectively published. These novelties were effectively published when they appeared in Portable Document Format (Ruck & al. in Notul. Alg. 10: 1–4. 17 Aug 2016), meeting the requirements of Art. 29.1.

ⓘ *Note 1.* The distribution before 1 January 2012 of electronic material does not constitute effective publication.

Ex. 3. Floristic accounts of the *Asteraceae* in *Flora of China* volume 20–21, containing numerous nomenclatural novelties, were published online in Portable Document Format on 25 October 2011. Because they were distributed before 1 January 2012 they were not effectively published. Effective publication occurred when the printed version of the same volume became available on 11 November 2011.

Ex. 4. The paper in which the diatom *"Tursiocola podocnemicola"* was first described was distributed online on 14 December 2011 as an "iFirst" PDF document (https://doi.org/10.1080/0269249X.2011.642498) available through the

Diatom Research website (ISSN 0269-249X, print; ISSN 2159-8347, online). Although the paper appeared online in an electronic publication with an ISSN and in Portable Document Format, it was distributed before 1 January 2012 and was not therefore effectively published. It did not become effectively published on 1 January 2012 merely by remaining available online. Effective publication occurred on 28 February 2012 upon distribution of the printed version of the journal in which the name *T. podocnemicola* C. E. Wetzel (in Diatom Res. 27: 2. 2012) was validly published.

29.2. For the purpose of Art. 29.1, "online" is defined as accessible electronically via the World Wide Web.

29.3. Should Portable Document Format (PDF) be succeeded, a successor international standard format communicated by the General Committee (see Div. III Prov. 7.10(1)) is acceptable.

ⓘ *Note 2.* Citation, for electronic material, of an inappropriate ISSN or ISBN (e.g. one that does not exist or that refers to a serial publication or book in which that electronic material is not included, not even as a declared supplement to an included item) does not result in effective publication under Art. 29.1.

Ex. 5. The paper by Meyer, Baquero, and Cameron in which *"Dracula trigonopetala"* was described as an intended new species was placed online as a PDF/A document on 1 March 2012. There was no mention of a journal or ISSN in the document itself, but, because it was made accessible through the homepage of *OrchideenJournal* (ISSN 1864-9459), it could be argued that it qualified as an "online publication with an International Standard Serial Number" (Art. 29.1). However, the content of the paper was not presented in a format suited for publication in the *OrchideenJournal* and was evidently not intended for inclusion in that journal. A new version of the paper, translated into German, appeared in print (in OrchideenJ. 19: 107–112) on 15 August 2012. Although this was effectively published, *"D. trigonopetala"* was not validly published there because no Latin or English description or diagnosis was provided. (The name was later validated as *D. trigonopetala* Gary Mey. & Baquero ex A. Doucette in Phytotaxa 74: 59. 9 December 2012.)

Recommendation 29A

29A.1. Electronic publication in Portable Document Format (PDF) should comply with the PDF/A archival standard (ISO 19005).

29A.2. Authors of electronic material should give preference to publications that are archived and curated, satisfying the following criteria as far as is practical (see also Rec. 29A.1):

(a) The material should be placed in multiple trusted online digital repositories, e.g. an ISO-certified repository.

(b) Digital repositories should be in more than one area of the world and preferably on different continents.

ARTICLE 30
FURTHER CONDITIONS OF EFFECTIVE PUBLICATION

30.1. Publication is not effected by communication of nomenclatural novelties at a public meeting, by the placing of names in collections or gardens open to the public, by the issue of microfilm made from manuscripts or typescripts or other unpublished material, or by distribution of electronic material other than as described in Art. 29.

> ***Ex. 1.*** Cusson announced his establishment of the genus *Physospermum* in a memoir read at the Société des Sciences de Montpellier in 1770, and later in 1782 or 1783 at the Société de Médecine de Paris, but its effective publication dates from 1787 (in Hist. Soc. Roy. Méd. 5(1): 279).

30.2. An electronic publication is not effectively published if there is evidence within or associated with the publication that its content is merely preliminary and was, or is to be, replaced by content that the publisher considers final, in which case only the version with that final content is effectively published.

> ***Ex. 2.*** *"Rodaucea"* was published in a paper first placed online on 12 January 2012 as a PDF document accessible through the website of the journal *Mycologia* (ISSN 0027-5514, print; ISSN 1557-2436, online). That document had a header stating "In Press", and on the journal website it was qualified as "Preliminary version", which is clear evidence that it was not considered by the publisher as final. Because the final version of the document appeared simultaneously online and in print, a correct citation of the name is: *Rodaucea* W. Rossi & Santam. in Mycologia 104 (print and online): 785. 11 Jun 2012.

> ***Ex. 3.*** *"Lycopinae"* appeared in a paper first placed online on 26 April 2012 as an "Advance Access" PDF document accessible through the website of the *American Journal of Botany* (ISSN 0002-9122, print; ISSN 1537-2197, online). Because the journal website stated (May 2012) "AJB Advance Access articles … have not yet been printed or posted online by issue" and "minor corrections may be made before the issue is released", this was evidently not considered as the final version by the publisher. The name *Lycopinae* B. T. Drew & Sytsma was validly published in Amer. J. Bot. 99: 945. 1 May 2012, when the printed volume containing it was effectively published.

> ***Ex. 4.*** The paper (in S. African J. Bot. 80: 63–66; ISSN 0254-6299, print; ISSN 1727-9321, online) in which the name *Nanobubon hypogaeum* Magee appeared

was effectively published online as a PDF document on 30 March 2012 in its "final and fully citable" form, before publication of the printed version (May 2012). Papers that appeared online in the same journal under the heading "In Press Corrected Proof" are not effectively published because the journal website clearly stated "Corrected proofs: articles that contain the authors' corrections. Final citation details, e.g. volume/issue number, publication year and page numbers, still need to be added and the text might change before final publication."

ⓘ **Note 1.** An electronic publication may be a final version even if details, e.g. volume, issue, article, or page numbers, are to be added or changed, provided that those details are not part of the content (see Art. 30.3).

30.3. Content of an electronic publication includes what is visible on the page, e.g. text, tables, illustrations, etc., but excludes volume, issue, article, and page numbers; it also excludes external sources accessed via a hyperlink or URL (Uniform Resource Locator).

Ex. 5. A paper describing the new genus *Partitatheca* and its four constituent species, accepted for the *Botanical Journal of the Linnean Society* (ISSN 0024-4074, print; ISSN 1095-8339, online), was placed online on 1 February 2012 as an "Early View" PDF document with preliminary pagination (1–29). This was evidently the version considered as final by the journal's publisher because, in the document itself, it was declared the "Version of Record" (an expression defined by the standard NISO-RP-8-2008). Later, in the otherwise identical electronic version published together with the printed version on 27 February 2012, the volume pagination (229–257) was added. A correct citation of the generic name is: *Partitatheca* D. Edwards & al. in Bot. J. Linn. Soc. 168 (online): [2 of 29], 230. 1 Feb 2012, or just "... 168 (online): 230. 1 Feb 2012".

Ex. 6. The new combination *Rhododendron aureodorsale* was made in a paper in *Nordic Journal of Botany* (ISSN 0107-055X, print; ISSN 1756-1051, online), first effectively published online on 13 March 2012 in "Early View", the "Online Version of Record published before inclusion in an issue", with a permanent Digital Object Identifier (DOI) but with preliminary pagination (1-EV to 3-EV). When the printed version was published on 20 April 2012, the pagination of the electronic version was changed to 184–186 and the date of the printed version was added. The combination can be cited as *Rhododendron aureodorsale* (W. P. Fang ex J. Q. Fu) Y. P. Ma & J. Nielsen in Nordic J. Bot. 30 (online): 184. 13 Mar 2012 (https://doi.org/10.1111/j.1756-1051.2011.01438.x).

Ex. 7. Two new *Echinops* species, including *E. antalyensis,* were described in *Annales Botanici Fennici* (ISSN 0003-3847, print; ISSN 1797-2442, online) in a paper effectively published in its definitive form on 13 March 2012 as an online PDF document, still with preliminary pagination ([1]–4) and the watermark "preprint". When the printed version was published on 26 April 2012, the online document was repaginated ([95]–98) and the watermark removed. A correct citation of the name is: *E. antalyensis* C. Vural in Ann. Bot. Fenn. 49 (online): 95. 13 Mar 2012.

30.4. The content of a particular electronic publication may not be altered after it is effectively published. Any such alterations are not themselves effectively published and have no effect on the original publication. Corrections or revisions must be issued separately to be effectively published. Electronic material that has been effectively published remains effectively published even if retracted.

> *Ex. 8. Bauhinia saksuwaniae* Mattapha & al. was effectively published in a paper first placed online on 11 December 2013 as a PDF document accessible through the website of the *Nordic Journal of Botany* (ISSN 1756-1051, online, https://doi.org/10.1111/j.1756-1051.2013.00102.x). That paper was later declared as "retracted" by the publisher and has not appeared in the printed version of the journal (ISSN 0107-055X, print). Despite the retraction, the paper remains effectively published under Art. 29 and 30 and the species name remains validly published.

30.5. Publication by indelible autograph before 1 January 1953 is effective. Indelible autograph produced on or after that date is not effectively published.

30.6. For the purpose of Art. 30.5, indelible autograph is handwritten material reproduced by some mechanical or graphic process (such as lithography, offset, or metallic etching).

> *Ex. 9.* Léveillé, *Flore du Kouy Tchéou* (1914–1915), is a work lithographed from a handwritten text.

> *Ex. 10. Catalogus plantarum hispanicarum … ab A. Blanco lectarum* (Webb & Heldreich, Paris, Jul 1850, folio) was effectively published as an indelible autograph catalogue.

> *Ex. 11.* The *Journal of the International Conifer Preservation Society,* vol. 5[1]. 1997 ("1998"), consists of duplicated sheets of typewritten text with handwritten additions and corrections in several places. The handwritten portions are not effectively published because they are indelible autograph published after 1 January 1953. Intended new combinations (e.g. *"Abies koreana* var. *yuanbaoshanensis",* p. 53) for which the basionym reference is handwritten are not validly published. The entirely handwritten account of a new taxon (p. 61: name, Latin description, statement of type) is not effectively published.

> *Ex. 12.* The generic designation *"Lindenia"* was handwritten in ink by Bentham in the margin of copies of a published but not yet distributed fascicle of the *Plantae hartwegianae* (p. 84. 1841) to replace the struck-out name *Siphonia* Benth., which he had discovered was a later homonym of *Siphonia* Rich. ex Schreb. (Gen. Pl.: 656. 1791). Although the fascicle was then distributed, the handwritten portion was not itself reproduced by mechanical or graphic process and is not therefore effectively published.

30.7. Publication on or after 1 January 1953 in trade catalogues or non-scientific newspapers, and on or after 1 January 1973 in seed-exchange lists, does not constitute effective publication.

30.8. The distribution on or after 1 January 1953 of printed matter accompanying specimens does not constitute effective publication.

ⓘ *Note 2.* If the printed matter is also distributed independently of the specimens, it is effectively published.

Ex. 13. The printed labels of Fuckel's *Fungi rhenani exsiccati* (1863–1874) are effectively published even though not independently issued. The labels antedate Fuckel's subsequent accounts (e.g. in Jahrb. Nassauischen Vereins Naturk. 23–24. 1870).

Ex. 14. Vězda's *Lichenes selecti exsiccati* (1960–1995) were issued with printed labels that were also distributed as printed fascicles; the latter are effectively published, and nomenclatural novelties appearing in Vězda's labels are to be cited from the fascicles.

30.9. Publication on or after 1 January 1953 of an independent non-serial work stated to be a thesis submitted to a university or other institute of education for the purpose of obtaining a degree does not constitute effective publication unless the work includes an explicit statement (referring to the requirements of the *Code* for effective publication) or other internal evidence that it is regarded as an effective publication by its author or publisher.

ⓘ *Note 3.* The presence of an International Standard Book Number (ISBN) or a statement of the name of the printer, publisher, or distributor in the original printed version is regarded as internal evidence that the work was intended to be effectively published.

Ex. 15. "*Meclatis* in *Clematis;* yellow flowering *Clematis* species – Systematic studies in *Clematis* L. *(Ranunculaceae),* inclusive of cultonomic aspects", a "Proefschrift ter verkrijging van de graad van doctor ... van Wageningen Universiteit [Dissertation to obtain the degree of doctor ... from Wageningen University]" by Brandenburg, was effectively published on 8 June 2000 because it has the ISBN 90-5808-237-7.

Ex. 16. The thesis "Comparative investigations on the life-histories and reproduction of some species in the siphoneous green algal genera *Bryopsis* and *Derbesia*" by Rietema, submitted to Rijksuniversiteit te Groningen in 1975, is stated to have been printed ("Druk") by Verenigde Reproduktie Bedrijven, Groningen and was therefore effectively published.

Ex. 17. The dissertation "Die Gattung *Mycena* s.l." by Rexer, submitted to the Eberhard-Karls-Universität Tübingen, was effectively published in 1994 because it includes the statement "Druck [Printing]: Zeeb-Druck, Tübingen 7 (Hagelloch)", referring to a commercial printer. The generic name *Roridomyces* Rexer and the names of new species in *Mycena,* such as *M. taiwanensis* Rexer, are therefore validly published.

Ex. 18. The thesis by Demoulin, "Le genre *Lycoperdon* en Europe et en Amérique du Nord" (1971), was not effectively published because it does not contain internal evidence that it is regarded as such. Even if photocopies of it can be found in some libraries, names of new species of *Lycoperdon,* e.g. *L. americanum* Demoulin, *L. cokeri* Demoulin, and *L. estonicum* Demoulin, introduced there, were validly published in the effectively published paper "Espèces nouvelles ou méconnues du genre *Lycoperdon* (Gastéromycètes)" (Demoulin in Lejeunia, ser. 2, 62: 1–28. 1972).

Ex. 19. The dissertation by Funk, "The Systematics of *Montanoa* Cerv. *(Asteraceae)*", submitted to the Ohio State University in 1980, was not effectively published because it does not contain internal evidence that it is regarded as such. The same applies to facsimile copies of the dissertation printed from microfiche and distributed, on demand, from 1980 onward, by University Microfilms, Ann Arbor. The name *Montanoa imbricata* V. A. Funk, introduced in the dissertation, was validly published in the effectively published paper "The systematics of *Montanoa (Asteraceae, Heliantheae)*" (Funk in Mem. New York Bot. Gard. 36: 1–133. 1982).

Ex. 20. The dissertation "Revision der südafrikanischen Astereengattungen *Mairia* und *Zyrphelis*" submitted in 1990 by Ursula Zinnecker-Wiegand to the Ludwig-Maximilians-Universität München (University of Munich) is not effectively published because it does not include an ISBN, the name of any printer or publisher or distributor, or any statement that it was intended to be effectively published under the *Code,* even though about 50 copies were distributed to other public libraries and all the other formalities for the publication of new taxa were met. The designations in the thesis became validly published names in the effectively published paper by Ortiz & Zinnecker-Wiegand (in Taxon 60: 1194–1198. 2011).

Recommendation 30A

30A.1. Preliminary and final versions of the same electronic publication should be clearly indicated as such when they are first issued. The phrase "Version of Record" should only be used to indicate a final version in which the content will not change.

30A.2. To facilitate citation, final versions of electronic publications should contain final pagination.

30A.3. Authors and editors are strongly recommended to include page numbers on the actual pages of publications, such that if electronic publications are printed, these page numbers are visible.

30A.4. It is strongly recommended that authors avoid publishing nomenclatural novelties in ephemeral printed matter of any kind, in particular printed matter that is multiplied in restricted and uncertain numbers, in which the permanence of the text may be limited, for which effective publication in terms of number of copies is not obvious, or that is unlikely to reach the general public. Authors should also avoid publishing nomenclatural novelties in popular periodicals, in abstracting journals, or on correction slips.

> *Ex. 1.* Kartesz provided an unpaginated printed insert titled "Nomenclatural innovations" to accompany the electronic version (1.0) of the *Synthesis of the North American flora* produced on compact disk (CD-ROM, which is not effectively published under Art. 30.1). This insert, which is effectively published under Art. 29–31, is the place of valid publication of 41 new combinations, which also appear on the disk, in an item authored by Kartesz: "A synonymized checklist and atlas with biological attributes for the vascular flora of the United States, Canada, and Greenland" (e.g. *Dichanthelium hirstii* (Swallen) Kartesz, Synth. N. Amer. Fl., Nomencl. Innov.: [1]. Aug 1999). Kartesz's procedure is not to be recommended, as the insert is unlikely to be permanently stored and catalogued in libraries and so reach the general public.

30A.5. To aid availability through time and place, authors publishing nomenclatural novelties should give preference to periodicals that regularly publish taxonomic work, or else they should send a copy of a publication (printed or electronic) to an indexing centre appropriate to the taxonomic group. When such publications exist only as printed matter, they should be deposited in at least ten, but preferably more, generally accessible libraries throughout the world.

30A.6. Authors and editors are encouraged to mention nomenclatural novelties in the summary or abstract, or list them in an index in the publication.

ARTICLE 31
DATE OF EFFECTIVE PUBLICATION

31.1. The date of effective publication is the date on which the printed matter or electronic material became available as defined in Art. 29 and 30. In the absence of proof establishing some other date, the one appearing in the printed matter or electronic material must be accepted as correct.

> *Ex. 1.* Individual parts of Willdenow's *Species plantarum* were published as follows: 1(1), Jun 1797; 1(2), Jul 1798; 2(1), Mar 1799; 2(2), Dec 1799; 3(1), 1800; 3(2), Nov 1802; 3(3), Apr–Dec 1803; 4(1), 1805; 4(2), 1806; these dates are presently accepted as the dates of effective publication (see Stafleu & Cowan in Regnum Veg. 116: 303. 1988).

Ex. 2. Fries first published *Lichenes arctoi* in 1860 as an independently paginated preprint, which antedates the identical content published in a journal (Nova Acta Regiae Soc. Sci. Upsal., ser. 3, 3: 103–398. 1861).

Ex. 3. *Diatom Research* 2(2) is dated December 1987. Nevertheless, Williams & Round, the authors of a paper in that issue, stated in a subsequent paper (in Diatom Res. 3: 265. 1988) that the actual date of publication had been 18 February 1988. Under Art. 31.1, their statement is acceptable as proof establishing another date of publication for issue 2(2) of the journal.

Ex. 4. The paper in which *Ceratocystis omanensis* Al-Subhi & al. is described was available online in final form on *Science Direct* on 7 November 2005 but was not effectively published (Art. 29 Note 1). It was distributed in print (in Mycol. Res. 110(2): 237–245) on 7 March 2006, which is the date of effective publication.

Ex. 5. On its last page, *Index secundus seminum, quae hortus botanicus Imperialis petropolitanus pro mutua commutatione offert [...],* a seed list co-authored by Fischer and Meyer and issued from Saint Petersburg, is dated "25 Декабря [December] 1835", which belongs to the Julian calendar then observed in Russia. In the Gregorian calendar, this date corresponds to 6 January 1836, which has been accepted as the date of effective publication (see Stafleu & Cowan in Regnum. Veg. 94: 836. 1976).

31.2. When a publication is issued in parallel as electronic material and printed matter, both must be treated as effectively published on the same date unless the dates of the versions are different as determined by Art. 31.1.

Ex. 6. The paper in which *Solanum baretiae* was validly published was placed on-line in final form, as a PDF document, on 3 January 2012 in the journal *PhytoKeys* (ISSN 1314-2003). The printed version (ISSN 1314-2011) of the corresponding issue of *PhytoKeys,* with identical pagination and content, is undated but demonstrably later because it includes a paper dated 6 January 2012. A correct citation of the name is: *S. baretiae* Tepe in PhytoKeys 8 (online): 39. 3 Jan 2012.

31.3. When separates from periodicals or other works placed on sale are issued in advance, the date on the separate is accepted as the date of effective publication unless there is evidence that it is erroneous.

Ex. 7. The names of the *Selaginella* species published by Hieronymus (in Hedwigia 51: 241–272. 1911) were effectively published on 15 October 1911 because the volume in which the paper appeared, though dated 1912, states (p. ii) that the separate appeared on that date.

Ex. 8. *Plantae novae thurberianae* […], a publication authored by Asa Gray and dated 1854, is a preprint from the journal *Memoirs of the American Academy of Arts and Sciences,* new series, volume 5, dated 1855 (see Stafleu & Cowan in Regnum Veg. 94: 991. 1976). The date appearing on the preprint is accepted as the date of effective publication.

Recommendation 31A

31A.1. The date on which the publisher or publisher's agent delivers printed matter to one of the usual carriers for distribution to the public should be accepted as its date of effective publication.

Recommendation 31B

31B.1. The date of effective publication should be clearly indicated as precisely as possible (day, month, year) within the printed matter or electronic material. In printed matter not already published as electronic material, the date should conform to Rec. 31A.1. When a publication is issued in parts, this date should be indicated in each part.

Recommendation 31C

31C.1. On reprints of papers published in a periodical, the name of the periodical, volume and part number, original pagination, and date (day, month, year) of publication should be indicated.

CHAPTER V
VALID PUBLICATION

SECTION 1
VALID PUBLICATION IN GENERAL

ARTICLE 32
GENERAL REQUIREMENTS FOR VALID PUBLICATION

32.1. To be validly published, a name of a taxon (autonyms excepted) must:

(a) be effectively published (Art. 29–31) on or after the starting-point date of the respective group (Art. 13.1 and F.1.1); and

(b) be composed only of letters of the Latin alphabet, except as provided in Art. 23.3, 60.4, 60.7, and 60.12–15; and

(c) have a form that complies with the provisions of Art. 16–27 (but see Art. 21.4 and 24.4) and Art. H.6 and H.7 (see also Art. 61).

ⓘ *Note 1.* The use of typographical signs, numerals, or letters of a non-Latin alphabet in the arrangement of taxa (such as Greek letters α, β, γ, etc. in the arrangement of varieties under a species) does not prevent valid publication because rank-denoting terms and devices are not part of the name.

32.2. Names above the rank of species are validly published even when they or their epithets were published with an improper Latin termination but otherwise in accordance with this *Code;* they are to be changed to accord with Art. 16–19 and 21, without change of authorship or date. Names of species or infraspecific taxa are validly published even when their epithets were published with an improper Latin or transcribed Greek termination but otherwise in accordance with this *Code;* they are to be changed to accord with Art. 23 and 24, without change of authorship or date (see also Art. 60.8).

> *Ex. 1.* The epithet in *Cassia* "*" *'Chamaecristae'* L. (Sp. Pl.: 379. 1753), the name of a subdivision of a genus, is a noun in the nominative plural, derived from *"Chamaecrista"*, a pre-Linnaean generic designation. Under Art. 21.2, however, this epithet must have the same form as a generic name, i.e. a noun in the nominative singular (Art. 20.1). The name is to be changed accordingly and is cited as *Cassia* [unranked] *Chamaecrista* L.

> ⓘ *Note 2.* Improper terminations of otherwise correctly formed names or epithets may result from the use of an inflectional form other than that required by Art. 32.2.

> *Ex. 2. Senecio* sect. *Synotii* Benth. (in Bentham & Hooker, Gen. Pl. 2: 448. 1873) was validly published with reference to certain species that constituted a section ("speciebus tamen nonnullis Asiaticis sectionem subdistinctam *(Synotios)* constituentibus [in some Asian species, however, constituting a subdistinct section *(Synotios)*]"). Although the sectional epithet was written as an adjective in the accusative plural (because it was a direct object), it is to be cited in the nominative plural, *S.* sect. *Synotii,* as required by Art. 21.2.

32.3. Autonyms (Art. 6.8) are accepted as validly published names, dating from the publication in which they were established (see Art. 22.3 and 26.3), whether or not they actually appear in that publication.

32.4. To be validly published, names of hybrids at specific or lower rank with Latin epithets must comply with the same rules as names of non-hybrid taxa at the same rank.

> *Ex. 3. "Nepeta ×faassenii"* (Bergmans, Vaste Pl. Rotsheesters, ed. 2: 544. 1939, with a description in Dutch; Lawrence in Gentes Herb. 8: 64. 1949, with a diagnosis in English) is not validly published because it is not accompanied by or associated with a Latin description or diagnosis (Art. 39.1). The name *Nepeta ×faassenii* Bergmans ex Stearn (in J. Roy. Hort. Soc. 75: 405. 1950) is validly published because it is accompanied by a Latin description.

Ex. 4. *"Rheum ×cultorum"* (Thorsrud & Reisaeter, Norske Plantenavn: 95. 1948) is a nomen nudum and is not therefore validly published (Art. 38.1(a)).

Ex. 5. *"Fumaria ×salmonii"* (Druce, List Brit. Pl.: 4. 1908) is not validly published (Art. 38.1(a)) because only the presumed parentage *(F. densiflora × F. officinalis)* was stated.

ⓘ **Note 3.** For names of hybrids at the rank of genus or of a subdivision of a genus, see Art. H.9.

ⓘ **Note 4.** For valid publication of names of organisms originally assigned to a group not covered by this *Code,* see Art. 45.

Recommendation 32A

32A.1. When publishing nomenclatural novelties, authors should indicate this by a phrase including the word "novus" or its abbreviation, e.g. genus novum (gen. nov., new genus), species nova (sp. nov., new species), combinatio nova (comb. nov., new combination), nomen novum (nom. nov., replacement name), or status novus (stat. nov., name at new rank).

ARTICLE 33
DATE OF VALID PUBLICATION

33.1. The date of a name is that of its valid publication. When the various conditions for valid publication are not simultaneously fulfilled, the date is that on which the last is fulfilled. However, the name must always be explicitly accepted in the place of its valid publication. When the various conditions for valid publication are not simultaneously fulfilled, a name is not validly published unless reference is given to the place(s) where these requirements were previously fulfilled. On or after 1 January 1973, this reference must be full and direct (Art. 41.5; see also Art. 41.7).

Ex. 1. *"Clypeola minor"* first appeared in the Linnaean thesis *Flora monspeliensis* (p. 21, 1756), in a list of names preceded by numerals but without an explanation of the meaning of these numerals and without any other descriptive matter; when the thesis was reprinted in vol. 4 of the *Amoenitates academicae* (1759), a statement was added (p. 475) explaining that the numbers referred to earlier descriptions published in Magnol's *Botanicum monspeliense* (1676). However, *"Clypeola minor"* was absent from the reprint and was not therefore validly published.

Ex. 2. When proposing *"Graphis meridionalis"* as a new species, Nakanishi (in J. Sci. Hiroshima Univ., Ser. B(2), 11: 75. 1966) provided a Latin description but did not designate a type. *Graphis meridionalis* M. Nakan. was validly published only when Nakanishi (l.c.: 265. 1967) designated the holotype of the name and provided a reference to his previous publication.

Ex. 3. *"Hypericum taygeteum"* (Quézel & Contandriopoulos in Naturalia Monspel., Sér. Bot. 16: 121. 1965) was not a validly published name because, although a Latin diagnosis was provided, no type was indicated. *Hypericum taygeteum* Quézel & Contandr. was validly published when the same authors (in Taxon 16: 240. 1967) indicated the type and referred to three of their previous publications but without specifying in which publication and on which page the description of *H. taygeteum* appears. Although this reference was not full and direct, before 1973 it was sufficient to validate the name under Art. 33.1.

Ex. 4. *"Passiflora salpoense"* (Leiva & Tantalean in Arnaldoa 22: 39. 2015) was not validly published because, although a single gathering, *S. Leiva & M. Leiva 5806,* was designated as "tipo", it was specified as being conserved in five herbaria, contrary to Art. 40.5. The name *P. salpoensis* S. Leiva & Tantalean (again as *'salpoense',* but correctable to *salpoensis* under Art. 23.5 and 32.2) was validly published only when the same authors (in Arnaldoa 23: 628. 2016) designated the same gathering as "lectotipo" in a single herbarium, HAO, with "isolectotipos" in CORD, F, MO, and HUT (correctable, respectively, to holotype and isotypes under Art. 9.10), while providing a full and direct reference to their previously published (l.c. 2015) validating English diagnosis of the species.

33.2. A correction of the original spelling of a name (see Art. 32.2 and 60) does not affect its date.

Ex. 5. The correction of the erroneous spelling of *Gluta 'benghas'* L. (Mant. Pl.: 293. 1771) to *G. renghas* L. does not affect the date of the name even though the correction dates from 1883 (Engler in Candolle & Candolle, Monogr. Phan. 4: 225).

ARTICLE 34
SUPPRESSED WORKS

34.1. New names at specified ranks included in publications or parts thereof listed as suppressed works (opera utique oppressa; App. I) are not validly published and no nomenclatural act[1] within the work associated with any name at the specified ranks is effective. Proposals for the addition of publications or parts thereof to App. I must be submitted to the General Committee, which will refer them for examination to the specialist committees for the various taxonomic groups (see

---

1 A nomenclatural act is an act requiring effective publication that results in a nomenclatural novelty (Art. 6 Note 4) or affects aspects of names such as typification (Art. 7.10, 7.11, and F.5.4), priority (Art. 11.5 and 53.5), orthography (Art. 61.3), or gender (Art. 62.3).

Rec. 34A.1, Div. III Prov. 2.2, 7.10(b), 7.11, and 8.13(a); see also Art. 14.12 and 56.2).

ⓘ **Note 1.** For the purpose of Art. 34.1, a "work" is normally a separately published book or a numbered part or supplement of a journal.

> **Ex. 1.** In the suppressed work (see App. I) of Motyka, *Porosty, Lecanoraceae* (3: 97. 1996), one of three specimens of *Lecanora dissipata* Nyl. (in Bull. Soc. Bot. France 13: 368. 1866) in Nylander's herbarium in H was designated as the lectotype for that name. This type designation is not effective and therefore has no nomenclatural status.

34.2. When a proposal for the suppression of a publication or part thereof has been approved by the General Committee after study by the specialist committees for the taxonomic groups concerned, suppression of that publication or part is authorized subject to the decision of a later International Botanical Congress (see also Art. 14.15, 38.5, 53.4, and 56.3) and takes retroactive effect.

Recommendation 34A

34A.1. When a proposal for the suppression of a publication or part thereof under Art. 34.1 has been referred to the appropriate specialist committees for study, authors should follow existing usage of names as far as possible pending the General Committee's recommendation on the proposal (see also Rec. 14A.1 and 56A.1).

34A.2. Individual papers or series of papers in regular issues of journals should not normally be considered as candidates for suppression.

ARTICLE 35
FURTHER REQUIREMENTS FOR VALID PUBLICATION

35.1. A name of a taxon below the rank of genus is not validly published unless the name of the genus or species to which it is assigned is validly published at the same time or was validly published previously (but see Art. 13.4).

> **Ex. 1.** Binary designations for six species of *"Suaeda"*, including *"S. baccata"* and *"S. vera"*, were published with descriptions and diagnoses by Forsskål (Fl. Aegypt.-Arab.: 69–71. 1775), but he provided no description or diagnosis for the genus; these were not therefore validly published names.

> **Ex. 2.** Müller (in Flora 63: 286. 1880) published the new genus *"Phlyctidia"* with the species *"P. hampeana* n. sp.", *"P. boliviensis"* (*Phlyctis boliviensis* Nyl.),

"P. sorediiformis" (*Phlyctis sorediiformis* Kremp.), *"P. brasiliensis"* (*Phlyctis brasiliensis* Nyl.), and *"P. andensis"* (*Phlyctis andensis* Nyl.). However, the intended new binomials were not validly published in this place because the intended generic name *"Phlyctidia"* was not validly published; Müller gave no generic description or diagnosis but only a description and a diagnosis for one additional species, *"P. hampeana",* and so did not validly publish *"Phlyctidia"* under Art. 38.6(b) because the genus was not monotypic (see Art. 38.7). Valid publication of the name *Phlyctidia* was by Müller (in Hedwigia 34: 141. 1895), who provided a short generic diagnosis and explicitly included only two species, the names of which, *P. ludoviciensis* Müll. Arg. and *P. boliviensis* (Nyl.) Müll. Arg., were also validly published in 1895.

ⓘ **Note 1.** Art. 35.1 applies also when specific and other epithets are published under words not to be regarded as names of genera or species (see Art. 20.4 and 23.7).

Ex. 3. The binary designation *"Anonymos aquatica"* (Walter, Fl. Carol.: 230. 1788) is not a validly published name (see Art. 20 Ex. 7). The first validly published name for the species concerned is *Planera aquatica* J. F. Gmel. (Syst. Nat. 2: 150. 1791). This name is not to be cited as *P. aquatica* "(Walter) J. F. Gmel."

Ex. 4. Despite the existence of the generic name *Scirpoides* Ség. (Pl. Veron. 3: 73. 1754), the binary designation *"S. paradoxus"* (Rottbøll, Descr. Pl. Rar.: 27. 1772) is not validly published because "Scirpoides" in Rottbøll's context was a word not intended as a generic name (see Art. 20 Ex. 8). The first validly published name for this species is *Fuirena umbellata* Rottb. (Descr. Icon. Rar. Pl.: 70. 1773).

35.2. A combination (autonyms excepted) is not validly published unless the author definitely associates the final epithet with the name of the genus or species, or with its abbreviation (see Art. 60.15).

ⓘ **Note 2.** This association can be achieved typographically by the position of the final epithet in the text, or by use of a symbol.

ⓘ **Note 3.** When an adjectival epithet is definitely associated with an abbreviation that can stand for two or more generic names differing in gender, the intended association can be ascertained through the gender of the epithet and result in a validly published new combination.

Ex. 5. Combinations validly published. In Linnaeus's *Species plantarum,* the placing of the epithet in the margin opposite the name of the genus clearly associates the epithet with the name of the genus. The same result is attained in Miller's *The gardeners dictionary,* ed. 8, by the inclusion of the epithet in parentheses immediately after the name of the genus, in Steudel's *Nomenclator botanicus* by the arrangement of the epithets in a list headed by the name of the genus, and in general by any typographical device that associates an epithet with a particular name of a genus or species.

Ex. 6. Combinations not validly published. Rafinesque's statement under *Blephilia* that "Le type de ce genre est la *Monarda ciliata* Linn. [The type of this genus is *Monarda ciliata* Linn.]" (in J. Phys. Chim. Hist. Nat. Arts 89: 98. 1819) does not constitute valid publication of the combination *B. ciliata* because Rafinesque did not definitely associate the epithet *ciliata* with the generic name *Blephilia*. Similarly, the combination *Eulophus peucedanoides* is not to be attributed to Bentham & Hooker (Gen. Pl. 1: 885. 1867) based on their listing of "*Cnidium peucedanoides,* H. B. et K." under *Eulophus.*

Ex. 7. *Erioderma polycarpum* subsp. *verruculosum* Vain. (in Acta Soc. Fauna Fl. Fenn. 7(1): 202. 1890) is validly published because Vainio clearly linked the sub-specific epithet to the specific epithet by an asterisk.

Ex. 8. When Tuckerman (in Proc. Amer. Acad. Arts 12: 168. 1877) described "*Erioderma velligerum,* sub-sp. nov.", he stated that his new subspecies was very near to *E. chilense,* from which he provided distinguishing features. However, because he did not definitely associate the subspecific epithet with that species name, he did not validly publish *"E. chilense* subsp. *velligerum".*

Ex. 9. *Andropogon brevifolius* Sw. was assigned to *Schizachyrium* Nees (in Martius, Fl. Bras. Enum. Pl. 2(1): 332. 1829) when Nees described that genus as new: "Hujusce generis species, praeter enumeratas, sunt et *Andropogon brevi-folius,* Sw. (*Pollinia* Spr.) … [Species of this genus, besides those enumerated, are also *Andropogon brevifolius,* Sw. (*Pollinia* Spr.) …]". However, Nees did not associate the final epithet of the species name with *Schizachyrium* and did not therefore validly publish a new combination. *Schizachyrium brevifolium* (Sw.) Nees ex Buse (in Miquel, Pl. Jungh.: 359. 1854) was validly published when Buse wrote "… a *Schiz. brevifolio* Nees (i.e. *Andr. brevifolio* Sw.) …", thereby referring to the basionym and definitely associating the final epithet with *Schizachyrium.*

Ex. 10. In "Chloris Novae Hollandiae", a catalogue of "hitherto published" names for Australian plants (in Ann. Bot. (König & Sims) 2: 504–532. 1806), species names are listed with their place of publication, preceded in several cases by the name of another genus in parentheses. For example, the listing includes "(*Myoporum*) *Pogonia debilis.* Andrews's reposit. 212" and a further statement "These plants belong to the genus *Myoporum* …" (p. 525). Nevertheless, none of the potential new combinations indicated in this way is validly published due to the lack of definite association of the parenthetical generic name with the specific epithet.

ARTICLE 36
NAMES NOT ACCEPTED BY THEIR AUTHORS, ALTERNATIVE NAMES

36.1. A name is not validly published when it is not accepted by its author(s) in the original publication, for example:

(a) when it is merely proposed in anticipation of the future accept-ance of the taxon concerned, or of a particular circumscription, position, or rank of the taxon (so-called provisional name); or

(b) when it is merely cited as a synonym.

These provisions do not apply to names published with a question mark or other indication of taxonomic doubt, but accepted by their author(s).

Ex. 1. *"Sebertia"*, proposed by Pierre (ms.) for a genus of one species, was not validly published by Baillon (in Bull. Mens. Soc. Linn. Paris 2: 945. 1891) because he did not accept the generic name. He gave a description of the species *"Sebertia acuminata* Pierre" but referred it to the genus *Sersalisia* R. Br., as *"Sersalisia?* *acuminata"*, which he thereby validly published as *Sersalisia acuminata* Baill., despite his use of a question mark. The name *Sebertia* was validly published by Engler (in Engler & Prantl, Nat. Pflanzenfam., Nachtr. 1: 280. 1897).

Ex. 2. The designations listed in the left-hand column of the Linnaean thesis *Herbarium amboinense* defended by Stickman (1754) were not names accepted by Linnaeus upon publication and are not validly published.

Ex. 3. The name *Coralloides gorgonina* Bory was validly published in a paper by Flörke (in Mag. Neuesten Entdeck. Gesammten Naturk. Ges. Naturf. Freunde Berlin 3: 125. 1809) even though Flörke did not accept the taxon as a new species. At Bory's request, Flörke included Bory's diagnosis (and species name) making Bory the publishing author as defined in Art. 46.6. The acceptance or otherwise of the name by Flörke is not therefore relevant for valid publication.

Ex. 4. Carrière (in Fl. Serres Jard. Eur. 8: 292. 1853) stated "On cultive encore dans les jardins, depuis plusieurs années, sous le nom de *Weigela splendens,* une autre espèce de *Diervilla,* voisine par le port du *D. canadensis,* Willd. [For several years now, another species of *Diervilla* has been cultivated in the gardens under the name of *Weigela splendens,* similar in appearance to *D. canadensis,* Willd.]" Because Carrière did not accept *"Weigela splendens"* and did not associate the specific epithet *splendens* with the generic name *Diervilla* (see Art. 35.2), neither *"Diervilla splendens"* nor *"Weigela splendens"* was validly published. Subsequently, *D.* ×*splendens* G. Kirchn. (in Petzold & Kirchner, Arbor. Muscav.: 442. 1864) was validly published.

Ex. 5. (a) The designation *"Conophyton"*, suggested by Haworth (Revis. Pl. Succ.: 82. 1821) for *Mesembryanthemum* sect. *Minima* Haw. (l.c.: 81. 1821) in the words "If this section proves to be a genus, the name of *Conophyton* would be apt", was not a validly published generic name because Haworth did not accept it but merely proposed *"Conophyton"* in anticipation of its future acceptance. The name was validly published as *Conophytum* N. E. Br. (in Gard. Chron., ser. 3, 71: 198. 1922).

Ex. 6. (a) *"Pteridospermaexylon"* and *"P. theresiae"* were published by Greguss (in Földt. Közl. 82: 171. 1952) for a genus and species of fossil wood. Because Greguss explicitly stated "Vorläufig benenne ich es mit dem Namen … [provisionally I designate it by the name …]", these are provisional names and as such are not validly published.

Ex. 7. (a) The designation *"Stereocaulon subdenudatum"* proposed by Havaas (in Univ. Bergen Årbok, Naturvidensk. Rekke. 1954(12): 13, 20. 1954) is not validly published, even though it was presented as a new species with a Latin diagnosis, because on both pages it was indicated to be "ad int. [ad interim, for the time being]".

Ex. 8. (b) "*Ornithogalum undulatum* Hort. Bouch." was not validly published by Kunth (Enum. Pl. 4: 348. 1843) when he cited it as a synonym under *Myogalum boucheanum* Kunth; the correct combination under *Ornithogalum* L. was validly published later: *O. boucheanum* (Kunth) Asch. (in Verh. Bot. Vereins Prov. Brandenburg 8: 165. 1866).

Ex. 9. (b) The intended new combination "*Henckelia membranacea* (Bedd.) Janeesha & Nampy comb. nov." was included by Janeesha & Nampy (in Rheedea 30: 77. 2020) in the synonymy of *H. missionis* (Wall. ex R. Br.) A. Weber & B. L. Burtt (in Beitr. Biol. Pflanzen 70: 350. 1998). "*Henckelia membranacea*" was not therefore validly published.

Ex. 10. *Besenna* A. Rich. and *B. anthelmintica* A. Rich. (Tent. Fl. Abyss. 1: 253. 1847) were simultaneously published by Richard, both with a question mark ("*Besenna* ?" and "*Besenna anthelmintica* ? Nob."). Richard's uncertainty was due to the absence of flowers or fruits for examination, but the names were nonetheless accepted by him, with *Besenna* listed as such (i.e. not italicized) in the index (p. [469]).

36.2. A name is not validly published by the mere mention of the subordinate taxa included in the taxon concerned.

Ex. 11. The family designation "*Rhaptopetalaceae*" was not validly published by Pierre (in Bull. Mens. Soc. Linn. Paris 2: 1296. May 1897), who merely mentioned the constituent genera, *Brazzeia* Baill., *Rhaptopetalum* Oliv., and "*Scytopetalum*", but gave no description or diagnosis; a description of the family was published under the name *Scytopetalaceae* Engl. (in Engler & Prantl, Nat. Pflanzenfam., Nachtr. 1: 242. Oct 1897).

Ex. 12. The generic designation "*Ganymedes*" was not validly published by Salisbury (in Trans. Hort. Soc. London 1: 353–355. 1812), who merely mentioned three included species but supplied no generic description or diagnosis.

36.3. When, on or after 1 January 1953, two or more different names based on the same type are accepted simultaneously for the same taxon by at least one author in common in the same publication (so-called alternative names), none of them, if new, is validly published. This rule does not apply in those cases where the same combination is simultaneously used at different ranks, either for infraspecific taxa or for subdivisions of a genus (see Rec. 22A.1, 22A.2, and 26A.1–3), nor where suprageneric names formed from the same generic name are simultaneously used at different ranks (see Rec. 19A.1 and 19A.2), nor to names provided for in Art. F.8.1.

Ex. 13. The species of *Brosimum* Sw. described by Ducke (in Arch. Jard. Bot. Rio de Janeiro 3: 23–29. 1922) were published with alternative names under *Piratinera* Aubl. added in a footnote (pp. 23–24), in which Ducke indicated acceptability of these names under the competing (alternative) *American Code*. The publication of both sets of names is valid because it occurred before 1 January 1953.

Ex. 14. *"Euphorbia jaroslavii"* (Poljakov in Bot. Mater. Gerb. Bot. Inst. Komarova Akad. Nauk SSSR 15: 155. 1953) was published with an alternative designation, *"Tithymalus jaroslavii"*. Neither was validly published. However, one name, *Euphorbia yaroslavii* (with a differently transcribed initial letter), was validly published by Poljakov (in Bot. Mater. Gerb. Bot. Inst. Komarova Akad. Nauk SSSR 21: 484. 1961), who provided a full and direct reference to the earlier publication and rejected the assignment to *Tithymalus*.

Ex. 15. Hitchcock (in Univ. Wash. Publ. Biol. 17(1): 507–508. 1969) used the name *Bromus inermis* subsp. *pumpellianus* (Scribn.) Wagnon and provided a full and direct reference to its basionym, *B. pumpellianus* Scribn. (in Bull. Torrey Bot. Club 15: 9. 1888). Within that subspecies, he recognized varieties, one of which he named *B. inermis* var. *pumpellianus* (without an author citation but clearly based on the same basionym and type). In so doing, he met the requirements for valid publication of *B. inermis* var. *pumpellianus* (Scribn.) C. L. Hitchc.

ARTICLE 37
REQUIREMENT OF INDICATION OF RANK

37.1. A name published on or after 1 January 1953 without a clear indication of the rank of the taxon concerned is not validly published.

37.2. For suprageneric names published on or after 1 January 1887, the use of one of the terminations[1] specified in Art. 16.3, 17.1, 18.1, 19.1, and 19.3 is accepted as an indication of the corresponding rank, unless this:

(a) would conflict with the explicitly designated rank of the taxon (which takes precedence); or

(b) would result in a rank sequence contrary to Art. 5 (in which case Art. 37.7 applies); or

---

1   The terminations specified in Art. 16.3, 17.1, 18.1, 19.1, and 19.3 are: *-phyta* (division or phylum in algae and plants), *-mycota* (division or phylum in fungi), *-phytina* (subdivision or subphylum in algae and plants), *-mycotina* (subdivision or subphylum in fungi), *-phyceae* (class in algae), *-mycetes* (class in fungi), *-opsida* (class in plants), *-phycidae* (subclass in algae), *-mycetidae* (subclass in fungi), *-idae* (subclass in plants), *-ales* (order), *-ineae* (suborder), *-aceae* (family), *-oideae* (subfamily), *-eae* (tribe), and *-inae* (subtribe).

(c) would result in a rank sequence in which the same rank-denoting term occurs at more than one hierarchical position.

Ex. 1. Jussieu (in Mém. Mus. Hist. Nat. 12: 497. 1827) proposed *Zanthoxyleae* without specifying the rank. Although he used the present termination for tribe *(-eae),* that name is unranked because it was published before 1887. *Zanthoxyleae* Dumort. (Anal. Fam. Pl.: 45. 1829), however, is the name of a tribe because Dumortier specified its rank.

Ex. 2. Nakai (Chosakuronbun Mokuroku [Ord. Fam. Trib. Nov.]. 1943) validly published the names *Parnassiales, Lophiolaceae, Ranzanioideae,* and *Urospatheae.* He indicated the respective ranks of order, family, subfamily, and tribe, by use of their terminations even though he did not mention these ranks explicitly.

37.3. A name published before 1 January 1953 without a clear indication of its rank is validly published provided that all other requirements for valid publication are fulfilled; it is, however, inoperative in questions of priority except for homonymy (see Art. 53.3). If it is the name of a new taxon, it may serve as a basionym or replaced synonym for subsequent new combinations, names at new rank, or replacement names at definite ranks.

Ex. 3. The unranked groups *"Soldanellae", "Sepincoli", "Occidentales",* etc., were published under *Convolvulus* L. by House (in Muhlenbergia 4: 50. 1908). The names *C.* [unranked] *Soldanellae* House, etc., are validly published but have no status in questions of priority except for purposes of homonymy under Art. 53.3.

Ex. 4. In *Carex* L., the epithet *Scirpinae* was used in the name of an unranked subdivision of a genus by Tuckerman (Enum. Meth. Caric.: 8. 1843); this taxon was assigned sectional rank by Kükenthal (in Engler, Pflanzenr. IV. 20 (Heft 38): 81. 1909) and its name is then cited as *Carex* sect. *Scirpinae* (Tuck.) Kük. (*C.* [unranked] *Scirpinae* Tuck.).

Ex. 5. Loesener published *Geranium andicola* "Var. vel forma α. *longipedicellata*" (in Bull. Herb. Boissier, ser. 2, 3: 93. 1903) with an ambiguous indication of infraspecific rank. The name is correctly cited as *G. andicola* [unranked] *longipedicellatum* Loes. The epithet was used in a subsequent combination, *G. longipedicellatum* (Loes.) R. Knuth (in Engler, Pflanzenr. IV. 129 (Heft 53): 171. 1912).

37.4. If in one whole publication (Art. 37.6), before 1 January 1890, only one infraspecific rank is admitted, it is considered to be that of variety unless this would be contrary to the author's statements in the same publication.

37.5. If in one whole publication (Art. 37.6) no general statement is made by the author(s) on the different infraspecific ranks used in that

publication, statements on ranks associated with individual infraspecific names can be used instead to assign rank throughout the publication as long as they do not result in misplaced terms contrary to Art.
5 (see Art. 37.7).

37.6. In questions of indication of rank, all publications appearing
under the same title and by the same author, such as different parts of
a flora issued at different times (but not different editions of the same
work), must be considered as a whole, and any statement made therein
designating the rank of taxa included in the work must be considered
as if it had been published together with the first instalment.

> *Ex. 6.* In Link's *Handbuch* (1829–1833) the rank-denoting term "O." (ordo) was
> used in all three volumes. These names of orders cannot be considered as hav-
> ing been published as names of families (Art. 18.2) because the term family was
> used for *Agaricaceae* and *Tremellaceae* under the order *Fungi* in vol. 3 (pp. 272,
> 337; see Art. 18 Note 3). This applies to all three volumes of the *Handbuch* even
> though vol. 3 was published later (Jul–29 Sep 1833) than vol. 1 and 2 (4–11 Jul
> 1829).

37.7. A name is not validly published if it is given to a taxon of which
the rank is at the same time denoted by a misplaced term, contrary to
Art. 5. Such misplacements include, e.g., forms divided into varieties,
species containing genera, and genera containing families or tribes
(but see Art. F.4.1).

37.8. Only those names published with rank-denoting terms that
must be removed to achieve a proper sequence are to be regarded as
not validly published. In cases where terms are switched, e.g. family-
order, and a proper sequence can be achieved by removing either or
both of the rank-denoting terms, names at neither rank are validly
published unless one is a secondary rank (Art. 4.1) and one is a principal rank (Art. 3.1), e.g. family-genus-tribe, in which case only names
published at the secondary rank are not validly published.

> *Ex. 7.* "Sectio *Orontiaceae*" (Brown, Prodr.: 337. 1810) is not a validly published
> name because Brown misapplied the term "sectio" to a rank higher than genus.

> *Ex. 8.* "Tribus *Involuta*" and "tribus *Brevipedunculata*" (Huth in Bot. Jahrb. Syst.
> 20: 365, 368. 1895) are not validly published names because Huth misapplied the
> term "tribus" to a rank lower than section within the genus *Delphinium*.

> *Ex. 9.* *Centaurea* "Ser. I. *Aplolepideae*" and *C.* "Sect. I. *Hyalaea*" (Candolle,
> Prodr. 6: 565. 1838) are not validly published names because Candolle misap-
> plied the term series to a rank higher than section within the genus *Centaurea*.

Consequently, *Hyalea* Jaub. & Spach (Ill. Pl. Orient. 3: 19. 1847), based on "*Centaureae* Sectio I" in Candolle (l.c. 1838), was published as the name of a new genus.

ⓘ **Note 1.** Consecutive use of the same rank-denoting term in a taxonomic sequence does not represent misplaced rank-denoting terms.

37.9. Situations where the same or equivalent rank-denoting term is used at more than one non-consecutive position in the taxonomic sequence represent informal usage of rank-denoting terms. Names published with such rank-denoting terms are treated as unranked (see Art. 37.1 and 37.3; see also Art. 16 Note 1).

Ex. 10. Names published with the term "series" by Bentham & Hooker (Gen. Pl. 1862–1883) are treated as unranked because this term was used at seven different hierarchical positions in the taxonomic sequence. Therefore, the sequence in *Rhynchospora* (l.c. 3: 1058–1060. 1883) of genus-"series"-section does not contain a misplaced rank-denoting term.

SECTION 2
VALID PUBLICATION OF NAMES OF NEW TAXA

ARTICLE 38
REQUIREMENT OF DESCRIPTION, DIAGNOSIS, OR ILLUSTRATION WITH ANALYSIS

38.1. To be validly published, a name of a new taxon (see Art. 6.9) must:

(a) be accompanied by a description or diagnosis of the taxon (see also Art. 38.8 and 38.9) or, if none is provided in the protologue, by a reference (see Art. 38.14) to a previously and effectively published description or diagnosis (except as provided in Art. 13.4 and H.9; see also Art. 14.9 and 14.14); and

(b) comply with the relevant provisions of Art. 32–45 and F.4–F.5.

ⓘ **Note 1.** An exception to Art. 38.1 is made for the generic names first published by Linnaeus in *Species plantarum,* ed. 1 (1753) and ed. 2 (1762–1763), which are treated as having been validly published in those works even though the validating descriptions were published later in *Genera plantarum,* ed. 5 (1754) and ed. 6 (1764), respectively (see Art. 13.4).

38.2. A diagnosis of a taxon is a statement of that which in the opinion of its author distinguishes the taxon from other taxa.

Ex. 1. *"Egeria"* (Néraud in Gaudichaud, Voy. Uranie, Bot.: 25, 28. 1826) was published without a description or a diagnosis or a reference to a former one (and is therefore a nomen nudum); it was not validly published.

Ex. 2. *"Loranthus macrosolen"* originally appeared without a description or diagnosis on the printed labels issued about the year 1843 with Sect. II, No. 529, 1288, of the herbarium specimens from Schimper's "Abyssinische Reise". The name *L. macrosolen* Steud. ex A. Rich. (Tent. Fl. Abyss. 1: 340. 1848) was validly published when Richard supplied a description.

*Ex. 3. In Don, *Sweet's Hortus britannicus,* ed. 3 (1839), for each listed species the flower colour, the duration of the plant, and a translation into English of the specific epithet are given in tabular form. In many genera the flower colour and duration may be identical for all species and clearly their mention is not intended as a validating description or diagnosis. Names of new taxa appearing in that work are not therefore validly published, except in some cases where reference is made to earlier descriptions or diagnoses.

Ex. 4. *"Crepis praemorsa* subsp. *tatrensis"* (Dvořák & Dadáková in Biológia (Bratislava) 32: 755. 1977) appeared with "a subsp. *praemorsa* karyotypo achaeniorumque longitudine praecipue differt [it differs from subsp. *praemorsa* principally by the karyotype and the length of the achenes]". This statement specifies the features in which the two taxa differ but not how these features differ and so it does not satisfy the requirement of Art. 38.1(a) for a "description or diagnosis".

Ex. 5. The generic name *Epilichen* Clem. (Gen. Fungi: 69, 174. 1909) is validly published by means of the key character "parasitic on lichens" (contrasting with "saprophytic" for *Karschia*) and the Latin diagnosis "Karschia lichenicola", referring to the ability of the included species formerly included in *Karschia* to grow on lichens. These statements, in the opinion of Clements, distinguished the genus from others, although provision of such a meagre diagnosis is not good practice.

Ex. 6. The protologue of *Iresine borschii* Zumaya & Flores Olv. (in Willdenowia 46: 166. 2016) includes both a morphological and a molecular diagnosis. Both are diagnoses because they indicate how the features of the new species, in the opinion of the authors, differ from those of other taxa.

38.3. The requirements of Art. 38.1(a) are not met by statements describing properties such as purely aesthetic features, economic, medicinal or culinary use, cultural significance, cultivation techniques, geographical origin, or geological age.

Ex. 7. *"Musa basjoo"* (Siebold in Verh. Bat. Genootsch. Kunsten 12: 18. 1830) appeared with "Ex insulis *Luikiu* introducta, vix asperitati hiemis resistens. Ex foliis linteum, praesertim in insulis *Luikiu* ac quibusdam insulis provinciae *Satzuma* conficitur. Est haud dubie linteum, quod Philippinis incolis audit

Nippis. [Introduced from the Ryukyu Islands, hardly withstands winter hardships. Linen is made from the leaves, principally in the Ryukyu Islands and in some islands of the Satsuma Province. This is doubtless the cloth called Nippis by the Philippine people.]" This statement gives information about the economic use (linen is made from the leaves), hardiness in cultivation (scarcely survives the winter), and geographical origin (introduced from the Ryukyu Islands). Because there is no explicit descriptive information (e.g. shape or texture) on the leaves, the only character mentioned, it does not satisfy the requirement of Art. 38.1(a) for a "description or diagnosis". *Musa basjoo* Siebold ex Miq. was later validly published by Miquel (in Ann. Mus. Bot. Lugduno-Batavi 3: 203. 1867) with a diagnosis in Latin.

38.4. A description is a statement explicitly describing one or more features of an individual taxon. A description need not be diagnostic (but see Rec. 38B.2).

38.5. When it is doubtful whether a descriptive statement satisfies the requirement of Art. 38.1(a) for a "description or diagnosis", a request for a binding decision may be submitted to the General Committee, which will refer it for examination to the specialist committee for the appropriate taxonomic group (see Div. III Prov. 2.2, 7.10(b), 7.11, and 8.13(a)). A General Committee recommendation as to whether or not the name concerned is validly published is to be treated as a binding decision subject to ratification by a later International Botanical Congress (see also Art. 14.15, 34.2, 53.4, and 56.3) and takes retroactive effect. These binding decisions are listed in App. VI.

Ex. 8. Ascomycota Caval.-Sm. (in Biol. Rev. (Cambridge) 73: 247. 1998, as "*Ascomycota* Berkeley 1857 stat. nov.") was published as the name of a phylum with the diagnosis "sporae intracellulares [spores intracellular]". Because Cavalier-Smith (l.c. 1998) did not provide a full and direct reference to Berkeley's publication (Intr. Crypt. Bot.: 270. 1857) of the name *Ascomycetes* (not *Ascomycota*), valid publication of *Ascomycota* is dependent on its meeting the requirements of Art. 38.1(a), and a request was made for a binding decision under Art. 38.5. The Nomenclature Committee for Fungi concluded that the requirements of Art. 38.1(a) were minimally fulfilled and recommended (in Taxon 59: 292. 2010) a binding decision that *Ascomycota* is validly published. This was approved by the General Committee (in Taxon 60: 1212. 2011) and ratified by the XVIII International Botanical Congress in Melbourne in 2011 (see App. VI).

Ex. 9. Brugmansia aurea Harrison (Floric. Cab. & Florist's Mag. 5: 144. 1837) was described in an account of a garden visit as comprising "plants about two feet high" with flowers "about the size of the *B. sanguinea,* but of fine rich golden yellow colour", and was compared with "an inferior kind … the flowers of which are of a dull buff colour". A binding decision has been made that the name is validly published (see App. VI).

38.6. The names of a genus and a species may be validly published simultaneously by provision of a single description (descriptio generico-specifica) or diagnosis, even though this may have been intended as either generic or specific, only if all of the following conditions are satisfied:

(a) the descriptio generico-specifica accompanies the names of the taxa described (reference instead to an earlier description or diagnosis is not acceptable); and

(b) the genus is at that time monotypic (see Art. 38.7); and

(c) no other names (at any rank) have previously been validly published based on the same type; and

(d) the names of the genus and species otherwise fulfil the requirements for valid publication.

> **Ex. 10.** (a) For *"Merremia convolvulacea"*, Dennstedt (Schlüssel Hortus Malab.: 12, 23, 34. 1818) did not provide a description or diagnosis but referred to an earlier Latin description and an illustration with analysis in Rheede (Hort. Malab. 8: 52 ["51"], t. 27. 1692). Therefore, under Art. 38.6(a), *"Merremia"* and *"Merremia convolvulacea"* were not validly published. The generic and species names were subsequently validated by Endlicher (Gen. Pl.: 1403. 1841) and Hallier (in Bot. Jahrb. Syst. 16: 552. 1893), respectively.

38.7. For the purpose of Art. 38.6, a monotypic genus is one for which a single binomial is validly published even though the author may indicate that other species are attributable to the genus.

> **Ex. 11.** Nylander (in Flora 62: 353. 1879) described the new species *"Anema nummariellum"* in a new genus *"Anema"* without providing a generic description or diagnosis. Because in the same publication (l.c.: 354. 1879) he wrote "Affine *Anemati nummulario* (DR.) Nyl., ... [related to *Anema nummularium* (DR.) Nyl., ...]", which was an attempted new combination in *"Anema"* based on *Collema nummularium* Dufour ex Durieu & Mont. (in Bory & Durieu, Expl. Sci. Algérie 1: 200. 1846), none of his designations was validly published. The names were later validly published by Forssell (Beitr. Gloeolich.: 40, 91, 93. 1885).

> **Ex. 12.** The names *Kedarnatha* P. K. Mukh. & Constance (in Brittonia 38: 147. 1986) and *K. sanctuarii* P. K. Mukh. & Constance, the latter designating the single, new species of the new genus, are both validly published although a Latin description was provided only under the generic name.

> **Ex. 13.** *Piptolepis phillyreoides* Benth. (Pl. Hartw.: 29. 1840) was a new species assigned to the monotypic new genus *Piptolepis*. Both names were validly published with a combined generic and specific description.

> **Ex. 14.** In publishing *"Phaelypea"* without a generic description or diagnosis, Browne (Civ. Nat. Hist. Jamaica: 269. 1756) included and described a single

species, but he gave the species a phrase name, not a validly published binomial. Art. 38.6 does not therefore apply and *"Phaelypea"* is not a validly published name.

38.8. For the purpose of Art. 38.6, before 1 January 1908, an illustration with analysis (see Art. 38.10 and 38.11) is acceptable in place of a written description or diagnosis.

Ex. 15. The generic name *Philgamia* Baill. (in Grandidier, Hist. Phys. Madagascar 35: t. 265. 1894) was validly published because it appeared on a plate with analysis of the only included species, *P. hibbertioides* Baill.

Ex. 16. The generic name *Torrentia* Vell. (Fl. Flumin. Icon. 8: t. 149. 1831) and that of its only included species, *T. quinquenervis* Vell., were validly published without a description or diagnosis but with an illustration with analysis showing details of the bracts, ray floret pappus, and stigma.

38.9. The name of a new species or infraspecific taxon published before 1 January 1908 may be validly published even if only accompanied by an illustration with analysis (see Art. 38.10 and 38.11).

Ex. 17. When *"Polypodium subulatum"* (Vellozo, Fl. Flumin. Icon. 11: ad t. 67. 1831) was published, only an illustration of part of a frond was presented, without analysis, hence this drawing does not fulfil the provisions of Art. 38.9 and the designation was not validly published there. The name *P. subulatum* Vell. was validly published when Vellozo's fern species descriptions appeared (in Arch. Mus. Nac. Rio de Janeiro 5: 447. 1881).

38.10. For the purposes of this *Code,* an analysis is a figure or group of figures, often presented separately from the main illustration of the organism (though usually on the same page or plate), showing details aiding identification, with or without a separate caption (see also Art. 38.11).

Ex. 18. *Panax nossibiensis* Drake (in Grandidier, Hist. Phys. Madagascar 35: t. 406. 1897) was validly published on a plate with analysis that includes details of flower structure.

38.11. For organisms other than vascular plants, single figures showing details aiding identification are considered as illustrations with analysis (see also Art. 38.10).

Ex. 19. *Eunotia gibbosa* Grunow (in Van Heurck, Syn. Diatom. Belgique: t. 35, fig. 13. 1881), a name of a diatom, was validly published by provision of a figure of a single valve.

38.12. For the purpose of valid publication of a name of a new taxon, reference to a previously and effectively published description or diagnosis is restricted as follows:

(a) For a name of a family or subdivision of a family, the earlier description or diagnosis must be that of a family or subdivision of a family.

(b) For a name of a genus or subdivision of a genus, the earlier description or diagnosis must be that of a genus or subdivision of a genus.

(c) For a name of a species or infraspecific taxon, the earlier description or diagnosis must be that of a species or infraspecific taxon (but see Art. 38.13).

Ex. 20. (a) "*Pseudoditrichaceae* fam. nov." (Steere & Iwatsuki in Canad. J. Bot. 52: 701. 1974) was not a validly published name of a family because there was no Latin description or diagnosis of the family, nor reference to either, but only mention of the single included genus and species (see Art. 36.2), as "*Pseudoditrichum mirabile* Steere et Iwatsuki, gen. et sp. nov.", the names of which were both validly published under Art. 38.6 by provision of a single Latin diagnosis.

Ex. 21. (b) *Scirpoides* Ség. (Pl. Veron. 3: 73. 1754) was published without a generic description or diagnosis. It was validly published by indirect reference (through the title of the book and a general statement in the preface) to the generic diagnosis and further direct references in Séguier (Pl. Veron. 1: 117. 1745).

Ex. 22. Because Art. 38.12 places no restriction on names at ranks higher than family, *Eucommiales* Němejc ex Cronquist (Integr. Syst. Class. Fl. Pl.: 182. 1981) was validly published by Cronquist, who provided a full and direct reference to the Latin description associated with the genus *Eucommia* Oliv. (in Hooker's Icon. Pl. 20: ad t. 1950. 1890).

38.13. A name of a new species may be validly published by reference (direct or indirect; see Art. 38.14 and 38.15) to a description or diagnosis of a genus, if the following conditions are satisfied:

(a) the name of the genus was previously and validly published simultaneously with its description or diagnosis; and

(b) neither the author of the name of the genus nor the author of the name of the species indicates that more than one species belongs to the genus in question.

Ex. 23. *Trilepisium* Thouars (Gen. Nov. Madagasc.: 22. 1806) was validated by a generic description but without mention of a name of a species. *Trilepisium madagascariense* DC. (Prodr. 2: 639. 1825) was subsequently proposed without a description or diagnosis of the species and with the generic name followed by a reference to Thouars. Neither author gave any indication that there was more than one species in the genus. Candolle's species name is therefore validly published. The type of *T. madagascariense* DC. is automatically the type of *Trilepisium* Thouars (see Art. 10.9).

38.14. For the purpose of valid publication of a name of a new taxon, reference to a previously and effectively published description or diagnosis may be direct or indirect (Art. 38.15). For names published on or after 1 January 1953, for which the requirement for a description or diagnosis was not previously fulfilled (see Art. 33.1), this reference must, however, be full and direct as specified in Art. 41.5.

38.15. An indirect reference is a clear (if cryptic) indication, by an author citation or in some other way, that a previously and effectively published description or diagnosis applies.

Ex. 24. *"Kratzmannia"* (Opiz in Berchtold & Opiz, Oekon.-Techn. Fl. Böhm. 1: 398. 1836) was published with a diagnosis but was not definitely accepted by the author and was not therefore validly published under Art. 36.1. *Kratzmannia* Opiz (Seznam: 56. 1852), lacking description or diagnosis, is however definitely accepted, and its citation as "*Kratzmannia* O." constitutes an indirect reference to Opiz's diagnosis published in 1836.

Ex. 25. The name *Statice tenoreana* was published by Gussone (Enum. Pl. Inarim.: 268. 1855) without any description or diagnosis but citing "*S. minuta.* Ten. Syll. p. 593 (non Lin.)". However, Tenore (Syll. Pl. Fl. Neapol.: 593. 1833) did not include any description of *S. minuta,* which was provided only later (Tenore, Fl. Napol. 5: 338. 1835–1838), along with a reference to page 593 of the *Sylloge* (Tenore, l.c. 1833). *Statice tenoreana* Guss., although lacking a description or diagnosis both in the protologue and in the directly cited *Sylloge* of Tenore, is validly published by the indirect reference to the description in Tenore's *Flora napolitana* (1835–1838).

Recommendation 38A

38A.1. A name of a new taxon should not be validated solely by a reference to a description or diagnosis published before 1753.

Recommendation 38B

38B.1. When a description is provided for valid publication of the name of a new taxon, a separate diagnosis should also be presented.

38B.2. Where no separate diagnosis is provided, the description of any new taxon should mention the points that distinguish the taxon from others.

Recommendation 38C

38C.1. When naming a new taxon, authors should not adopt a name that has been previously but not validly published for a different taxon.

Recommendation 38D

38D.1. In describing or diagnosing new taxa, authors should, when possible, supply figures with details of structure as an aid to identification.

38D.2. In the explanation of figures, authors should indicate the specimen(s) on which they are based (see also Rec. 8A.2).

38D.3. Authors should indicate clearly and precisely the scale of the figures that they publish.

Recommendation 38E

38E.1. Descriptions or diagnoses of new taxa of parasitic organisms, especially fungi, should always be followed by indication of the hosts. The hosts should be designated by their scientific names and not solely by names in modern languages, the application of which is often doubtful.

ARTICLE 39
LANGUAGE OF VALIDATING DESCRIPTION OR DIAGNOSIS

39.1. To be validly published, a name of a new taxon (algae and fossils excepted) published between 1 January 1935 and 31 December 2011, inclusive, must be accompanied by a Latin description or diagnosis or by a reference (see Art. 38.14) to a previously and effectively published Latin description or diagnosis (but see Art. H.9; for fossils see Art. 43.1; for algae see Art. 44.1).

Ex. 1. Arabis "Sekt. *Brassicoturritis* O. E. Schulz" and *A.* "Sekt. *Brassicarabis* O. E. Schulz" (in Engler & Prantl, Nat. Pflanzenfam., ed. 2, 17b: 543–544. 1936), published with German but no Latin descriptions or diagnoses, are not validly published names.

Ex. 2. "Schiedea gregoriana" (Degener, Fl. Hawaiiensis, fam. 119. 9 Apr 1936) was accompanied by an English but no Latin description and is not therefore a validly published name. *Schiedea kealiae* Caum & Hosaka (in Occas. Pap. Bernice Pauahi Bishop Mus. 11(23): 3. 10 Apr 1936), the type of which is part of the material used by Degener, is provided with a Latin description and is validly published.

Ex. 3. Alyssum flahaultianum Emb., first published without a Latin description or diagnosis (in Bull. Soc. Sci. Nat. Maroc 15: 199. 1935), was validly published posthumously when a Latin translation of Emberger's original French description was provided (in Willdenowia 15: 62–63. 1985).

39.2. To be validly published, a name of a new taxon published on or after 1 January 2012 must be accompanied by a Latin or English description or diagnosis or by a reference (see Art. 38.14) to a previously and effectively published Latin or English description or diagnosis (for fossils see also Art. 43.1).

Recommendation 39A

39A.1. Authors publishing names of new taxa should give or cite a full description in Latin or English in addition to the diagnosis.

ARTICLE 40
REQUIREMENT OF INDICATION OF TYPE

40.1. Publication on or after 1 January 1958 of the name of a new taxon at the rank of genus or below is valid only when the type of the name is indicated (see Art. 7–10; but see Art. H.9 Note 1 for the names of certain hybrids).

ⓘ **Note 1.** When elements that cannot on their own serve as types as defined under Art. 8 and 40 are cited as part of the type indication (e.g. living organisms cited contrary to Art. 8.4 or illustrations cited contrary to Art. 8.5 and 40.6), they do not affect valid publication of the name and are not relevant for the purpose of Art. 40.1.

40.2. For the name of a new genus or subdivision of a genus, reference (direct or indirect) to a single species name, or citation of a single element that is the type of a previously or simultaneously published species name, even if that element is not explicitly designated as type, is acceptable as indication of the type (see also Art. 10.8; but see Art. 40.4).

40.3. For the name of a new species or infraspecific taxon published on or after 1 January 1958, mention of a single specimen, a single gathering or a part thereof, or an illustration is acceptable as indication of the type, even if that element is not explicitly designated as type (but see Art. 40.4) or if it consists of two or more specimens as defined in Art. 8 (but see Art. 40.5).

Ex. 1. When Cheng described *"Gnetum cleistostachyum"* (in Acta Phytotax. Sin. 13(4): 89. 1975) the name was not validly published because two gatherings were designated as types: *K. H. Tsai 142* (as "♀ Typus") and *X. Jiang 127* (as "♂ Typus").

Ex. 2. *"Baloghia pininsularis"* was published by Guillaumin (in Mém. Mus. Natl. Hist. Nat., B, Bot. 8: 260. 1962) with two cited gatherings: *Baumann 13813* and *Baumann 13823*. Because the author did not designate one of them as the type, the designation was not validly published. Valid publication of the name *B. pininsularis* Guillaumin was achieved when McPherson & Tirel (Fl. Nouv.-Calédonie & Dépend. 14: 58. 1987) wrote "Lectotype (désigné ici): *Baumann-Bodenheim 13823* (P!; iso-, Z)" while providing a full and direct reference to Guillaumin's Latin description (Art. 33.1; see Art. 46 Ex. 22); McPherson & Tirel's use of "lectotype" is correctable to "holotype" under Art. 9.10.

ⓘ **Note 2.** Mere citation of a locality does not constitute mention of a single specimen or gathering. Concrete reference to some detail relating to the actual type is required, such as the collector's name, collecting number or date, or unique specimen identifier.

ⓘ **Note 3.** When the type is indicated by mention of an entire gathering, or a part thereof, consisting of more than one specimen, those specimens are syntypes (see Art. 9.6).

Ex. 3. The protologue of *Laurentia frontidentata* E. Wimm. (in Engler, Pflanzenr. IV. 276 (Heft 108): 855. 1968) includes the type statement "*E. Esterhuysen No. 17070!* Typus – Pret., Bol." The name is validly published because a single gathering is cited, despite the mention of duplicate specimens (syntypes) in two different herbaria, and Art. 40.5 does not apply.

Ex. 4. Radcliffe-Smith (Gen. Croton Madag. Comoro: 169. 2016) indicated the type of *Croton nitidulus* var. *acuminatus* Radcl.-Sm. as "*Cours 4871* (holotypus P)". In the herbarium P there are four duplicates of *Cours 4871*. The name is validly published because a single gathering in a single herbarium was indicated as type. These specimens are syntypes, and one of them was subsequently designated as the lectotype by Berry & al. (in PhytoKeys 90: 69. 2017).

ⓘ **Note 4.** Cultures of algae and fungi preserved in a metabolically inactive state are acceptable as types (Art. 8.4; see also Rec. 8B, Art. 40.7, and Rec. F.11A).

40.4. For the name of a new taxon at the rank of genus or below published on or after 1 January 1990, indication of the type must include one of the words "typus" or "holotypus", its equivalent in a modern language, or abbreviations of these (see also Rec. 40A.3). This requirement is also satisfied by use of one of the words "lectotypus" or "neotypus" (or its equivalent in a modern language, or abbreviations of these), which are to be treated as errors to be corrected under Art. 9.10. In the case of the name of a monotypic (as defined in Art. 38.7) new genus or subdivision of a genus with the simultaneously published name of a new species, indication of the type of the species name is sufficient.

Ex. 5. *"Sedum mucizonia* (Ortega) Raym.-Hamet subsp. *urceolatum"* was described by Stephenson (in Cact. Succ. J. (Los Angeles) 64: 234. 1992) but was not validly published because an indication of "typus" or "holotypus", its equivalent in a modern language, or an abbreviation of one of these was not provided.

Ex. 6. *"Dendrobium sibuyanense"* (see Art. 8 Ex. 11) was described with a living plant indicated as "Type specimen" and was not therefore validly published. *Dendrobium sibuyanense* Lubag-Arquiza & al. (in Orchid Digest 70: 174. 2006) was validly published later, when Lubag-Arquiza & Christenson designated a published drawing as "lectotype", which under Art. 9.10 and 40.4 is treated as an error to be corrected to "holotype".

40.5. For the name of a new species or infraspecific taxon published on or after 1 January 1990 of which the type is a specimen or unpublished illustration, the single herbarium, collection, or institution in which the type is conserved must be specified (see also Rec. 40A.4 and 40A.5).

Ex. 7. In the protologue of *Setaria excurrens* var. *leviflora* Keng ex S. L. Chen (in Bull. Nanjing Bot. Gard. 1988–1989: 3. 1990) the gathering *Guangxi Team 4088* was indicated as "模式 [type]" and the herbarium where the type is conserved was specified as "中国科学院植物研究所标本室 [Herbarium, Institute of Botany, The Chinese Academy of Sciences]", i.e. PE.

Ex. 8. *"Sedum eriocarpum* subsp. *spathulifolium"* was described by 't Hart (in Ot 2(2): 7. 1995) without specification of herbarium, collection, or institution in which the holotype specimen was conserved and therefore was not validly published. Valid publication was achieved when 't Hart (in Strid & Tan, Fl. Hellen. 2: 325. 2002) wrote "Type … 't Hart HRT-27104 … (U)" and provided a full and direct reference to his previously published Latin diagnosis (Art. 33.1).

Ex. 9. *"Rhodotorula portillonensis"* (Laich & al. in Int. J. Syst. Evol. Microbiol. 63: 3889. 2013) was introduced with the statement "Typus stirpis Pi2^T (=CBS 12733^T=CECT 13081^T)", where Pi2^T was a strain designation. Because the type was cited as conserved in more than one culture collection (CBS and CECT), the name was not validly published. The name was validly published as *R. portillonensis* Laich & al. (in Laich, Index Fungorum 361: 1. 2018) with the type designated as "Holotype CBS 12733".

ⓘ **Note 5.** Specification of the herbarium, collection, or institution may be made in an abbreviated form, e.g. as given in Index Herbariorum or in the

Culture Collections Information Worldwide database of the World Data Centre for Microorganisms.

40.6. For the name of a new species or infraspecific taxon published on or after 1 January 2007, the type indicated in accordance with Art. 40 may not be an illustration (for fossils see also Art. 8.5). An exception is permitted for names of non-fossil microscopic algae and non-fossil microfungi, for which the type may be an effectively published illustration if there are technical difficulties of specimen preservation or if it is impossible to preserve a specimen that would show the features attributed to the taxon by the author of the name.

> ***Ex. 10.*** Lücking & Moncada (in Fungal Diversity 84: 119–138. 2017) introduced *"Lawreymyces"* and seven intended microfungal species names using representations of diagnostic sequences of bases of DNA from the Internal Transcribed Spacer (ITS) region as intended types. These representations are not illustrations under Art. 6.1 footnote because they are not depictions of features of the organisms, and consequently the intended names were not validly published.

40.7. For the name of a new species or infraspecific taxon published on or after 1 January 2019 of which the type is a culture, the protologue must include a statement that the culture is preserved in a metabolically inactive state.

> ***Ex. 11.*** *"Trichoderma botryosum"* (Rodríguez & al. in Sci. Rep. 11(5671): 12. 11 Mar 2021) was not validly published because the holotype was cited as a culture, "H.C. Evans, K. Belachew & R.W. Barreto. Ex-type culture: COAD 2422", without a statement that it is preserved in a metabolically inactive state. Subsequently, Rodríguez & al. (in Sci. Rep. 11(19229): 1. 22 Sep 2021) fulfilled the requirement of Art. 40.7 by citing the type as "H.C. Evans, K. Belachew & R.W. Barreto. VIC 47493 (dried metabolically inactive culture). Ex-type culture: COAD 2422" and thus validly published the name *T. botryosum* M. C. H. Rodr. & al. The validation cited a new identifier (MB840985) as required by Art. F.5.7.

40.8. For the name of a new fossil-species or infraspecific fossil-taxon published on or after 1 January 2026, the protologue must clearly indicate where the holotype specimen (see Art. 8.6) is located within the rock, sediment, or preparation.

Recommendation 40A

40A.1. Authors proposing names of new families or subdivisions of families are urged to ensure that the generic name from which the new name is formed is itself effectively typified (see Art. 7 and 10), if necessary by designating a type for that generic name under the relevant provisions of Art. 7 and 10 (see also Rec. 40A.2).

40A.2. For the name of a new genus or subdivision of a genus, authors should cite the type of the species name (see Art. 7–9) that provides the type (Art. 10.1) of the new name and, if necessary, designate the type for that species name under the relevant provisions of Art. 7 and 9.

40A.3. Details of the type specimen of the name of a new species or infraspecific taxon should be published in the Latin alphabet.

40A.4. Specification of the herbarium, collection, or institution of deposition should be followed by any available number permanently and unambiguously identifying the holotype specimen.

> **Ex. 1.** The type of *Lycianthes lucens* S. Knapp (in PhytoKeys 209: 65. 2022) was designated as "*O. Gideon LAE-57196* (holotype: LAE [acc. # 256314]; isotypes: K [K000922490], L [L.2882045])", where "LAE [acc. # 256314]" is the accession number on the holotype in the herbarium of the Papua New Guinea Forest Research Institute (LAE).

40A.5. Citation of the herbarium, collection, or institution of deposition should use one of the standards mentioned in Art. 40 Note 5 or, when those standards give no abbreviated form, the herbarium, collection, or institution should be cited in full with its location.

SECTION 3
NEW COMBINATIONS, NAMES AT NEW RANK, AND REPLACEMENT NAMES

ARTICLE 41

41.1. To be validly published, a new combination, name at new rank, or replacement name must be accompanied by a reference to the basionym or replaced synonym (see Art. 6.10 and 6.11).

41.2. For the purpose of valid publication of a new combination, name at new rank, or replacement name, the following restrictions apply:

(a) For a name of a family or subdivision of a family, the basionym or replaced synonym must be a name of a family or subdivision of a family.

(b) For a name of a genus or subdivision of a genus, the basionym or replaced synonym must be a name of a genus or subdivision of a genus.

(c) For a name of a species or infraspecific taxon, the basionym or replaced synonym must be a name of a species or infraspecific taxon.

Ex. 1. (a) Carl Presl did not validly publish *"Cuscuteae"* (in Presl & Presl, Delic. Prag.: 87. 1822) as the name of a family (see "Praemonenda", pp. [3–4]) based on *Cuscutales* Bercht. & J. Presl (Přir. Rostlin: 247. 1820, *'Cuscuteae'*) because the latter is the name of an order (see Art. 18 *Ex. 6).

Ex. 2. (b) *Thuspeinanta* T. Durand (Index Gen. Phan.: 703. 1888) is a replacement name for *Tapeinanthus* Boiss. ex Benth. (in Candolle, Prodr. 12: 436. 1848) non Herb. (Amaryllidaceae: 190. 1837); *Aspalathoides* (DC.) K. Koch (Hort. Dendrol.: 242. 1853) is based on *Anthyllis* sect. *Aspalathoides* DC. (Prodr. 2: 169. 1825).

41.3. An indirect reference (see Art. 38.15) to a basionym or replaced synonym is sufficient for valid publication of a new combination, name at new rank, or replacement name published before 1 January 1953. Therefore, errors in the citation of the basionym or replaced synonym, or in author citation (Art. 46), do not affect valid publication of such names.

Ex. 3. Gross (in Bot. Jahrb. Syst. 49: 275. 1913) ascribed the name *Persicaria runcinata* to "(Ham.)" without giving further information. Previously, the name *Polygonum runcinatum* had been validly published by Don (Prodr. Fl. Nepal.: 73. 1825) and ascribed there to "Hamilton MSS." The mention of "Ham." by Gross is regarded as an indirect reference to the basionym published by Don, and thus the new combination *Persicaria runcinata* (Buch.-Ham. ex D. Don) H. Gross was validly published.

Ex. 4. Opiz validly published the name at new rank *Hemisphace* (Benth.) Opiz (Seznam: 50. 1852) by writing *"Hemisphace* Benth."*, which is regarded as an indirect reference to the basionym *Salvia* sect. *Hemisphace* Benth. (Labiat. Gen. Spec.: 193, 207, 310. 1833).

Ex. 5. The new combination *Cymbopogon martini* (Roxb.) Will. Watson (in Gaz. N.-W. Prov. India 10: 392. 1882) is validly published through the cryptic notation "309", which, as explained at the top of the same page, is the running-number of the species (*Andropogon martini* Roxb.) in Steudel (Syn. Pl. Glumac. 1: 388. 1854). Although the reference to the basionym *A. martini* is indirect, it is unambiguous (but see Art. 33 Ex. 1; see also Rec. 60C.1).

Ex. 6. Miller (1768), in the preface to *The gardeners dictionary,* ed. 8, stated that he had "now applied Linnaeus's method entirely, except in such particulars ...", of which he gave examples. In the main text, he often referred to Linnaean genera under his own generic headings, e.g. to *Cactus* L. (pro parte) under *Opuntia* Mill. Therefore, an implicit reference to a Linnaean binomial may be assumed when this is appropriate, and Miller's binomials are accepted as new combinations (e.g. *O. ficus-indica* (L.) Mill., based on *C. ficus-indica* L.) or replacement names (e.g. *O. vulgaris* Mill., based on *C. opuntia* L.: both names have the reference to *"Opuntia vulgo herbariorum"* of Bauhin & Cherler in common).

Ex. 7. When Haines (Forest Fl. Chota Nagpur: 530. 1910) published the name *Dioscorea belophylla,* he attributed the name to "Voight". Previously, Prain (Bengal Pl.: 1065, 1067. 1903) had validly published *D. nummularia* var. *belophylla* Prain, citing "Voigt (sp.)", an apparent reference to the nomen nudum *"Dioscorea belophylla"* (Voigt, Hort. Suburb. Calcutt.: 653. 1845). The mention by Haines of "Voight" is regarded as an indirect reference to Prain's varietal name, and thus *D. belophylla* (Prain) Voigt ex Haines was validly published as a new combination and name at new rank.

Ex. 8. Cortinarius collinitus var. *trivialis* (J. E. Lange) A. H. Sm. (in Lloydia 7: 175. 1944) was validly published as a new combination based on *C. trivialis* J. E. Lange (Fl. Agaric. Danic. 5(Taxon. Consp.): iii. 1940), even though Smith referred to the basionym as "*C. trivialis* Lange 'Studies,' pt. 10: 24. 1935", where that name was not validly published because Lange did not provide a Latin description or diagnosis.

41.4. If, for a name of a genus or lower-ranked taxon published before 1 January 1953, no reference to a basionym is given but the conditions for its valid publication as the name of a new taxon or a replacement name are fulfilled, that name is nevertheless treated as a new combination or name at new rank when this was the author's presumed intent and a potential basionym (Art. 6.10) applying to the same taxon exists.

Ex. 9. In Kummer's *Der Führer in die Pilzkunde* (1871) the note (p. 12) explaining that the author intended to adopt at generic rank the subdivisions of *Agaricus* then in use, which at the time were those of Fries, and the general arrangement of the work, which faithfully follows that of Fries, have been considered to provide indirect reference to Fries's earlier names of "tribes" as basionyms (see Art. F.4.1). Even though this was Kummer's presumed intent, he did not actually mention Fries. Nevertheless, even when Art. 41.3 is not considered to apply, because Kummer provided diagnoses in a key and thus fulfilled the conditions for valid publication of names of new taxa, Art. 41.4 rules that names such as *Hypholoma* (Fr.) P. Kumm. and *H. fasciculare* (Huds.) P. Kumm. are to be accepted as new combinations or names at new rank based on the corresponding Friesian names (here: *A.* "tribus" [unranked] *Hypholoma* Fr., nom. sanct. and *A. fascicularis* Huds., nom. sanct.).

Ex. 10. Scaevola taccada was validly published by Roxburgh (Hort. Bengal.: 15. 1814) solely by reference to an illustration in Rheede (Hort. Malab. 4: t. 59. 1683) that is associated with a description of a species. Because the same illustration was cited in the protologue of the earlier name *Lobelia taccada* Gaertn. (Fruct. Sem. Pl. 1: 119. 1788) and the two names apply to the same species, *S. taccada* is treated as a new combination, *S. taccada* (Gaertn.) Roxb., not as the name of a new species, even though in Roxburgh's protologue there is no reference, either direct or indirect, to *L. taccada*.

Ex. 11. When Moench (Methodus: 272. 1794) described *Chamaecrista,* he did not refer to *Cassia* [unranked] *Chamaecrista* L. (Sp. Pl.: 379. 1753; see Art. 32 Ex. 1) but used its epithet as the generic name and included its type, *Cassia chamaecrista* L. (cited in synonymy). Therefore, he published a name at new rank, *Chamaecrista* (L.) Moench, and not a name of a new genus.

Ex. 12. *Cololejeunea* was published by Stephani (in Hedwigia 30: 208. 1891) for a taxon that had previously been described as *Lejeunea* subg. *Cololejeunea* Spruce (in Trans. & Proc. Bot. Soc. Edinburgh 15: 79, 291. 1884) but without even an indirect reference to Spruce's earlier publication. Because Stephani provided a description of *C. elegans* Steph. that under Art. 38.6 is acceptable as a descriptio generico-specifica, he fulfilled the requirements for valid publication of *Cololejeunea* as the name of a new monotypic genus. Under Art. 41.4, *Cololejeunea* is therefore to be treated as a name at new rank, *Cololejeunea* (Spruce) Steph., based on Spruce's subgeneric name.

Ex. 13. When Sampaio published "*Psoroma murale* Samp." (in Bol. Real Soc. Esp. Hist. Nat. 27: 142. 1927), he adopted the epithet of *Lichen muralis* Schreb. (Spic. Fl. Lips.: 130. 1771), a name applied to the same taxon, without referring to that name either directly or indirectly. He cited in synonymy *Lecanora saxicola* (Pollich) Ach. (Lichenogr. Universalis: 431. 1810), which is based on *Lichen saxicola* Pollich (Hist. Pl. Palat. 3: 225. 1777). Under Art. 41.4, *Psoroma murale* (Schreb.) Samp. is treated as a new combination based on *Lichen muralis;* otherwise, it would be a validly published but illegitimate replacement name for *Lichen saxicola.*

41.5. A new combination, name at new rank, or replacement name published on or after 1 January 1953 is not validly published unless its basionym or replaced synonym is clearly indicated and a full and direct reference given to its author and place of valid publication, with page or plate reference and date (but see Art. 41.6 and 41.8). In addition, a new combination, name at new rank, or replacement name published on or after 1 January 2007 is not validly published unless its basionym or replaced synonym is cited.

Ex. 14. In transferring *Ectocarpus mucronatus* D. A. Saunders to *Giffordia,* Kjeldsen & Phinney (in Madroño 22: 90. 27 Apr 1973) cited the basionym and its author but without reference to its place of valid publication. They later (in Madroño 22: 154. 2 Jul 1973) validly published the new combination *G. mucronata* (D. A. Saunders) Kjeldsen & H. K. Phinney by giving a full and direct reference to the place of valid publication of the basionym.

Ex. 15. The new combination *Conophytum marginatum* subsp. *littlewoodii* (L. Bolus) S. A. Hammer (Dumpling & His Wife: New Views Gen. Conophytum: 181. 2002), because it was made before 1 January 2007, was validly published even though Hammer did not cite the basionym (*C. littlewoodii* L. Bolus) but only indicated it by giving a full and direct reference to its place of valid publication.

i **Note 1.** For the purpose of Art. 41.5, a page reference (for publications with a consecutive pagination) is a reference to the page or pages on which the basionym or replaced synonym was validly published or on which the protologue appears, but not to the pagination of the whole publication unless it is coextensive with that of the protologue.

Ex. 16. When proposing *"Cylindrocladium infestans"*, Peerally (in Mycotaxon 40: 337. 1991) cited the basionym as *"Cylindrocladiella infestans* Boesew., Can. J. Bot. 60: 2288–2294. 1982". Because this refers to the pagination of Boesewinkel's entire paper, not of the protologue of the intended basionym alone, the combination was not validly published by Peerally.

i **Note 2.** For the purpose of Art. 41.5, a virtual page reference can be achieved for publications lacking page numbers, e.g. by:

(a) citing an assumed page number when there is continuous pagination; or
(b) citing the page number automatically generated within the PDF of an electronic publication; or
(c) using the words "without page number", "sine pagina", "s.p." or similar; or
(d) including any indication that refers to the exact page, for example citing the species number, or the words "addition" or "supplement" if indicated as such on the page.

The citation of a DOI or URL is not by itself sufficient for page indication (see also Rec. 41A.2).

41.6. For names published on or after 1 January 1953, errors in the citation of the basionym or replaced synonym, including incorrect author citation (Art. 46), but not omissions (Art. 41.5), do not preclude valid publication of a new combination, name at new rank, or replacement name.

Ex. 17. *Aronia arbutifolia* var. *nigra* (Willd.) F. Seym. (Fl. New England: 308. 1969) was published as a new combination "Based on *Mespilus arbutifolia* L. var. *nigra* Willd., in Sp. Pl. 2: 1013. 1800." Willdenow treated these plants in the genus *Pyrus,* not *Mespilus,* and publication was in 1799, not 1800; these errors of citation do not preclude valid publication of the new combination.

Ex. 18. The name at new rank *Agropyron desertorum* var. *pilosiusculum* (Melderis) H. L. Yang (in Kuo, Fl. Reipubl. Popularis Sin. 9(3): 113. 1987) was inadvertently but validly published by Yang, who wrote *"Agropyron desertorum* ... var.

pilosiusculum Meld. in Norlindh, Fl. Mong. Steppe. 1: 121. 1949", which constitutes a full and direct reference to the basionym, *A. desertorum* f. *pilosiusculum* Melderis, despite the error in citing the rank-denoting term.

Ex. 19. *Nekemias grossedentata* (Hand.-Mazz.) J. Wen & Z. L. Nie (in PhytoKeys 42: 16. 2014) was published as a new combination, with the basionym cited as "*Ampelopsis cantoniensis* var. *grossedentata* Hand.-Mazz., Sitzungsber. Kaiserl. Akad. Wiss., Math.-Naturwiss. Cl., Abt. 1, 59: 105. 1877". The actual place of publication of the cited basionym was in Anz. Akad. Wiss. Wien, Math.-Naturwiss. Kl. 59: 105. 1922. These errors of citation (name of the journal and date) do not prevent valid publication of the new combination.

Ex. 20. The basionym of *Eremogone hookeri* (Nutt.) W. A. Weber (in Brittonia 33: 326. 1981) was cited as "*Arenaria hookeri* Nutt. ex T. & G. Fl. N. Amer. 1: 178. 1838." The author of the basionym, however, is Nuttall alone (see Art. 46 Ex. 6). Weber's incorrect authorship citation "Nutt. ex T. & G." (Nuttall ex Torrey & A. Gray) does not prevent valid publication of the new combination.

41.7. Mere reference to the *Index kewensis,* the *Index of fungi,* or any work other than that in which the name was validly published does not constitute a full and direct reference to the place of publication of a name (but see Art. 41.8).

Ex. 21. "*Leptosiphon croceus* (Eastw.) J. M. Porter & L. A. Johnson, comb. nov." (in Aliso 19: 80. 2000) was published with the basionym citation "*Linanthus croceus* Eastw., Pl. hartw. p. 325. 1849." Because the actual place of publication of *Linanthus croceus* was in Bot. Gaz. 37: 442–443. 1904, Porter & Johnson's combination was not validly published.

Ex. 22. Ciferri (in Mycopathol. Mycol. Appl. 7: 86–89. 1954), in proposing 142 intended new combinations in *Meliola,* omitted references to places of publication of basionyms, stating that they could be found in Petrak's lists or in the *Index of fungi;* none of these combinations was validly published. Similarly, Grummann (Cat. Lich. Germ.: 18. 1963) introduced a new combination in the form *Lecanora campestris* f. "*pseudistera* (Nyl.) Grumm. c.n. – *L. p.* Nyl., Z 5: 521", in which "Z 5" referred to Zahlbruckner (Cat. Lich. Univ. 5: 521. 1928), who gave the full citation of the basionym, *Lecanora pseudistera* Nyl.; Grummann's combination was not validly published.

🛈 **Note 3.** For the purpose of Art. 41.7, an unpaginated or independently paginated electronic publication and a later version with definitive pagination are not considered to be different publications (see Art. 30 Note 1).

🛈 **Note 4.** A new name published for a taxon previously known under a misapplied name is always the name of a new taxon and must therefore meet all relevant requirements of Art. 32–45 and F.4–F.5 for valid publication of such a name. This procedure is not the same as publishing a replacement name for a validly published but illegitimate name (Art. 58.1), the type of which is necessarily that of the replaced synonym (Art. 7.4).

Ex. 23. *Sadleria hillebrandii* W. J. Rob. (in Bull. Torrey Bot. Club 40: 226. 1913) was introduced as a "nom. nov." for "*Sadleria pallida* Hilleb. Fl. Haw. Is. 582. 1888. Not Hook. & Arn. Bot. Beech. 75. 1832." Because the requirements for valid publication were satisfied (before 1935, simple reference to a previous description or diagnosis in any language was sufficient), *S. hillebrandii* is the name of a new species validated by Hillebrand's description of the taxon to which he misapplied the name *S. pallida* Hook. & Arn., not a replacement name as stated by Robinson (see Art. 6.14).

Ex. 24. *"Juncus bufonius* var. *occidentalis"* (Hermann in Gen. Techn. Rep. R. M., U.S. Forest Serv. 18: 14. 1975) was published as a "nom. et stat. nov." for *J. sphaerocarpus* "auct. Am., non Nees". Because there is no Latin description or diagnosis, indication of type, or reference to any previous publication providing these requirements, this is not a validly published name.

41.8. In any of the following cases, a full and direct reference to a work other than that in which the basionym or replaced synonym was validly published is treated as an error to be corrected, not affecting the valid publication of a new combination, name at new rank, or replacement name published on or after 1 January 1953:

(a) when the actual basionym or replaced synonym was validly published earlier than the name or later isonym cited as such, but in the cited publication, in which all conditions for valid publication of the name as cited are fulfilled, there is no reference, in association with that name, to the place of valid publication of the actual basionym or replaced synonym (but see Art. 41.5 second sentence); or

(b) when the failure to cite the place of valid publication of the basionym or replaced synonym is explained by the later nomenclatural starting-point for the group concerned (Art. 13.1), or by the backward shift of the starting date for some fungi; or

(c) when the resulting new combination or name at new rank would otherwise be validly published as a (legitimate or illegitimate) replacement name; or

(d) when the resulting new combination, name at new rank, or replacement name would otherwise be the validly published name of a new taxon.

Ex. 25. (a) The new combination *Trichipteris kalbreyeri* was proposed by Tryon (in Contr. Gray Herb. 200: 45. 1970) with a full and direct reference to "*Alsophila Kalbreyeri* C. Chr. Ind. Fil. 44. 1905". This, however, is not the place of valid publication of the intended basionym, which had previously been published, with the same type, by Baker (1892; see Art. 6 Ex. 2). Because Christensen provided no reference to Baker's earlier publication, Tryon's error of citation does not affect the valid publication of his new combination, which is cited as *T. kalbreyeri* (Baker) R. M. Tryon.

Ex. 26. (a) The intended new combination *"Machaerina iridifolia"* was proposed by Koyama (in Bot. Mag. (Tokyo) 69: 64. 1956) with a full and direct reference to "*Cladium iridifolium* Baker, Flor. Maurit. 424 (1877)". However, *C. iridifolium* had been proposed by Baker as a new combination based on *Scirpus iridifolius* Bory (Voy. Îles Afrique 2: 94. 1804). Because Baker provided an explicit reference to Bory, Art. 41.8(a) does not apply and the combination under *Machaerina* was not validly published by Koyama.

Ex. 27. (b) The combination *Lasiobelonium corticale* was proposed by Raitviir (in Scripta Mycol. 9: 106. 1980) with a full and direct reference to *Peziza corticalis* in Fries (Syst. Mycol. 2: 96. 1822). This, however, is not the place of valid publication of the basionym, which, under the *Code* operating in 1980, was in Mérat (Nouv. Fl. Env. Paris, ed. 2, 1: 22. 1821), and under the current *Code* is in Persoon (Observ. Mycol. 1: 28. 1796). Raitviir's error of citation is partly explained by the backward shift of the starting date for some fungi and partly by the absence of a reference to Mérat in Fries's work and does not therefore prevent valid publication of the new combination, which is cited as *L. corticale* (Pers.) Raitv.

Ex. 28. (b) *Malvidae* C. Y. Wu (in Acta Phytotax. Sin. 40: 306. 2002) was validly published as a name at new rank based on *Malvaceae* Juss. (Gen. Pl.: 271. 1789), even though Wu cited as the basionym *"Malvaceae"* (Adanson, Fam. Pl. 2: 390. 1763). Wu's error of citation, explained by the later nomenclatural starting-point for suprageneric names of *Spermatophyta* and *Pteridophyta* (Art. 13.1(a)), does not prevent valid publication of the name at new rank.

Ex. 29. (c) The new combination *Mirabilis laevis* subsp. *glutinosa* was proposed by Murray (in Kalmia 13: 32. 1983) with a full and direct reference to "*Mirabilis glutinosa* A. Nels., Proc. Biol. Soc. Wash. 17: 92 (1904)" as the intended basionym. This, however, cannot be a basionym because it is an illegitimate later homonym of *M. glutinosa* Kuntze (Revis. Gen. Pl. 3: 265. 1898); it is also the replaced synonym of *Hesperonia glutinosa* Standl. (in Contr. U. S. Natl. Herb. 12: 365. 1909). Under Art. 41.8(c), Murray validly published a new combination based on *H. glutinosa*, because otherwise he would have published a replacement name for *M. glutinosa* A. Nelson. The name is therefore to be cited as *M. laevis* subsp. *glutinosa* (Standl.) A. E. Murray.

Ex. 30. (c) The new combination *Tillandsia barclayana* var. *minor* was proposed by Butcher (in Bromeliaceae 43(6): 5. 2009) with a reference, but not a full and direct one, to *Vriesea barclayana* var. *minor* Gilmartin (in Phytologia 16: 164. 1968). Butcher also provided a full and direct reference to *T. lateritia* André ("BASIONYM: *Tillandsia lateritia* Andre, Enum. Bromel. 6. 13 Dec 1888; Revue Hort. 60: 566. 16 Dec 1888"), which is the replaced synonym of *V. barclayana* var. *minor*. Under Art. 41.8(c), *T. barclayana* var. *minor* (Gilmartin) Butcher was validly published as a new combination based on *V. barclayana* var. *minor* because it would otherwise have been published as a replacement name for *T. lateritia*.

Ex. 31. (d) When Koyama published the new combination *Carex henryi* (C. B. Clarke) T. Koyama (in Jap. J. Bot. 15: 175. 1956), he cited the basionym, *C. longicruris* var. *henryi* C. B. Clarke (in J. Linn. Soc., Bot. 36: 295. 1903), with a full and direct reference not to the work in which that name was validly published, but to a later work (Kükenthal in Engler, Pflanzenr. IV. 20 (Heft 38): 603. 1909), in which the name was accompanied by a Latin diagnosis. Koyama's reference to Kükenthal is treated as an error to be corrected, not affecting the valid publication of the new combination *C. henryi,* because otherwise that name would be validly published as the name of a new species by direct reference to Kükenthal's Latin diagnosis (Art. 38.1(a)).

Recommendation 41A

41A.1. The full and direct reference to the place of publication of the basionym or replaced synonym should immediately follow a proposed new combination, name at new rank, or replacement name. It should not be provided by mere cross-reference to a bibliography at the end of the publication or to other parts of the same publication, e.g. by use of the abbreviations "loc. cit." or "op. cit."

41A.2. In the absence of established tradition, if publications are not paginated, page numbers indicated according to Art. 41 Note 2(a), (b), or (c) should be enclosed in square brackets.

Ex. 1. The name *Crocus antalyensioides* Rukšāns was published electronically in *International Rock Gardener* (ISSN 2053-7557), Volume 64, April 2015, in Portable Document Format (PDF), without page numbers included on the actual pages of the publication. The reference should be cited as Int. Rock Gard. 64: [6]. 2015.

SECTION 4
REGISTRATION OF NAMES AND NOMENCLATURAL ACTS

ARTICLE 42

42.1. Interested institutions, in particular those with expertise in nomenclatural indexing, may apply for recognition as nomenclatural repositories under this *Code*. A nomenclatural repository takes charge, for specified categories of organisms, of registering nomenclatural novelties (Art. 6 Note 4) and/or any nomenclatural act (Art. 34.1 footnote).

42.2. Applications for recognition as nomenclatural repositories for organisms other than those treated as fungi are to be addressed to the General Committee, which will refer the applications to the Registration Committee (see Div. III Prov. 7.15) and act upon its recommendation (for organisms treated as fungi see Art. F.5.1). Prior to such a recommendation, mechanisms and modalities of registration, and definition of coverage, will be developed in consultations among the applicant(s), the Registration Committee, and the Permanent Nomenclature Committee(s) for the group(s) concerned, and be widely publicized in the taxonomic community; a public trial run of at least one year must have shown that the procedure works efficiently and sustainably. The General Committee has the power to suspend or revoke a granted recognition.

42.3. The General Committee has the power to:

(a) Appoint one or more localized or decentralized, open and accessible, electronic repositories (recognized repositories) to accession the information required by Art. 42.5 and 42.6 and issue identifiers.
(b) Cancel such appointment at its discretion.
(c) Set aside the requirements of Art. 42.5 and 42.6, if the repository mechanism, or essential parts of it, cease to function.

Decisions made by this Committee under these powers are subject to ratification by a later International Botanical Congress.

42.4. Registration may be proactive and/or synchronous and/or retrospective; that is, it may occur before and/or simultaneously with and/or after the valid publication of a nomenclatural novelty (Art. 6 Note 4) or the effective publication of any nomenclatural act (Art. 34.1 footnote).

*ⓘ **Note 1.*** In contrast to mandatory registration of fungal names, registration of algal and plant names is voluntary.

*ⓘ **Note 2.*** For ways in which proactive registration of nomenclatural novelties functions, see Art. F.5.2 and F.5.3, relevant for names of organisms treated as fungi, including fossil fungi and lichen-forming fungi.

42.5. For an identifier to be issued by a recognized repository for a nomenclatural novelty (Art. 6 Note 4) applied to an organism treated as an alga or plant under this *Code*, the minimum elements of information that must be provided to the repository by either the author(s) or other user(s) of these scientific names are proof of effective publication (Art. 29–31) of the name itself along with those elements required for valid publication under Art. 38.1(a) and 39.2 (validating description or diagnosis) and Art. 40.1 and 40.5 (type) or Art. 41.5 (reference to the basionym or replaced synonym) and for algae Art. 44.2 (illustration or figure).

42.6. For an identifier to be issued by a recognized repository for the purpose of specifying the designation of a type of the name of an organism treated as an alga or plant under this *Code*, the minimum elements of information that must be provided to the repository by either the author(s) or other user(s) of these type designations are proof of effective publication (Art. 29–31) of the name being typified, along with the author(s) designating the type and those elements required by Art. 9.21, 9.22, and 9.23.

i **Note 3.** Proof of effective and/or valid publication for the purpose of issuing an identifier may be provided to the repository in a variety of forms including (but not limited to) PDFs, scanned page images, and/or URLs/ DOIs that lead to free and publicly accessible websites where these may be obtained.

i **Note 4.** For organisms treated as algae or plants under this *Code*, issuance of an identifier by a recognized repository has no effect on valid publication of a name (Art. 32–45) or effective type designation (Art. 7.8–7.11). Instead, these simply serve as stable digital references to the information present in the actual place of publication.

42.7. In the interests of fully exploring best practices for the registration of nomenclatural novelties, the General Committee has the power to appoint one or more repositories (as specified under Art. 42.3) to register nomenclatural novelties in a manner other than the protocol described in Art. 42.5, for example, those that follow the well-established practices for fungal names (Art. F.5.3). Regardless of the protocol employed, registration for algae and plants will remain voluntary. Such appointments made by this Committee under these powers are subject to the decision of a later International Botanical Congress.

Recommendation 42A

42A.1. Following the effective and valid publication of new names and no-menclatural acts, authors should report them to a recognized nomenclatural repository (Art. 42.1) for indexing and assignment of a unique identifier (Art. 42.5 and 42.6). While it is in the interest of the authors to do this promptly, any other party may present proof of effective and valid publication to these repositories for the purposes of both indexing and identifier assignment. In groups where proactive registration is mandatory (see Art. F.5), additional publication details (e.g. final pagination, precise date of publication, etc.) should be provided to the repository at this time.

42A.2. Specification of names and nomenclatural acts for the purpose of ex-change of digital information should include the unique identifier for those entities as established by a recognized nomenclatural repository (Art. 42.1).

SECTION 5
NAMES IN PARTICULAR GROUPS

ARTICLE 43
NAMES OF FOSSIL-TAXA

43.1. To be validly published, a name of a new fossil-taxon published on or after 1 January 1996 must be accompanied by a Latin or English description or diagnosis or by a reference (see Art. 38.14) to a previously and effectively published Latin or English description or diagnosis.

🛈 *Note 1.* Because Art. 39.1 does not apply to names of fossil-taxa, a validating description or diagnosis (see Art. 38) in any language is acceptable for them before 1996.

43.2. A name of a new fossil-genus or lower-ranked fossil-taxon published on or after 1 January 1912 is not validly published unless it is accompanied by an illustration or figure showing the essential characters or by a reference to a previously and effectively published such illustration or figure. For this purpose, in the case of a name of a fossil-genus or subdivision of a fossil-genus, citation of, or reference (direct or indirect) to, a name of a fossil-species validly published on or after 1 January 1912 will suffice.

Ex. 1. *"Laconiella"* when published by Krasser (in Sitzungsber. Akad. Wiss. Wien, Math.-Naturwiss. Kl. Abt. 1, 129: 16. 1920) included only one species, the intended name of which, *"Laconiella sardinica",* was not validly published as no illustration or figure or reference to a previously and effectively published illustration or figure was provided. *"Laconiella"* is not therefore a validly published generic name.

Ex. 2. *Batodendron* Chachlov (in Izv. Sibirsk. Otd. Geol. Komiteta 2(5): 9, fig. 23–25. 1921) was published with a description and illustrations. Even though the new fossil-genus did not include any named species, its name is validly published (albeit as an illegitimate later homonym of the non-fossil generic name *Batodendron* Nutt. in Trans. Amer. Philos. Soc., ser. 2, 8: 261. 1842).

43.3. A name of a new fossil-species or infraspecific fossil-taxon published on or after 1 January 2001 is not validly published unless at least one of the validating illustrations is identified as representing the type specimen (see also Art. 9.15).

Ex. 3. In the protologue of *Eophysaloides inflata* Cam. Martínez & Deanna (in New Phytol. 238: 2688. 2023), the authors designated the holotype as "STRI-SGC 36163. (Fig. 2a–c)" and wrote in the caption of figure 2 "(a–c) Holotype of *E. inflata,* STRI-SGC 36163" thus clearly identifying the validating illustrations as representing the holotype specimen.

ⓘ **Note 2.** To be validly published, a nomenclatural novelty applied to a fungal fossil-taxon and published on or after 1 January 2013 must comply with Art. F.5.2 and F.5.3.

ARTICLE 44
NAMES OF ALGAE

44.1. To be validly published, a name of a new taxon of non-fossil algae published between 1 January 1958 and 31 December 2011, inclusive, must be accompanied by a Latin description or diagnosis or by a reference (see Art. 38.14) to a previously and effectively published Latin description or diagnosis.

ⓘ **Note 1.** Because Art. 39.1 does not apply to names of algal taxa, a validating description or diagnosis (see Art. 38) in any language is acceptable for them before 1958.

Ex. 1. Although *Neoptilota* Kylin (Gatt. Rhodophyc.: 392. 1956) was accompanied only by a description in German, it is a validly published name because it applies to an alga and was published before 1958.

Ex. 2. *"Skeletonemopsis"* (Sims in Diatom Res. 9: 408. 1995) was not validly published because it was not accompanied by a Latin description or diagnosis. The designation was applied to a genus of fossil diatoms, which are nevertheless treated as non-fossil algae (see Art. 1.1 and 13.3), to which Art. 44.1 applies.

Skeletonemopsis P. A. Sims was validly published in App. III of the *Melbourne Code* (in Regnum Veg. 157: 50. 2015) by a full and direct reference to Sims's previously published English description of *"Skeletonemopsis"* (see Art. 39.2) and by indicating the type as *Skeletonema barbadense* Grev.

44.2. A name of a new species or infraspecific taxon of non-fossil algae published on or after 1 January 1958 is not validly published unless it is accompanied by an illustration or figure showing the distinctive morphological features, or by a reference to a previously and effectively published such illustration or figure.

Recommendation 44A

44A.1. The illustration or figure required by Art. 44.2 should be prepared from actual specimens, preferably including the holotype.

ARTICLE 45
NAMES IN GROUPS COVERED BY OTHER *CODES*

45.1. If a taxon originally assigned to a group not covered by this *Code* is treated as belonging to the algae or fungi, any of its names need satisfy only the requirements of the relevant other *Code* that the author was using for status equivalent to valid publication under this *Code* (but see Art. 54 and F.6.1, regarding homonymy). The *Code* used by the author is determined through internal evidence, regardless of any claim by the author as to the group of organisms to which the taxon is assigned. However, a name generated in zoological nomenclature in accordance with the Principle of Coordination (*ICZN* Article 46) is not validly published under this *Code* unless and until it actually appears in a publication as the accepted name of a taxon.

Ex. 1. *Amphiprora* Ehrenb. (in Abh. Königl. Akad. Wiss. Berlin 1841: 401, t. II(VI), fig. 28. 1843), available[1] under the *International Code of Zoological Nomenclature* as the name of a genus of animals, was first treated as belonging to the algae by Kützing (Kieselschal. Bacill.: 107. 1844). Under the *International Code of Nomenclature for algae, fungi, and plants*, *Amphiprora* is validly published and dates from 1843, not 1844.

Ex. 2. *Petalodinium* Cachon & Cachon-Enj. (in Protistologica 5: 16. 1969) is available under the *International Code of Zoological Nomenclature* as the name of a

---

1 The term "available" (when applied to a name) in the *International Code of Zoological Nomenclature* is equivalent to "validly published" in the *International Code of Nomenclature for algae, fungi, and plants*.

genus of dinoflagellates. When the taxon is treated as belonging to the algae, its name is validly published and retains its original authorship and date even though the original publication lacked a Latin description or diagnosis (Art. 44.1).

Ex. 3. *Prochlorothrix hollandica* Burger-Wiersma & al. (in Int. J. Syst. Bacteriol. 39: 256. 1989) was published according to the *International Code of Nomenclature of Prokaryotes*. When the taxon is treated as an alga, its name is validly published and retains its original authorship and date even though it was based on a living culture (Art. 8.4) and the original publication lacked a Latin description or diagnosis (Art. 44.1).

Ex. 4. *Labyrinthodictyon* Valkanov (in Progr. Protozool. 3: 373. 1969, '*Labyrinthodyction*') is available under the *International Code of Zoological Nomenclature* as the name of a genus of rhizopods. When the taxon is treated as belonging to the fungi, its name is validly published and retains its original authorship and date even though the original publication lacked a Latin description or diagnosis (Art. 39.1).

Ex. 5. *Protodiniferaceae* Kof. & Swezy (in Mem. Univ. Calif. 5: 111. 1921, '*Protodiniferidae*'), available under the *International Code of Zoological Nomenclature,* is validly published as a name of a family of algae and retains its original authorship and date but with the original termination changed in accordance with Art. 18.4 and 32.2.

Ex. 6. *Pneumocystis* P. Delanoë & Delanoë (in Compt. Rend. Hebd. Séances Acad. Sci. 155: 660. 1912) was published for a "protozoan" genus with a description expressing doubt as to its generic status, "Si celui-ci doit constituer un genre nouveau, nous proposons de lui donner le nom de *Pneumocystis Carinii.* [If this is to constitute a new genus, we propose to give it the name *Pneumocystis carinii.*]" Under Art. 36.1(a) *Pneumocystis* would not be validly published, but Article 11.5.1 of the *International Code of Zoological Nomenclature* allows for such qualified publication before 1961. Therefore, *Pneumocystis,* because it is an available name under the *ICZN,* is validly published under Art. 45.1.

Ex. 7. *Pneumocystis jirovecii* Frenkel (in Natl. Cancer Inst. Monogr. 43: 16. 1976, '*jiroveci*'), treated as a protozoan, was published with only an English description and without designation of a type, but the former condition is no obstacle to availability under the *International Code of Zoological Nomenclature* (see *ICZN* Recommendation 13B) and the latter condition was no obstacle under that *Code* until after 1999 (*ICZN* Article 72.3). Therefore, when considered to be the name of a fungus, *P. jirovecii,* with corrected termination (Art. 60.8), is validly published under Art. 45.1. Subsequent publication of a Latin diagnosis and indication of type by Frenkel (in J. Eukaryot. Microbiol. 46: 91S. 1999), who treated the species as a fungus, was necessary for valid publication under the edition of the *International Code of Botanical Nomenclature* in operation at that time but is no longer so; *P. jirovecii* dates from 1976, not 1999.

ⓘ ***Note 1.*** Names of *Microsporidia* are not covered by this *Code* (see Pre. 8 and Art. F.1.1) even when *Microsporidia* are considered as fungi.

131

ⓘ **Note 2.** If a taxon originally assigned to a group not covered by this *Code* is treated as belonging to the plants (i.e. not the algae or fungi), the authorship and date of any of its names are determined by the first publication that satisfies the relevant requirements of Art. 32–45 for valid publication.

CHAPTER VI
CITATION OF NAMES

ARTICLE 46
AUTHOR CITATIONS

46.1. In publications, particularly those dealing with taxonomy and nomenclature, it may be desirable, even when no bibliographic reference to the protologue is made, to cite the author(s) of the name concerned (see also Art. 22.1 and 26.1). In so doing, the following rules of Art. 46 apply.

> *Ex. 1. Rosaceae* Juss. (Gen. Pl.: 334. 1789), *Rosa* L. (Sp. Pl.: 491. 1753), *Rosa gallica* L. (l.c.: 492. 1753), *Rosa gallica* var. *versicolor* L. (Sp. Pl., ed. 2: 704. 1762), *Rosa gallica* L. var. *gallica*.

ⓘ **Note 1.** A name of a taxon is attributed to the author(s) of the publication in which it appears (see Art. 46.5) unless one or more of the provisions of Art. 46 rule otherwise.

> *Ex. 2.* Wallich (Pl. Asiat. Rar. 3: 66. 15 Aug 1832) ascribed the name *Aikinia brunonis* to himself ("Wall.") and, although he ascribed both the diagnosis and description to "Brown", the correct attribution is *A. brunonis* Wall. because Wallich is the author of the publication and the name is not ascribed to anyone else.

46.2. A name of a new taxon is attributed to the author(s) to whom the name was ascribed when the validating description or diagnosis was simultaneously ascribed to or unequivocally associated with the same author(s), even when authorship of the publication is different. A new combination, name at new rank, or replacement name is attributed to the author(s) to whom it was ascribed when, in the publication in which it appears, it is explicitly stated that the same author(s) contributed in some way to that publication. Despite Art. 46.5, authorship of a nomenclatural novelty is always accepted as ascribed, even when it differs from authorship of the publication, when at least one author is common to both.

Ex. 3. The name *Pinus longaeva* was published in a paper by Bailey (in Ann. Missouri Bot. Gard. 57: 243. 1971) and was ascribed to "D. K. Bailey". The validating description is unequivocally associated with Bailey because he is the author of the publication (see Art. 46 Note 5). The name is therefore cited as *P. longaeva* D. K. Bailey (see also Art. 46 Note 1).

Ex. 4. The name *Viburnum ternatum* was published in Sargent (Trees & Shrubs 2: 37. 1907). It was ascribed to "Rehd.", and the account of the species has "Alfred Rehder" at the end. The name is therefore cited as *V. ternatum* Rehder.

Ex. 5. In a paper by Hilliard & Burtt (in Notes Roy. Bot. Gard. Edinburgh 43: 365. 1986) names of new species of *Schoenoxiphium,* including *S. altum,* were ascribed to Kukkonen, preceded by a statement "The following diagnostic descriptions of new species have been supplied by Dr. I. Kukkonen in order to make the names available for use." The name is therefore cited as *S. altum* Kukkonen.

Ex. 6. In Torrey & Gray (Fl. N. Amer. 1: 198. 1838) the names *Calyptridium* and *C. monandrum* were ascribed to "Nutt. mss.", and the descriptions were enclosed in double quotes indicating that Nuttall wrote them, as acknowledged in the preface. The names are therefore cited as *Calyptridium* Nutt. and *C. monandrum* Nutt.

Ex. 7. When publishing *Eucryphiaceae* (in Bot. Zeitung (Berlin) 6: 130. 1848) the otherwise unnamed author "W.", in a review of Gay's *Flora chilena* (1845–1854), wrote "wird die Gattung *Eucryphia* als Typus einer neuen Familie, der *Eucryphiaceae,* angesehen [the genus *Eucryphia* is considered as type of a new family, the *Eucryphiaceae*]", thus ascribing both the name and its validating description to Gay (Fl. Chil. 1: 348. 1846), who had used the designation "Eucrifiáceas" (see Art. 18.4). The name is therefore cited as *Eucryphiaceae* Gay.

Ex. 8. When Candolle (Essai Propr. Méd. Pl., ed. 2: 87. 1816) wrote "*Elaeocarpeae.* Juss., Ann. Mus. 11, p. 233" he ascribed the name to Jussieu and, to validate it, used Jussieu's diagnosis of an unnamed family (in Ann. Mus. Hist. Nat. 11: 233. 1808). The name is therefore cited as *Elaeocarpaceae* Juss., nom. cons. (see App. IIB), not *Elaeocarpaceae* "Juss. ex DC."

Ex. 9. Green (Census Vasc. Pl. W. Australia, ed. 2: 6. 1985) ascribed the new combination *Neotysonia phyllostegia* to Wilson and elsewhere in the same publication acknowledged his assistance. The name is therefore cited as *N. phyllostegia* (F. Muell.) Paul G. Wilson.

Ex. 10. The authorship of *Sophora tomentosa* subsp. *occidentalis* (L.) Brummitt (in Kirkia 5: 265. 1966) is accepted as originally ascribed, although the new combination was published in a paper authored jointly by Brummitt & Gillett.

ⓘ **Note 2.** When authorship of a name differs from authorship of the publication in which it was validly published, both are sometimes cited, connected by the word "in". In such a case, "in" and what follows are part of a bibliographic citation and are better omitted unless the place of publication is being cited.

Ex. 11. The name and original description of *Verrucaria aethiobola* Wahlenb. (in Acharius, Methodus, Suppl.: 17. 1803) were published in a single paragraph followed by an ascription to "Wahlenb. Msc." The name is therefore cited as *V. aethiobola* Wahlenb., not "Wahlenb. ex Ach." nor "Wahlenb. in Ach." (unless a full bibliographic citation is given), regardless of the accompanying description provided by Acharius.

Ex. 12. The new combination *Crepis lyrata* was published in Candolle's *Prodromus systematis naturalis regni vegetabilis* (7: 170. 1838), as "*C. lyrata* (Froel. in litt. 1837)", and in a footnote on p. 160 Candolle acknowledged Froelich as having authored the account of the relevant section of *Crepis:* "Sectiones generis iv, v et vi, à cl. Froelich elaboratae sunt [Sections iv, v and vi of the genus are provided by the famous Froelich]". The name is therefore cited as *C. lyrata* (L.) Froel. or *C. lyrata* (L.) Froel. in Candolle (followed by a bibliographic citation of the place of publication), but not *C. lyrata* "(L.) Froel. ex DC."

Ex. 13. The name *Physma arnoldianum* was published in a paper authored by Arnold (in Flora 41: 94. 1858). Arnold introduced the name as "*Ph. Arnoldianum* Hepp. lit. 12. Decbr. 1857", and the description is immediately followed by the phrase "Hepp. in lit." The name is therefore cited as *P. arnoldianum* Hepp, not *P. arnoldianum* "Hepp ex Arnold". Because Arnold is the author of the paper, not of the whole work (the journal *Flora*), his name is not required even in a full bibliographic citation.

ⓘ Note 3. The authorship of a descriptive name (Art. 16.1(b)) is not changed if the name is used at a rank different from that at which it was first validly published because it is not a name at new rank (see Art. 6 Note 3; see also Art. 49.2).

Ex. 14. *Streptophyta* Caval.-Sm. (in Lewin, Origins of Plastids: 340. 1993) was originally published as a name at the rank of infrakingdom (used as a rank between subkingdom and phylum). When the name is used at the rank of phylum, it is still cited as *Streptophyta* Caval.-Sm. (1993).

46.3. For the purposes of Art. 46, ascription is the direct association of the name of a person or persons with a new name or description or diagnosis of a taxon. An author citation associated with a synonym does not constitute ascription of the accepted name, nor does reference to a basionym or a replaced synonym (regardless of bibliographic accuracy) or reference to a homonym.

Ex. 15. The name *Atropa sideroxyloides* was published in Roemer & Schultes (Syst. Veg. 4: 686. 1819), with the name and diagnosis in a single paragraph followed by an ascription to "Reliq. Willd. MS." As this represents direct association of Willdenow with both the name and the diagnosis, the name is cited as *A. sideroxyloides* Willd., not *A. sideroxyloides* "Roem. & Schult." nor *A. sideroxyloides* "Willd. ex Roem. & Schult."

Ex. 16. *Sicyos triqueter* Moc. & Sessé ex Ser. (in Candolle, Prodr. 3: 309. 1830) was ascribed to Mociño and Sessé by Seringe's writing "*S. triqueter* (Moc. & Sessé, fl. mex. mss.)". However, *Malpighia emarginata* DC. (Prodr. 1: 578. 1824) was not ascribed to these authors by Candolle's writing "*M. emarginata* (fl. mex. ic. ined.)".

Ex. 17. *Lichen debilis* Sm. (in Smith & Sowerby, Engl. Bot. 35: t. 2462. 1812) was not ascribed to Turner and Borrer by Smith's citing "*Calicium debile*. Turn. and Borr. Mss." as a synonym.

Ex. 18. When Opiz (1852) wrote "*Hemisphace* Benth." he did not ascribe the generic name to Bentham but provided an indirect reference to the basionym, *Salvia* sect. *Hemisphace* Benth. (see Art. 41 Ex. 4).

Ex. 19. When Brotherus (in Engler & Prantl, Nat. Pflanzenfam. 1(3): 875. 1907) published "*Dichelodontium nitidum* Hook. fil. et Wils." he provided an indirect reference to the basionym, *Leucodon nitidus* Hook. f. & Wilson, and did not ascribe the new combination to Hooker and Wilson. He did, however, ascribe to them the simultaneously published name of his new genus, *Dichelodontium* Hook. f. & Wilson ex Broth.

Ex. 20. When Sheh & Watson (in Wu & al., Fl. China 14: 72. 2005) wrote "*Bupleurum hamiltonii* var. *paucefulcrans* C. Y. Wu ex R. H. Shan & Yin Li, Acta Phytotax. Sin. 12: 291. 1974" they did not ascribe the new combination to any of those authors but provided a full and direct reference to the basionym, *B. tenue* var. *paucefulcrans* C. Y. Wu ex R. H. Shan & Yin Li.

Ex. 21. When Sirodot (1872) wrote "*Lemanea* Bory" he published a later homonym because he excluded the nomenclatural type of *Lemanea* Bory (see Art. 48 Ex. 1). His reference to Bory's earlier homonym is not therefore ascription of the later homonym, *Lemanea* Sirodot, to Bory.

ⓘ **Note 4.** When the name of a new taxon is validly published by reference to a previously and effectively published description or diagnosis (Art. 38.1(a)), the name of the author of that description or diagnosis, even if not explicitly mentioned, is unequivocally associated with that description or diagnosis.

Ex. 22. The appropriate author citation for *Baloghia pininsularis* (see Art. 40 Ex. 2) is Guillaumin, and not McPherson & Tirel, because in the protologue the name was ascribed to Guillaumin and a full and direct reference was given to Guillaumin's earlier Latin description. Even though McPherson & Tirel did not explicitly ascribe the validating description to its author, Guillaumin, he is "unequivocally associated" with it.

Ex. 23. "*Pancheria humboldtiana*" was published by Guillaumin (in Mém. Mus. Natl. Hist. Nat., B, Bot. 15: 47. 1964), but not validly so because no type was indicated. Valid publication was achieved by Hopkins & Bradford (in Adansonia 31: 119. 2009), who designated "*Baumann-Bodenheim 15515* (P! P00143076)" as the holotype, ascribed the name to Guillaumin, and by citing "*Pancheria humboldtiana* Guillaumin, *Mémoires du Muséum national d'Histoire naturelle,* sér. B, botanique 15: 47 (1964), nom. inval.", provided a full and direct reference to a validating description that is unequivocally associated with Guillaumin. Despite Art. 46.10, the name is therefore attributed to Guillaumin, not "Guillaumin ex H. C. Hopkins & J. Bradford" as given by Hopkins & Bradford.

ⓘ Note 5. A name or its validating description or diagnosis is treated as though ascribed to the author(s) of the publication (as defined in Art. 46.6) when there is no ascription to or unequivocal association with a different author or different authors.

Ex. 24. The name *Asperococcus pusillus* was published in Hooker (Brit. Fl., ed. 4, 2(1): 277. 1833), with the name and diagnosis ascribed simultaneously, at the end of the paragraph, to "Carm. MSS." followed by a description ascribed similarly to Carmichael. Direct association of Carmichael with both the name and the diagnosis is evident, and the name must be cited as *A. pusillus* Carmich. However, the paragraph containing the name and the diagnosis of *A. castaneus,* published by Hooker on the same page of the same work, ends with "*Scytosiphon castaneus,* Carm. MSS." Because Carmichael is directly associated with "*S. castaneus*" and not *A. castaneus,* the latter name is correctly cited as *A. castaneus* Hook., the author of the publication, even though the description is ascribed to Carmichael.

Ex. 25. Brown is accepted as the author of the treatments of genera and species appearing under his name in Aiton's *Hortus kewensis,* ed. 2 (1810–1813), even when names of new taxa or the descriptions validating them are not explicitly ascribed to him. In a postscript to that work (5: 532. 1813), Aiton wrote: "Much new matter has been added by this gentleman, and some without reference to his name [Robert Brown]; but the greater part of his able improvements are distinguished by the signature of *Brown mss.*" The latter phrase is therefore a statement of authorship not merely an ascription. For example, the new combination *Oncidium triquetrum,* based by indirect reference on *Epidendrum triquetrum* Sw. (Prodr.: 122. 1788), is cited as *O. triquetrum* (Sw.) R. Br. (in Aiton, Hort. Kew., ed. 2, 5: 216. 1813) and is not attributed to "R. Br. ex W. T. Aiton", nor to Aiton alone, because in the generic heading Brown is credited with authorship of the treatment of *Oncidium.*

46.4. When a validly published name or its final epithet is taken up from and attributed to the author of a different "name" that has not been validly published, or one at a different rank likewise not validly published, only the author of the validly published name is cited (except as provided in Art. 46.7).

Ex. 26. When publishing the new generic name *Anoplon,* Reichenbach (Consp. Regn. Veg.: 212b. 1828–1829) attributed the name to Wallroth and referred to the designation published by Wallroth (Orobanches Gen. Diask.: 25, 66. 1825) as *Orobanche* "Tribus III. *Anoplon*", which was not validly published under Art. 37.7 because its rank was denoted by a misplaced term (tribe between genus and species). The generic name is cited as *Anoplon* Rchb., not *Anoplon* "Wallr. ex Rchb."

Ex. 27. When publishing *Andropogon drummondii,* Steudel (Syn. Pl. Glumac. 1: 393. 1854) attributed the name to "Nees. (mpt. sub: *Sorghum.*)". This reference to the unpublished binary designation "*Sorghum drummondii* Nees" is not ascription of *A. drummondii* to Nees, and the name is cited as *A. drummondii* Steud., not *A. drummondii* "Nees ex Steud."

Ex. 28. "*Porphyra yezoensis* f. *narawaensis*" was published by Miura (in J. Tokyo Univ. Fish. 71: 6. 1984), but two gatherings (from the same place but on different dates) were cited as "holotype" and the designation was not therefore validly published. Kikuchi & al. (in J. Jap. Bot. 90: 381. 2015), using Miura's description and designating a single specimen as the holotype, validly published the name *Pyropia yezoensis* f. *narawaensis* N. Kikuchi & al., which is not to be cited as *P. yezoensis* f. *narawaensis* "A. Miura ex N. Kikuchi & al."

46.5. A name of a new taxon is attributed to the author(s) of the publication in which it appears when the name was ascribed to a different author or different authors but the validating description or diagnosis was neither ascribed to nor unequivocally associated with that author or those authors. A new combination, name at new rank, or replacement name is attributed to the author(s) of the publication in which it appears, although it was ascribed to a different author or different authors, when no separate statement was made that one or more of those authors contributed in some way to that publication. However, in both cases, authorship as ascribed, followed by "ex", may be inserted before the name(s) of the publishing author(s).

Ex. 29. Henry (in Bull. Trimestriel Soc. Mycol. France 74: 303. 1958) published the designation "*Cortinarius balteatotomentosus*" with a Latin description and a locality citation but without indicating a type (Art. 40 Note 2). He later (in Bull. Trimestriel Soc. Mycol. France 101: 4. 1985) validated the name by designating a holotype and providing a full and direct reference to his earlier description (see Art. 33.1). The description is therefore unequivocally associated with Henry (Art. 46 Note 4) and the name, although not explicitly ascribed, is treated as ascribed to Henry because he was the author of the publication (Art. 46 Note 5). Liimatainen & al. (in Persoonia 33: 118. 2014) cited the authorship as *C. balteatotomentosus* "Rob. Henry ex Rob. Henry", but Art. 46.5 does not apply because Henry did not ascribe the name to a different author. Under Art. 46.2, the name is correctly cited as *C. balteatotomentosus* Rob. Henry.

Ex. 30. *Lilium tianschanicum* was described by Grubov (in Grubov & Egorova, Rast. Tsent. Azii, Mater. Bot. Inst. Komarova 7: 70. 1977) as a new species, with its name ascribed to Ivanova; because there is no indication that Ivanova provided the validating description, the name is cited as either *L. tianschanicum* N. A. Ivanova ex Grubov or *L. tianschanicum* Grubov.

Ex. 31. In a paper by Boufford, Tsi & Wang (in J. Arnold Arbor. 71: 123. 1990) the name *Rubus fanjingshanensis* was ascribed to Lu with no indication that Lu provided the description; the name is attributed to either L. T. Lu ex Boufford & al. or Boufford & al.

Ex. 32. Seemann (Fl. Vit.: 22. 1865) published *Gossypium tomentosum* "Nutt. mss.", followed by a validating description not ascribed to Nuttall; the name is cited as either *G. tomentosum* Nutt. ex Seem. or *G. tomentosum* Seem.

Ex. 33. Rudolphi published *Pinaceae* (Syst. Orb. Veg.: 35. 1830) as "*Pineae.* Spreng.", followed by a validating diagnosis not ascribed to Sprengel; the name is cited as either *Pinaceae* Spreng. ex F. Rudolphi or *Pinaceae* F. Rudolphi.

Ex. 34. Green (Census Vasc. Pl. W. Australia, ed. 2: 6. 1985) ascribed the new combination *Tersonia cyathiflora* to "(Fenzl) A. S. George"; because Green nowhere mentioned that George had contributed in any way, the name is cited as either *T. cyathiflora* (Fenzl) A. S. George ex J. W. Green or *T. cyathiflora* (Fenzl) J. W. Green.

46.6. For the purposes of Art. 46, the authorship of a publication is the authorship of that part of a publication in which a name appears regardless of the authorship or editorship of the publication as a whole.

Ex. 35. *Pittosporum buxifolium* was described as a new species, with its name ascribed to Feng, in Wu & Li, *Flora yunnanica,* vol. 3 (1983). The account of *Pittosporaceae* in that flora was authored by Yin, while the whole volume was edited by Wu & Li. The author of the publication (including the validating diagnosis) was Yin. The name is therefore cited as either *P. buxifolium* K. M. Feng ex W. Q. Yin or *P. buxifolium* W. Q. Yin, but not *P. buxifolium* "K. M. Feng ex C. Y. Wu & H. W. Li" nor *P. buxifolium* "C. Y. Wu & H. W. Li".

Ex. 36. *Vicia amurensis* f. *sanneensis,* ascribed to Jiang & Fu, was published in Ma & al. (editors), *Flora intramongolica,* ed. 2, vol. 3 (1989). The author of the account of *Vicia* in that flora is Jiang, one of the persons to whom the name was ascribed (see Art. 46.2 last sentence). The name is therefore cited as *V. amurensis* f. *sanneensis* Y. C. Jiang & S. M. Fu, not *V. amurensis* f. *sanneensis* "Y. C. Jiang & S. M. Fu ex Ma & al."

Ex. 37. *Centaurea funkii* var. *xeranthemoides* "Lge. ined." was described in *Prodromus florae hispanicae,* which was authored as a whole by Willkomm & Lange, although the different family treatments are by individual authors, and Fam. 63 *Compositae* has a footnote "Auctore Willkomm". Because the validating description was not ascribed to Lange, the name is cited as *C. funkii* var. *xeranthemoides* Lange ex Willk. Its full bibliographic citation is *C. funkii* var. *xeranthemoides* Lange ex Willk. in Willkomm & Lange, Prodr. Fl. Hispan. 2: 154. 1865.

Ex. 38. The name *Solanum dasypus* was published in a work of Candolle (Prodr. 13(1): 161. 1852), in which the account of *Solanaceae* was authored by Dunal. Dunal introduced the name as "*S. dasypus* (Drège, n. 1933, in h. DC)" thereby ascribing it to Drège. The name is therefore cited as either *S. dasypus* Drège ex Dunal or *S. dasypus* Dunal.

Ex. 39. Schultes & Schultes (Mant. 3: 526. 1827), in a note, published a new classification of the traditional genera *Avena* and *Trisetum,* which they had received
from "Besser in litt." The publishing author of that text, in which the new genera
Acrospelion Bess., *Helictotrichon* Bess., and *Heterochaeta* Bess. were described,
is Besser. The new names are validly published, authored by Besser alone, regardless of whether or not the volume authors, Schultes & Schultes, accepted
them (see also Art. 36 Ex. 3).

46.7. When a name has been ascribed by its author to a pre-starting-
point author, the latter may be included in the author citation, followed
by "ex". For groups with a starting-point later than 1753, when a taxon
of a pre-starting-point author was changed in rank or taxonomic position upon valid publication of its name, that pre-starting-point author
may be cited in parentheses, followed by "ex".

Ex. 40. Linnaeus (Gen. Pl., ed. 5: 322. 1754) ascribed the name *Lupinus* to the pre-
starting-point author Tournefort; the name is cited as either *Lupinus* Tourn. ex L.
(Sp. Pl.: 751. 1753) or *Lupinus* L. (see Art. 13.4).

Ex. 41. *"Lyngbya glutinosa"* (Agardh, Syst. Alg.: 73. 1824) was taken up as
Hydrocoleum glutinosum by Gomont in the publication that marks the starting-
point of the *"Nostocaceae homocysteae"* (in Ann. Sci. Nat., Bot., ser. 7, 15:
339. 1892). The name is cited as either *H. glutinosum* (C. Agardh) ex Gomont or
H. glutinosum Gomont.

Ex. 42. Designations of desmids published before their starting-point (see Art.
13.1(e)) may be cited according to their validation in Ralfs (Brit. Desmid. 1848) as
follows: *"Closterium dianae"* (Ehrenberg, Infusionsthierchen: 92. 1838), cited as
C. dianae Ehrenb. ex Ralfs (l.c.: 168. 1848); *"Euastrum pinnatifidum"* (Kützing,
Phycol. Germ.: 134. 1845), cited as *Micrasterias pinnatifida* (Kütz.) ex Ralfs (l.c.:
77. 1848).

46.8. In determining the correct author citation, only internal evidence in the publication as a whole (as defined in Art. 37.6) where the
name was validly published is to be accepted, including ascription of
the name, statements in the introduction, title, or acknowledgements,
and typographical or stylistic distinctions in the text.

Ex. 43. Although the descriptions in Aiton's *Hortus kewensis* (1789) are generally considered to have been written by Solander or Dryander, the names of new
taxa published there are attributed to Aiton, the stated author of the work, except
where a name and description were both ascribed in that work to someone else.

Ex. 44. The name *Andreaea angustata* was published in a work of Limpricht
(Laubm. Deutschl. 1: 144. 1885) with the ascription "nov. sp. Lindb. in litt. ad
Breidler 1884 [new species of Lindberg in a letter to Breidler in 1884]", but there
is no internal evidence that Lindberg had supplied the validating description.
Authorship is therefore cited as either Lindb. ex Limpr. or Limpr., but not "Lindb."

46.9. External evidence may be used to determine authorship of nomenclatural novelties included in a publication for which there is no internal evidence of authorship.

Ex. 45. If no internal or external evidence of authorship of effectively and validly published names can be determined, the standard form "Anon." (for Anonymous) may be used, e.g. *Ficus cooperi* Anon. (in Proc. Roy. Hort. Soc. London 2: 374. 1862) or *Nymphaea gigantea* f. *hudsonii* (Anon.) K. C. Landon (in Phytologia 40: 439. 1978).

Ex. 46. No authorship appears anywhere in the work known as "Cat. Pl. Upper Louisiana. 1813", a catalogue of plants available from the Fraser Brothers Nursery. Based on external evidence (cf. Stafleu & Cowan in Regnum Veg. 105: 785. 1981), authorship of the document, and of included nomenclatural novelties such as *Oenothera macrocarpa,* is attributed to Thomas Nuttall.

Ex. 47. The book that appeared under the title *Vollständiges systematisches Verzeichniß aller Gewächse Teutschlandes …* (Leipzig 1782) has no explicit authorship but is attributed "einem Mitgliede der [to a member of the] Gesellschaft Naturforschender Freunde". External evidence may be used to determine that G. A. Honckeny is the author of the work and of the nomenclatural novelties that appear in it (e.g. *Poa vallesiana* Honck., *Phleum hirsutum* Honck.; see also Art. 23 Ex. 24), as was done by Pritzel (Thes. Lit. Bot.: 123. 1847).

46.10. Authors publishing nomenclatural novelties and wishing other persons' names followed by "ex" to precede theirs in author citation may adopt the "ex" citation in the protologue.

Ex. 48. In validly publishing the name *Nothotsuga,* Page (in Notes Roy. Bot. Gard. Edinburgh 45: 390. 1989) ascribed it to "H.-H. Hu ex C. N. Page", noting that in 1951 Hu had published it as a nomen nudum; the name is attributed to either Hu ex C. N. Page or C. N. Page.

Ex. 49. Atwood (in Selbyana 5: 302. 1981) ascribed the name of a new species, *Maxillaria mombachoensis,* to "Heller ex Atwood", with a note stating that it was originally named by Heller, then deceased; the name is attributed to either A. H. Heller ex J. T. Atwood or J. T. Atwood.

Recommendation 46A

46A.1. For the purpose of author citation, prefixes indicating ennoblement (see Rec. 60C.4(d) and (e)) should be suppressed unless they are an inseparable part of the name.

Ex. 1. Lam. for J. B. P. A. Monet Chevalier de Lamarck, but De Wild. for E. De Wildeman.

46A.2. When a name in an author citation is abbreviated, the abbreviation should be long enough to be distinctive, and should normally end with a consonant that, in the full name, precedes a vowel. The first letters should be given without any omission, but one of the last characteristic consonants of the name may be added when this is customary.

> *Ex. 2.* L. for Linnaeus; Fr. for Fries; Juss. for Jussieu; Rich. for Richard; Bertol. for Bertoloni, to be distinct from Bertero; Michx. for Michaux, to be distinct from Micheli.

46A.3. Given names or accessory designations serving to distinguish two authors of the same name should be abridged in the same way.

> *Ex. 3.* R. Br. for Robert Brown; A. Juss. for Adrien de Jussieu; Burm. f. for Burman filius; J. F. Gmel. for Johann Friedrich Gmelin, J. G. Gmel. for Johann Georg Gmelin, C. C. Gmel. for Carl Christian Gmelin, S. G. Gmel. for Samuel Gottlieb Gmelin; Müll. Arg. for Jean Müller argoviensis (of Aargau).

46A.4. When it is a well-established custom to abridge a name in another manner, it is advisable to conform to custom.

> *Ex. 4.* DC. for A.-P. de Candolle; A. St.-Hil. for A. F. C. P. de Saint-Hilaire; Rchb. for H. G. L. Reichenbach.

ⓘ **Note 1.** Brummitt & Powell's *Authors of plant names* (1992) provides unambiguous standard forms for many authors of names of organisms in conformity with this Recommendation. These standard forms, updated as necessary from the International Plant Names Index and Index Fungorum, have been used for author citations throughout this *Code,* although with additional spacing.

Recommendation 46B

46B.1. In citing the author of the scientific name of a taxon, the romanization of the author's name given in the original publication should normally be accepted. Where an author did not give a romanization, or where an author has at different times used different romanizations, then the romanization known to be preferred by the author or that most frequently adopted by the author should be accepted. In the absence of such information, the author's name should be romanized in accordance with an internationally available standard.

46B.2. Authors of scientific names whose personal names are not written in the Latin alphabet should romanize their names, preferably (but not necessarily) in accordance with an internationally recognized standard and, as a matter of typographical convenience, without diacritical signs. Once authors have selected the romanization of their personal names, they should use it consistently. Whenever possible, authors should not permit editors or publishers to change the romanization of their personal names.

Recommendation 46C

46C.1. After a name published jointly by two authors, both authors should be cited, linked by an ampersand (&) or by the word "et".

> *Ex. 1.* *Didymopanax gleasonii* Britton & P. Wilson or *D. gleasonii* Britton et P. Wilson.

46C.2. After a name published jointly by more than two authors, the citation should be restricted to the first author followed by "& al." or "et al.", except in the original publication.

> *Ex. 2.* *Lapeirousia erythrantha* var. *welwitschii* (Baker) Geerinck, Lisowski, Malaisse & Symoens (in Bull. Soc. Roy. Bot. Belgique 105: 336. 1972) should be cited as *L. erythrantha* var. *welwitschii* (Baker) Geerinck & al. or *L. erythrantha* var. *welwitschii* (Baker) Geerinck et al.

Recommendation 46D

46D.1. Authors should cite themselves by name after each nomenclatural novelty they publish rather than refer to themselves by expressions such as "nobis" (nob.) or "mihi" (m.).

ARTICLE 47
NAMES OF ALTERED TAXA WITHOUT EXCLUSION OF TYPE

47.1. An alteration of the diagnostic characters or of the circumscription of a taxon without the exclusion of the type does not warrant a change of authorship of the name of the taxon.

> *Ex. 1.* When the original material of *Arabis beckwithii* S. Watson (in Proc. Amer. Acad. Arts 22: 467. 1887) is attributed to two different species, as by Munz (in Bull. S. Calif. Acad. Sci. 31: 62. 1932), the species not including the lectotype must have a different name (*A. shockleyi* Munz) but the other species is still named *A. beckwithii* S. Watson.

> *Ex. 2.* *Myosotis* as revised by Brown differs from the genus as originally circumscribed by Linnaeus, but the generic name remains *Myosotis* L. because the type of the name is still included in Brown's circumscription of the genus (it may be cited as *Myosotis* L. emend. R. Br.: see Rec. 47A).

> *Ex. 3.* The variously defined species that includes the types of *Centaurea jacea* L. (Sp. Pl.: 914. 1753), *C. amara* L. (Sp. Pl., ed. 2: 1292. 1763), and a variable number of other species names is still called *C. jacea* L. (or, as the case may be, *C. jacea* L. emend. Coss. & Germ., *C. jacea* L. emend. Vis., or *C. jacea* L. emend. Godr.: see Rec. 47A).

Recommendation 47A

47A.1. When an alteration as mentioned in Art. 47.1 has been considerable, the nature of the change may be indicated by adding such words, abbreviated where suitable, as "emendavit" (emend.) followed by the name of the author responsible for the change, "mutatis characteribus" (mut. char.), "pro parte" (p. p.), "excluso genere" or "exclusis generibus" (excl. gen.), "exclusa specie" or "exclusis speciebus" (excl. sp.), "exclusa varietate" or "exclusis varietatibus" (excl. var.), "sensu amplo" (s. ampl.), "sensu lato" (s. l.), "sensu stricto" (s. str.), etc.

> *Ex. 1.* *Phyllanthus* L. emend. Müll. Arg.; *Globularia cordifolia* L. excl. var. (emend. Lam.).

ARTICLE 48
NAMES OF ALTERED TAXA WITH EXCLUSION OF TYPE

48.1. The application of an existing name to a different taxon without exclusion of the type is considered to be a misapplication that has no nomenclatural status (but see Art. 57.1; see also Rec. 50D). However, if an author applies an existing name but definitely excludes its type, a later homonym that must be attributed solely to that author is considered to have been published. Similarly, when an author who adopts a name refers to an apparent basionym or replaced synonym but explicitly excludes its type, the name of a new taxon is considered to have been published that must be attributed solely to that author. Exclusion can be achieved by simultaneous explicit inclusion of the type in a different taxon by the same author.

> *Ex. 1.* Sirodot included *Lemanea corallina* Bory, the type of *Lemanea* Bory (in Ann. Mus. Hist. Nat. 12: 178, 183. 1808), in his new genus *Sacheria* Sirodot (in Ann. Sci. Nat., Bot., ser. 5, 16: 69. 1872), the name of which was therefore illegitimate (Art. 52.1). As a result, *Lemanea,* as treated by Sirodot (l.c. 1872), is cited as *Lemanea* Sirodot non Bory, and not as *Lemanea* "Bory emend. Sirodot".

> *Ex. 2.* In the protologue of *Peltophorum brasiliense* Urb. (in Symb. Antill. 2: 285. 1900), Urban cited in synonymy *"Caesalpinia brasiliensis* Linn. Spec. I ed. vol. I (1753) p. 380 (p. p.)", but on p. 279 he also accepted *C. brasiliensis* L. (Sp. Pl.: 380. 1753) as a distinct species. He thereby definitely excluded the type of *C. brasiliensis* from *P. brasiliense,* which was therefore published not as a new combination but as the legitimate name of a new species.

> *Ex. 3.* The type of *Myginda* sect. *Gyminda* Griseb. (Cat. Pl. Cub.: 55. 1866) is *M. integrifolia* Poir. even though Grisebach misapplied the latter name. When Sargent raised the section to the rank of genus, he named the species described by Grisebach *G. grisebachii* and explicitly excluded *M. integrifolia* from the genus.

Gyminda Sarg. (in Gard. & Forest 4: 4. 1891) is therefore the name of a new genus, typified by *G. grisebachii* Sarg., not a name at new rank based on *M.* sect. *Gyminda*.

ⓘ **Note 1.** Misapplication of a new combination, name at new rank, or replacement name to a different taxon, but without explicit exclusion of the type of the basionym or replaced synonym, is dealt with under Art. 7.3 and 7.4.

ⓘ **Note 2.** Retention of a name in a sense that excludes its original type, or its type designated under Art. 7–10, can be achieved only by conservation (see Art. 14.9).

48.2. For the purpose of Art. 48.1, definite exclusion or inclusion of the type of a name means exclusion or inclusion of:

(a) the holotype under Art. 9.1 or the original type under Art. 10; or
(b) all syntypes under Art. 9.6 or all elements eligible as types under Art. 10.2; or
(c) the type previously designated under Art. 9.11–9.13 or 10.2; or
(d) the type previously conserved under Art. 14.9.

Exclusion of the type is also achieved by:

(e) explicit exclusion of the name itself or any name homotypic at that time, unless the type is at the same time included either explicitly or by implication.

Ex. 4. The name *Chusquea quila* was published by Kunth (Révis. Gramin.: 138. 1829) with reference to "*Arundo quila* Poir., excl. Syn." The only synonym cited by Poiret (in Lamarck & al., Encycl. 6: 274. 1804) was the phrase name for *A. quila* Molina (Sag. Stor. Nat. Chili: 154, 155, 349. 1782). *Chusquea quila* Kunth is the name of a new taxon validated by Poiret's description because Kunth explicitly excluded its apparent basionym *A. quila* Molina.

ⓘ **Note 3.** For the purpose of Art. 48.1, the inclusion of an apparent basionym with an expression of doubt, or in a sense that excludes one or more but not all of its potential type elements, does not by itself constitute exclusion of its type.

Ex. 5. The name *Meum segetum* was published by Gussone (Fl. Sicul. Prodr. 1: 346. 1827) with citation of "*Anethum segetum.* Lin. mant. 219?" in synonymy. Because Gussone's expression of doubt did not exclude the type of *A. segetum* L. (Mant. Pl.: 219. 1771), he published the new combination *M. segetum* (L.) Guss., not the name of a new taxon.

Ex. 6. The name *Amorphophallus campanulatus* was published by Decaisne (in Nouv. Ann. Mus. Hist. Nat. 3: 366. 1834) with citation of *Arum campanulatum* Roxb. (Pl. Coromandel 3: 68. 1820) in synonymy, but with exclusion of certain elements included by Roxburgh ("Excl. syn. Hort. malab. nec non t. 112. Herb. Amb. V."). Because Decaisne did not explicitly exclude the type of *A. campanulatum,* which in 1834 had no holotype, syntypes, previously designated lectotype, or previously conserved type, he published the new combination *Amorphophallus campanulatus* (Roxb.) Decne., not the name of a new taxon.

ARTICLE 49
PARENTHETICAL AUTHOR CITATIONS

49.1. Author citation for a name at the rank of genus or below that has a basionym (Art. 6.10) comprises the author(s) of the basionym cited in parentheses followed by the author(s) of the name itself (see also Art. 46.7).

Ex. 1. Medicago polymorpha var. *orbicularis* L. (Sp. Pl.: 779. 1753) when raised to the rank of species is cited as *M. orbicularis* (L.) Bartal. (Cat. Piante Siena: 60. 1776).

Ex. 2. Anthyllis sect. *Aspalathoides* DC. (Prodr. 2: 169. 1825) raised to generic rank, retaining the epithet *Aspalathoides* as its name, is cited as *Aspalathoides* (DC.) K. Koch (Hort. Dendrol.: 242. 1853).

Ex. 3. Cineraria sect. *Eriopappus* Dumort. (Fl. Belg.: 65. 1827) when transferred to *Tephroseris* (Rchb.) Rchb. is cited as *T.* sect. *Eriopappus* (Dumort.) Holub (in Folia Geobot. Phytotax. 8: 173. 1973).

Ex. 4. Cistus aegyptiacus L. (Sp. Pl.: 527. 1753) when transferred to *Helianthemum* Mill. is cited as *H. aegyptiacum* (L.) Mill. (Gard. Dict., ed. 8: *Helianthemum* No. 23. 1768).

Ex. 5. Fumaria bulbosa var. *solida* L. (Sp. Pl.: 699. 1753) was raised to specific rank as *F. solida* (L.) Mill. (Gard. Dict. Abr., ed. 6: *Fumaria* No. 8. 1771). The name of this species when transferred to *Corydalis* DC. is cited as *C. solida* (L.) Clairv. (Man. Herbor. Suisse: 371. 1811), not *C. solida* "(Mill.) Clairv."

Ex. 6. Pulsatilla montana var. *serbica* W. Zimm. (in Feddes Repert. Spec. Nov. Regni Veg. 61: 95. 1958), originally placed under *P. montana* subsp. *australis* (Heuff.) Zämelis, retains its authorship when placed under *P. montana* subsp. *dacica* Rummelsp. (see Art. 24.1) and is not to be cited as var. *serbica* "(W. Zimm.) Rummelsp." (in Feddes Repert. 71: 29. 1965).

Ex. 7. Salix subsect. *Myrtilloides* C. K. Schneid. (Ill. Handb. Laubholzk. 1: 63. 1904), originally placed under *S.* sect. *Argenteae* W. D. J. Koch, retains its authorship when placed under *S.* sect. *Glaucae* Pax (see Art. 21.1) and is not to be cited as *S.* subsect. *Myrtilloides* "(C. K. Schneid.) Dorn" (in Canad. J. Bot. 54: 2777. 1976).

Ex. 8. The name *Lithocarpus polystachyus* published by Rehder (in J. Arnold Arbor. 1: 130. 1919) was based on *Quercus polystachya* A. DC. (Prodr. 16(2): 107.

1864), ascribed by Candolle to "Wall.! list n. 2789" (a nomen nudum); Rehder's combination is cited as either *L. polystachyus* (Wall. ex A. DC.) Rehder or *L. polystachyus* (A. DC.) Rehder (see Art. 46.5).

ⓘ Note 1. Author citation for a replacement name (Art. 6.11) comprises only the author(s) of the name itself, not those of the replaced synonym.

> **Ex. 9.** *Mycena coccineoides,* a replacement name for *Omphalina coccinea* Murrill (see Art. 6 Ex. 15), is cited as *M. coccineoides* Grgur., not *M. coccineoides* "(Murrill) Grgur." (see also Art. 58 Ex. 1, 3, and 4).

ⓘ Note 2. Art. 46.7 provides for the use of parenthetical author citations preceding the word "ex" after some names in groups with a starting-point later than 1753.

49.2. Parenthetical author citations are not used for suprageneric names.

> **Ex. 10.** Even though *Illiciaceae* A. C. Sm. (in Sargentia 7: 8. 1947) was validly published by reference to *Illicieae* DC. (Prodr. 1: 77. 1824) it is not to be cited as *Illiciaceae* "(DC.) A. C. Sm."

ARTICLE 50
TRANSFER BETWEEN THE HYBRID AND NON-HYBRID CATEGORY

50.1. When a taxon at the rank of species or below is transferred from the non-hybrid category to the hybrid category at the same rank (Art. H.10 Note 1), or vice versa, the authorship remains unchanged but may be followed by an indication in parentheses of the original category.

> **Ex. 1.** *Stachys ambigua* Sm. (in Smith & Sowerby, Engl. Bot. 30: t. 2089. 1809) was published as the name of a species. If regarded as applying to a hybrid, it may be cited as *S. ×ambigua* Sm. (pro sp.).

> **Ex. 2.** *Salix ×glaucops* Andersson (in Candolle, Prodr. 16(2): 281. 1868) was published as the name of a hybrid. Later, Rydberg (in Bull. New York Bot. Gard. 1: 270. 1899) considered the taxon to be a species. If this view is accepted, the name may be cited as *S. glaucops* Andersson (pro hybr.).

GENERAL RECOMMENDATIONS ON CITATION

Recommendation 50A

50A.1. In the citation of a designation that is not validly published because it was merely cited as a synonym (Art. 36.1(b)), the words "as synonym" or "pro syn." should be added.

Recommendation 50B

50B.1. In the citation of a nomen nudum, its status should be indicated by adding the words "nomen nudum" or "nom. nud."

> *Ex. 1. "Carex bebbii"* (Olney, Carices Bor.-Amer. 2: 12. 1871), published without a description or diagnosis, should be cited as *Carex bebbii* Olney, nomen nudum (or nom. nud.).

Recommendation 50C

50C.1. The citation of a later homonym should be followed by the name of the author of the earlier homonym preceded by the word "non", preferably with the date of publication added. In some instances, it is advisable to also cite any other homonyms, preceded by the word "nec".

> *Ex. 1. Ulmus racemosa* Thomas in Amer. J. Sci. Arts 19: 170. 1831, non Borkh. 1800.

> *Ex. 2. Lindera* Thunb., Nov. Gen. Pl.: 64. 1783, nom. cons., non Adans. 1763.

> *Ex. 3. Bartlingia* Brongn. in Ann. Sci. Nat. (Paris) 10: 373. 1827, non Rchb. 1824 nec F. Muell. 1882.

Recommendation 50D

50D.1. Misidentifications should not be included in synonymies but added after them. A misapplied name should be indicated by the words "auct. non" followed by the name(s) of the original author(s) and the bibliographic reference of the misidentification.

> *Ex. 1. Ficus stortophylla* Warb. in Ann. Mus. Congo Belge, Bot., ser. 4, 1: 32. 1904. *F. irumuensis* De Wild., Pl. Bequaert. 1: 341. 1922. *F. exasperata* auct. non Vahl: De Wildeman & Durand in Ann. Mus. Congo Belge, Bot., ser. 2, 1: 54. 1899; De Wildeman, Miss. Ém. Laurent: 26. 1905; Durand & Durand, Syll. Fl. Congol.: 505. 1909.

Recommendation 50E

50E.1. After a conserved name (nomen conservandum; see Art. 14 and App. II–IV) the abbreviation "nom. cons." or, in the case of a conserved spelling, "orth. cons." (orthographia conservanda) should be added in a formal citation.

> *Ex. 1. Protea* L., Mant. Pl.: 187. 1771, nom. cons., non L. 1753.

> *Ex. 2. Combretum* Loefl. 1758, nom. cons.

> *Ex. 3. Glechoma* L. 1753, orth. cons., *'Glecoma'*.

50E.2. After a name rejected under Art. 56 (nomen utique rejiciendum, suppressed name; see App. V) the abbreviation "nom. rej." should be added in a formal citation.

> *Ex. 4. Betula alba* L. 1753, nom. rej.

ⓘ **Note 1.** Rec. 50E.2 also applies to any combination based on a nomen utique rejiciendum (suppressed name; see Art. 56.1).

> *Ex. 5. Dryobalanops sumatrensis* (J. F. Gmel.) Kosterm. in Blumea 33: 346. 1988, nom. rej.

Recommendation 50F

50F.1. If a name is cited with alterations from the form as originally published, it is desirable that in full citations the exact original form should be added, preferably between single or double quotation marks.

> *Ex. 1. Pyrus calleryana* Decne. *(P. mairei* H. Lév. in Repert. Spec. Nov. Regni Veg. 12: 189. 1913, *'Pirus').*

> *Ex. 2. Zanthoxylum cribrosum* Spreng., Syst. Veg. 1: 946. 1824, *'Xanthoxylon' (Z. caribaeum* var. *floridanum* (Nutt.) A. Gray in Proc. Amer. Acad. Arts 23: 225. 1888, *'Xanthoxylum').*

> *Ex. 3. Spathiphyllum solomonense* Nicolson in Amer. J. Bot. 54: 496. 1967, *'solomonensis'.*

Recommendation 50G

50G.1. Authors should avoid mentioning in their publications previously unpublished names that they do not accept, especially if the persons responsible for these unpublished names have not formally authorized their publication (see Rec. 23A.3(i)).

CHAPTER VII
REJECTION OF NAMES

ARTICLE 51
LIMITATION OF REJECTION

51.1. A legitimate name must not be rejected merely because it, or its epithet, is inappropriate or disagreeable, or because another is preferable or better known (but see Art. 14, 56.1, and F.7.1), or because it has lost its original meaning.

Ex. 1. Changes such as the following are contrary to Art. 51.1: *Mentha* to *Minthe*, *Staphylea* to *Staphylis, Tamus* to *Tamnus, Thamnos,* or *Thamnus, Tillaea* to *Tillia, Vincetoxicum* to *Alexitoxicon;* and *Orobanche artemisiae* to *O. artemisiepiphyta, O. columbariae* to *O. columbarihaerens, O. rapum-genistae* to *O. rapum* or *O. sarothamnophyta.*

Ex. 2. Ardisia quinquegona Blume (Bijdr. Fl. Ned. Ind. 13: 689. 1825) is not to be rejected in favour of *A. pentagona* A. DC. (in Trans. Linn. Soc. London 17: 124. 1834) merely because the specific epithet *quinquegona* is a hybrid word (Latin and Greek) (contrary to Rec. 23A.3(c)).

Ex. 3. The name *Scilla peruviana* L. (Sp. Pl.: 309. 1753) is not to be rejected merely because the species does not grow in Peru.

Ex. 4. The name *Petrosimonia oppositifolia* (Pall.) Litv. (Sched. Herb. Fl. Ross. 7: 13. 1911), based on *Polycnemum oppositifolium* Pall. (Reise Russ. Reich. 1: 484. 1771), is not to be rejected merely because the species has leaves only partly opposite, and partly alternate, although there is another closely related species, *Petrosimonia brachiata* (Pall.) Bunge, that has all its leaves opposite.

Ex. 5. Richardia L. (Sp. Pl.: 330. 1753) is not to be rejected in favour of *Richardsonia,* as was done by Kunth (in Mém. Mus. Hist. Nat. 4: 430. 1818), merely because the name was originally dedicated to Richardson.

51.2. Despite Art. 51.1, a legitimate name of a new taxon or a replacement name published on or after 1 January 2026 may be rejected (under Art. 56.1) because it, or its epithet, is derogatory to a group of people.

Recommendation 51A

51A.1. Authors are strongly advised to avoid publishing names of new taxa or replacement names that could be viewed as inappropriate, disagreeable, offensive, or unacceptable by any national, ethnic, cultural, or other groups.

ARTICLE 52
NOMENCLATURALLY SUPERFLUOUS NAMES

52.1. A name, unless conserved (Art. 14), protected (Art. F.2), or sanctioned (Art. F.3), is illegitimate and is to be rejected if it was nomenclaturally superfluous when published, i.e. if the taxon to which it was applied, as circumscribed by its author, definitely included the type (as qualified in Art. 52.2) of a name that ought to have been adopted, or of which the epithet ought to have been adopted, under Art. 11 (but see Art. 52.4 and F.8.1).

52.2. For the purpose of Art. 52.1, definite inclusion of the type of a name is achieved by citation of:

(a) the holotype under Art. 9.1 or the original type under Art. 10; or
(b) all syntypes under Art. 9.6 or all elements eligible as types under Art. 10.2; or
(c) the type previously designated under Art. 9.11–9.13 or 10.2; or
(d) the type previously conserved under Art. 14.9; or
(e) the name itself or any name homotypic at that time, unless the type is at the same time excluded either explicitly or by implication.

For this purpose, citation of an illustration of a specimen is treated as citation of the specimen.

Ex. 1. The generic name *Cainito* Adans. (Fam. Pl. 2: 166. 1763) is illegitimate because it was a superfluous name for *Chrysophyllum* L. (Sp. Pl.: 192. 1753), which Adanson cited as a synonym.

Ex. 2. *Picea excelsa* Link (in Linnaea 15: 517. 1841) is illegitimate because it is based on *Pinus excelsa* Lam. (Fl. Franç. 2: 202. 1779), a superfluous name for *Pinus abies* L. (Sp. Pl.: 1002. 1753). Under *Picea* the correct name is *Picea abies* (L.) H. Karst. (Deut. Fl.: 324. 1881).

Ex. 3. *Salix myrsinifolia* Salisb. (Prodr. Stirp. Chap. Allerton: 394. 1796) is legitimate because it is explicitly based on *"S. myrsinites"* of Hoffmann (Hist. Salic. Ill.: 71. 1787), a misapplication of *S. myrsinites* L. (Sp. Pl.: 1018. 1753), a name that Salisbury excluded by implication by not citing Linnaeus as he did under each of the other 14 species of *Salix*.

Ex. 4. *Cucubalus latifolius* Mill. and *C. angustifolius* Mill. are not illegitimate names, although Miller's species are now united with the species previously named *C. behen* L. (Sp. Pl.: 414. 1753): *C. latifolius* and *C. angustifolius* as circumscribed by Miller (Gard. Dict., ed. 8: *Cucubalus* No. 2, 3. 1768) did not include the type of *C. behen* L., a name that he adopted for another species.

Ex. 5. Explicit exclusion of type. When publishing the name *Galium tricornutum,* Dandy (in Watsonia 4: 47. 1957) cited *G. tricorne* Stokes (Bot. Arr. Brit. Pl., ed. 2, 1: 153. 1787) pro parte as a synonym while explicitly excluding its type.

Ex. 6. Exclusion of type by implication. *Tmesipteris elongata* P. A. Dang. (in Botaniste 2: 213. 1891) was published as a new species but *Psilotum truncatum* R. Br. was cited as a synonym. However, on the following page, *T. truncata* (R. Br.) Desv. is recognized as a different species and two pages later both are distinguished in a key, thus showing that the meaning of the cited synonym was either *"P. truncatum* R. Br. pro parte" or *"P. truncatum* auct. non R. Br."

Ex. 7. Under *Bauhinia semla* Wunderlin (in Taxon 25: 362. 1976), the name *B. retusa* Roxb. ex DC. (Prodr. 2: 515. 1825) non Poir. (in Lamarck, Encycl. Suppl. 1: 599. 1811), was cited as the replaced synonym while *B. emarginata* Roxb. ex G. Don (Gen. Hist. 2: 462. 1832) non Mill. (Gard. Dict., ed. 8: *Bauhinia* No. 5.

1768), was also cited in synonymy, and hence the types of the two synonyms were definitely included. However, *B. roxburghiana* Voigt (Hort. Suburb. Calcutt.: 254. 1845), which was published as a replacement name (Art. 6.12) for *B. emarginata* Roxb. ex G. Don, is necessarily homotypic with it (Art. 7.4) and should have been adopted by Wunderlin. Therefore, *B. semla* is an illegitimate superfluous name but is typified by the type of its replaced synonym, *B. retusa* Roxb. ex DC. (see Art. 7 Ex. 5).

Ex. 8. Both *Apios americana* Medik. (in Vorles. Churpfälz. Phys.-Ökon. Ges. 2: 355. 1787) and *A. tuberosa* Moench (Methodus: 165. 1794) are replacement names for the legitimate *Glycine apios* L. (Sp. Pl.: 753. 1753), the epithet of which in combination with *Apios* would form a tautonym (Art. 23.4) and would not therefore be validly published (Art. 32.1(c)). *Apios tuberosa* was nomenclaturally superfluous when published, and is therefore illegitimate, because Moench cited in synonymy *G. apios,* which was then, as now, homotypic with *A. americana,* the name that has priority and that Moench should have adopted.

Ex. 9. *Welwitschia* Rchb. (Handb. Nat. Pfl.-Syst.: 194. 1837) was based on *Hugelia* Benth. (Edwards's Bot. Reg. 19: t. 1622. 1833), non *Huegelia* Rchb. (in Mitth. Geb. Fl. Pomona 1829(13): 50. 1829). *Welwitschia* Hook. f. (in Gard. Chron. 1862: 71. 1862) was conserved against *Welwitschia* Rchb., becoming effective on 18 May 1910 (see Art. 14 Note 4(b)). *Eriastrum* Wooton & Standl. (in Contr. U. S. Natl. Herb. 16: 160. 1913), also based on *Hugelia* Benth., was not therefore nomenclaturally superfluous when published because *Welwitschia* Rchb. was no longer available for use.

ⓘ *Note 1.* The inclusion, with an expression of doubt, of an element in a new taxon, e.g. the citation of a name with a question mark, or in a sense that excludes one or more of its potential type elements, does not make the name of the new taxon nomenclaturally superfluous.

Ex. 10. The protologue of *Blandfordia grandiflora* R. Br. (Prodr.: 296. 1810) includes, in synonymy, "*Aletris punicea.* Labill. nov. holl. 1. p. 85. t. 111 ?", indicating that the new species might be the same as *A. punicea* Labill. (Nov. Holl. Pl. 1: 85. 1805). *Blandfordia grandiflora* is nevertheless a legitimate name.

ⓘ *Note 2.* The inclusion, in a new taxon, of an element that was subsequently designated as the type of a name that, so typified, ought to have been adopted, or of which the epithet ought to have been adopted, does not by itself make the name of the new taxon illegitimate.

Ex. 11. *Leccinum* Gray (Nat. Arr. Brit. Pl. 1: 646. 1821) does not include any of the elements eligible as types of the then untypified *Boletus* L. (Sp. Pl.: 1176. 1753), nom. cons., and is not therefore illegitimate even though it included, as *L. edule* (Bull.) Gray, the subsequently conserved type of *Boletus, B. edulis* Bull., nom. sanct.

Ex. 12. The protologue of *Capparis baducca* L. (Sp. Pl.: 504. 1753), nom. rej., included American and Indian elements. Candolle (Prodr. 1: 246. 1824) published *C. rheedei* DC., for which he included only the Indian element ("Badukka" in

Rheede, Hort. Malab. 6: t. 57. 1686) and excluded the American element. Jacobs (in Blumea 12: 435. 1965) lectotypified the Linnaean name on the Indian element. The name *C. rheedei* is legitimate even though it includes the subsequently designated lectotype of *C. baducca* (see App. V).

52.3. For the purpose of Art. 52.2(e), citation of a name can be achieved by a direct and unambiguous reference to it, e.g. by citation of its original sequential number or exact diagnostic phrase name (Linnaean "nomen specificum legitimum") rather than its epithet.

> *Ex. 13.* In publishing the name *Matricaria suaveolens* (Fl. Suec., ed. 2: 297. 1755), Linnaeus adopted the phrase name and included all the synonyms of *M. recutita* L. (Sp. Pl.: 891. 1753), but did not explicitly cite *M. recutita*. Because in 1755 *M. recutita* had no holotype, no syntypes, and no designated lectotype or conserved type, the provisions of Art. 52.2 alone do not make *M. suaveolens* illegitimate. However, because the exact diagnostic phrase name (nomen specificum legitimum) of *M. recutita* was that provided for *M. suaveolens,* the latter name is illegitimate under Art. 52.3.

> *Ex. 14.* *Cyperus involucratus* Rottb. (Descr. Pl. Rar.: 22. 1772) was validly published with the species number "57" and a short description (diagnostic phrase name). Subsequently, Rottbøll (Descr. Icon. Rar. Pl.: 42. 1773) published *C. flabelliformis* Rottb., referred to his previous place of publication, and repeated the earlier diagnostic phrase name almost verbatim, but added as a remark "Ob formam involucri nomen triviale mutavi. [I have changed the trivial name (i.e. the epithet) because of the shape of the involucre.]" Although Rottbøll did not cite the name *C. involucratus* as a synonym, his direct and unambiguous reference to it and his remark about the change of the epithet make the name *C. flabelliformis* superfluous and illegitimate.

ⓘ **Note 3.** For the purpose of Art. 52.2(e), citation of a later isonym is equivalent to citation of the name itself if the citing author does not normally cite the primary source, or if the name is usually not cited from its primary source in contemporary literature. However, if it is possible to imply that the isonym is cited "in the sense of" the later author or "as used in" the later source, its inclusion does not by itself cause illegitimacy.

52.4. A name that was nomenclaturally superfluous when published is not illegitimate on account of its superfluity if it has a basionym (which is necessarily legitimate; see Art. 6.10), or if it is formed from a legitimate generic name. When published it is incorrect, but it may become correct later.

> *Ex. 15.* *Chloris radiata* (L.) Sw. (Prodr.: 26. 1788) was nomenclaturally superfluous when published because Swartz cited the legitimate *Andropogon fasciculatus* L. (Sp. Pl.: 1047. 1753) as a synonym. However, it is not illegitimate because it has a basionym, *Agrostis radiata* L. (Syst. Nat., ed. 10: 873. 1759). *Chloris radiata*

is the correct name in the genus *Chloris* for *Agrostis radiata* when *Andropogon fasciculatus* is treated as a different species, as was done by Hackel (in Candolle & Candolle, Monogr. Phan. 6: 177. 1889).

Ex. 16. *Juglans major* (Torr.) A. Heller (in Muhlenbergia 1: 50. 1904), based on *J. rupestris* var. *major* Torr. (in Sitgreaves, Rep. Exped. Zuni & Colorado Rivers: 171. 1853), was nomenclaturally superfluous when published because Heller cited the legitimate *J. californica* S. Watson (in Proc. Amer. Acad. Arts 10: 349. 1875) as a synonym. Nevertheless, *J. major* is legitimate because it has a basionym, and it may be correct when treated as taxonomically distinct from *J. californica*.

Ex. 17. The generic name *Hordelymus* (Jess.) Harz (Landw. Samenk.: 1147. 1885) was nomenclaturally superfluous when published because its type, *Elymus europaeus* L., is also the type of *Cuviera* Koeler (Descr. Gram.: 328. 1802). However, it is not illegitimate because it has a basionym, *Hordeum* [unranked] *Hordelymus* Jess. (Deutschl. Gräser: 202. 1863). *Cuviera* Koeler has since been rejected in favour of its later homonym *Cuviera* DC., and *Hordelymus* can now be used as the correct name for a segregate genus containing *E. europaeus* L.

Ex. 18. *Carpinaceae* Vest (Anleit. Stud. Bot.: 265, 280. 1818) was nomenclaturally superfluous when published because of the inclusion of *Salix* L., the type of *Salicaceae* Mirb. (Elém. Physiol. Vég. Bot. 2: 905. 1815). However, it is not illegitimate because it is formed from a legitimate generic name, *Carpinus* L.

Ex. 19. *Wormia suffruticosa* Griff. ex Hook. f. & Thomson (in Hooker, Fl. Brit. India 1: 35. 1872), nom. cons., was nomenclaturally superfluous when published because of the inclusion of *W. subsessilis* Miq. (Fl. Ned. Ind., Eerste Bijv.: 619. 1861), nom. rej. With conservation, the previously illegitimate *W. suffruticosa* became available to serve as basionym of *Dillenia suffruticosa* (Griff. ex Hook. f. & Thomson) Martelli (in Malesia 3: 163. 1886), which thereby also became legitimate (see Art. 6.4), although it too was nomenclaturally superfluous when published because of the inclusion of *W. subsessilis*.

ⓘ ***Note 4.*** In no case does a statement of parentage accompanying the publication of a name for a hybrid make the name illegitimate (see Art. H.4 and H.5).

Ex. 20. The name *Polypodium* ×*shivasiae* Rothm. (in Kulturpflanze, Beih. 3: 245. 1962) was proposed for hybrids between *P. australe* Fée and *P. vulgare* subsp. *prionodes* (Asch.) Rothm., while in the same publication (l.c. 1962) the author accepted *P.* ×*font-queri* Rothm. (in Cadevall y Diars & Font Quer, Fl. Catalun. 6: 353. 1937) for hybrids between *P. australe* and *P. vulgare* L. subsp. *vulgare*. Under Art. H.4.1, *P.* ×*shivasiae* is a synonym of *P.* ×*font-queri;* nevertheless, it is not an illegitimate name.

ⓘ ***Note 5.*** Nothogeneric names, because they do not have types (see Art. H.9 Note 1), do not cause superfluity.

Ex. 21. In the protologue of *Majovskya* Sennikov & Kurtto (in Memoranda Soc. Fauna Fl. Fenn. 93: 63. 2017), the authors cited ×*Chamaearia* Mezhenskyj (in Mezhenskyj & al., Netraditsīni Plodoy Kul'turi: 27. 2012) as a taxonomic synonym. Because ×*Chamaearia* is a nothogeneric name that has no type, and *Majovskya* did not include the type of a name that ought to have been adopted, *Majovskya* was not nomenclaturally superfluous when published and is not an illegitimate name.

ARTICLE 53
HOMONYMS

53.1. A name of a family, genus, or species, unless conserved (Art. 14), protected (Art. F.2), or sanctioned (Art. F.3), is illegitimate if it is a later homonym, that is, if it is spelled exactly like a name based on a different type that was previously and validly published for a taxon at the same rank (see also Art. 53.3 and F.3.3).

Ex. 1. Tapeinanthus Boiss. ex Benth. (in Candolle, Prodr. 12: 436. 1848) *(Labiatae)* is a later homonym of *Tapeinanthus* Herb. (Amaryllidaceae: 190. 1837) *(Amaryllidaceae). Tapeinanthus* Boiss. ex Benth. is therefore illegitimate and unavailable for use; it was replaced by *Thuspeinanta* T. Durand (Index Gen. Phan.: 703. 1888).

Ex. 2. Torreya Arn. (in Ann. Nat. Hist. 1: 130. 1838) is a conserved name and is therefore available for use despite the existence of the earlier homonym *Torreya* Raf. (in Amer. Monthly Mag. & Crit. Rev. 3: 356. 1818).

Ex. 3. Astragalus rhizanthus Boiss. (Diagn. Pl. Orient., ser. 1, 2: 83. 1843) is a later homonym of the validly published name *A. rhizanthus* Royle ex Benth. (in Royle, Ill. Bot. Himal. Mts.: 200. 1835) and is therefore illegitimate; it was replaced by *A. cariensis* Boiss. (Diagn. Pl. Orient., ser. 1, 9: 56. 1849).

Ex. 4. Both *Molina* Ruiz & Pav. (Fl. Peruv. Prodr.: 111. 1794) and *M. racemosa* Ruiz & Pav. (Syst. Veg. Fl. Peruv. Chil. 1: 209. 1798) *(Compositae)* are illegitimate later homonyms of *Molina* Cav. (Diss. 9: 435. 1790) and *M. racemosa* Cav. (l.c. 1790) *(Malpighiaceae)*, respectively, even though Cavanilles's species name is itself illegitimate under Art. 52.1 (Art. 53 Note 4).

Ex. 5. Moreae Britton & Rose (in Britton, N. Amer. Fl. 23: 201, 217. 1930), formed from *Mora* Benth. (in Trans. Linn. Soc. London 18: 210. 1839), although a later homonym of *Moreae* Dumort. (Anal. Fam. Pl.: 17. 1829), formed from *Morus* L. (Sp. Pl.: 986. 1753), is not illegitimate because the provisions on homonymy do not apply to subdivisions of families.

🛈 *Note 1.* Simultaneously published homonyms are not illegitimate on account of their homonymy unless an earlier homonym exists (see also Art. 53.5).

🛈 *Note 2.* Nothogeneric names, even though they do not have types (see Art. H.9 Note 1), can be homonyms (Art. H.3.3).

154

ⓘ **Note 3.** Later homonyms are illegitimate regardless of whether the type is fossil or non-fossil.

Ex. 6. *Endolepis* Torr. (in Pacif. Railr. Rep. 12(2, 2): 47. 1860–1861), based on a non-fossil type, is an illegitimate later homonym of *Endolepis* Schleid. (in Schmid & Schleiden, Geognos. Verhältnisse Saalthales Jena: 72. 1846), based on a fossil type.

Ex. 7. *Cornus paucinervis* Hance (in J. Bot. 19: 216. 1881), based on a non-fossil type, is an illegitimate later homonym of *C. paucinervis* Heer (Fl. Tert. Helv. 3: 289. 1859), based on a fossil type.

Ex. 8. *Ficus crassipes* F. M. Bailey (Rep. Pl. Prelim. Gen. Rep. Bot. Meston's Exped. Bellenden-Ker Range: 2. 1889), *F. tiliifolia* Baker (in J. Linn. Soc., Bot. 21: 443. 1885), and *F. tremula* Warb. (in Bot. Jahrb. Syst. 20: 171. 1894), each based on a non-fossil type, were illegitimate later homonyms of, respectively, *F. crassipes* (Heer) Heer (Fl. Foss. Arct. 6(2): 70. 1882), *F. tiliifolia* (A. Braun) Heer (Fl. Tert. Helv. 2: 68. 1856), and *F. tremula* Heer (in Abh. Schweiz. Paläontol. Ges. 1: 11. 1874), each based on a fossil type. The three names with non-fossil types have been conserved against their earlier homonyms in order to maintain their use (see App. IV).

ⓘ **Note 4.** A validly published earlier homonym, even if illegitimate, rejected under Art. 56 or F.7, or otherwise generally treated as a synonym, causes illegitimacy of any later homonym that is not conserved, protected, or sanctioned (but see Art. F.3.3).

Ex. 9. *Zingiber truncatum* S. Q. Tong (in Acta Phytotax. Sin. 25: 147. 1987) is illegitimate because it is a later homonym of the validly published *Z. truncatum* Stokes (Bot. Mat. Med. 1: 68. 1812), even though the latter name is itself illegitimate under Art. 52.1; *Z. truncatum* S. Q. Tong was replaced by *Z. neotruncatum* T. L. Wu & al. (in Novon 10: 91. 2000).

Ex. 10. *Amblyanthera* Müll. Arg. (in Martius, Fl. Bras. 6(1): 141. 1860) is a later homonym of the validly published *Amblyanthera* Blume (Mus. Bot. 1: 50. 1849) and is therefore illegitimate, even when *Amblyanthera* Blume is treated as a synonym of *Osbeckia* L. (Sp. Pl.: 345. 1753).

53.2. When two or more names of genera or species based on different types are so similar that they are likely to be confused (because they are applied to related taxa or for any other reason) they are to be treated as homonyms (see also Art. 61.5). If established practice has been to treat two similar names as homonyms, this practice is to be continued if it is in the interest of nomenclatural stability.

*Ex. 11. Names treated as homonyms: *Asterostemma* Decne. (in Ann. Sci. Nat., Bot., ser. 2, 9: 271. 1838) and *Astrostemma* Benth. (in Hooker's Icon. Pl. 14: 7. 1880); *Pleuropetalum* Hook. f. (in London J. Bot. 5: 108. 1846) and *Pleuripetalum* T. Durand (Index Gen. Phan.: 493. 1888); *Eschweilera* DC. (Prodr. 3: 293. 1828) and *Eschweileria* Boerl. (in Ann. Jard. Bot. Buitenzorg 6: 106, 112. 1887); *Skytanthus* Meyen (Reise 1: 376. 1834) and *Scytanthus* Hook. (in Icon. Pl. 7: ad t. 605–606. 1844).

Ex. 12. *Bradlea* Adans. (Fam. Pl. 2: 324, 527. 1763), *Bradleja* Banks ex Gaertn. (Fruct. Sem. Pl. 2: 127. 1790), and *Braddleya* Vell. (Fl. Flumin.: 93. 1829), all commemorating Richard Bradley, are treated as homonyms because only one can be used without serious risk of confusion.

Ex. 13. *Acanthoica* Lohmann (in Wiss. Meeresuntersuch., Abt. Kiel 7: 68. 1902) and *Acanthoeca* W. N. Ellis (in Ann. Soc. Roy. Zool. Belgique 60: 77. 1930), both applied to flagellates, are sufficiently alike to be considered as homonyms (Voss in Taxon 22: 313. 1973).

Ex. 14. Epithets so similar that they are likely to be confused if combined under the same name of a genus or species: *ceylanicus* and *zeylanicus; chinensis* and *sinensis; heteropodus* and *heteropus; macrocarpon* and *macrocarpum; macrostachys* and *macrostachyus; napaulensis, nepalensis,* and *nipalensis; poikilantha* and *poikilanthes; polyanthemos* and *polyanthemus; pteroides* and *pteroideus; thibetanus* and *tibetanus; thibetensis* and *tibetensis; thibeticus* and *tibeticus; trachycaulon* and *trachycaulum; trinervis* and *trinervius.*

Ex. 15. Names not likely to be confused: *Desmostachys* Miers (in Ann. Mag. Nat. Hist., ser. 2, 9: 399. 1852) and *Desmostachya* (Stapf) Stapf (in Thiselton-Dyer, Fl. Cap. 7: 316. 1898); *Euphorbia peplis* L. (Sp. Pl.: 455. 1753) and *E. peplus* L. (l.c.: 456. 1753); *Gerrardina* Oliv. (in Hooker's Icon. Pl. 11: 60. 1870) and *Gerardiina* Engl. (in Bot. Jahrb. Syst. 23: 507. 1897); *Iris* L. (l.c.: 38. 1753) and *Iria* (Pers.) R. Hedw. (Gen. Pl.: 360. 1806); *Lysimachia hemsleyana* Oliv. (in Hooker's Icon. Pl. 20: ad t. 1980. 1891) and *L. hemsleyi* Franch. (in J. Bot. (Morot) 9: 461. 1895) (but see Rec. 23A.2); *Monochaetum* (DC.) Naudin (in Ann. Sci. Nat., Bot., ser. 3, 4: 48. 1845) and *Monochaete* Döll (in Martius, Fl. Bras. 2(3): 78. 1875); *Peltophorus* Desv. (in Nouv. Bull. Sci. Soc. Philom. Paris 2: 188. 1810) and *Peltophorum* (Vogel) Benth. (in J. Bot. (Hooker) 2: 75. 1840); *Peponia* Grev. (in Trans. Microscop. Soc. London, n.s., 11: 75. 1863) and *Peponium* Engl. (in Engler & Prantl, Nat. Pflanzenfam., Nachtr. 1: 318. 1897); *Rubia* L. (l.c.: 109. 1753) and *Rubus* L. (l.c.: 492. 1753); *Senecio napaeifolius* (DC.) Sch. Bip. (in Flora 28: 498. 1845, *'napeaefolius';* see Art. 60 Ex. 45) and *S. napifolius* MacOwan (in J. Linn. Soc., Bot. 25: 388. 1890; the epithets derived, respectively, from *Napaea* L. and *Brassica napus* L.); *Symphyostemon* Miers (in Proc. Linn. Soc. London 1: 123. 1841) and *Symphostemon* Hiern (Cat. Afr. Pl. 1: 867. 1900); *Urvillea* Kunth (in Humboldt & al., Nov. Gen. Sp. 5, ed. qu.: 105; ed. fol.: 81. 1821) and *Durvillaea* Bory (Dict. Class. Hist. Nat. 9: 192. 1826).

Ex. 16. Names conserved against earlier names treated as homonyms (see App. III): *Cephalotus* Labill. (against *Cephalotos* Adans.); *Columellia* Ruiz & Pav. (against *Columella* Lour., both commemorating Columella, the Roman writer on agriculture); *Lyngbya* Gomont (against *Lyngbyea* Sommerf.); *Simarouba* Aubl. (against *Simaruba* Boehm.).

53.3. The names of two subdivisions of the same genus, or of two infraspecific taxa within the same species, even if they are at different ranks, are homonyms if they are not based on the same type and have the same final epithet or are treated as homonyms if they have a confusingly similar final epithet. The later name is illegitimate.

Ex. 17. *Andropogon sorghum* subsp. *halepensis* (L.) Hack. (in Candolle & Candolle, Monogr. Phan. 6: 501. 1889) and *A. sorghum* var. *halepensis* (L.) Hack. (l.c.: 502. 1889) are legitimate because both have the same type (see also Rec. 26A.1).

Ex. 18. *Anagallis arvensis* subsp. *caerulea* Hartm. (Sv. Norsk Exc.-Fl.: 32. 1846), based on the later homonym *A. caerulea* Schreb. (Spic. Fl. Lips.: 5. 1771), is illegitimate because it is itself a later homonym of *A. arvensis* var. *caerulea* (L.) Gouan (Fl. Monsp.: 30. 1765), based on *A. caerulea* L. (Amoen. Acad. 4: 479. 1759).

Ex. 19. *Scenedesmus armatus* var. *brevicaudatus* (Hortob.) Pankow (in Arch. Protistenk. 132: 153. 1986), based on *S. carinatus* var. *brevicaudatus* Hortob. (in Acta Bot. Acad. Sci. Hung. 26: 318. 1981), is a later homonym of *S. armatus* f. *brevicaudatus* L. Ş. Péterfi (in Stud. Cercet. Biol. (Bucharest), Ser. Biol. Veg. 15: 25. 1963) even though the two names apply to taxa at different infraspecific ranks. However, *S. armatus* var. *brevicaudatus* (L. Ş. Péterfi) E. H. Hegew. (in Arch. Hydrobiol. Suppl. 60: 393. 1982) is not a later homonym because it is based on the same type as *S. armatus* f. *brevicaudatus* L. Ş. Péterfi.

ⓘ *Note 5.* The same final epithet may be used in the names of subdivisions of different genera and in the names of infraspecific taxa within different species.

Ex. 20. *Verbascum* sect. *Aulacosperma* Murb. (Monogr. Verbascum: 34, 593. 1933) is permissible, although there is an earlier *Celsia* sect. *Aulacospermae* Murb. (Monogr. Celsia: 34, 56. 1926). This, however, is not an example to be followed because it is contrary to Rec. 21B.3 second sentence.

53.4. When it is doubtful whether names or their epithets are sufficiently alike to be confused, a request for a binding decision may be submitted to the General Committee, which will refer it for examination to the specialist committee(s) for the appropriate taxonomic group(s) (see Div. III Prov. 2.2, 7.10(b), 7.11, and 8.13(a)). A General Committee recommendation as to whether or not to treat the names concerned as homonyms is to be treated as a binding decision subject to ratification by a later International Botanical Congress (see also Art. 14.15, 34.2, 38.5, and 56.3) and takes retroactive effect. These binding decisions are listed in App. VII.

Ex. 21. Gilmania Coville (in J. Wash. Acad. Sci. 26: 210. 1936) was published as a replacement name for *Phyllogonum* Coville (in Contr. U. S. Natl. Herb. 4: 190. 1893) because the author considered the latter to be a later homonym of *Phyllogonium* Brid. (Bryol. Univ. 2: 671. 1827). Although treating *Phyllogonum* Coville and *Phyllogonium* Brid. as homonyms had become accepted, e.g. in *Index Nominum Genericorum,* a binding decision was requested under Art. 53.4. The Nomenclature Committee for Spermatophyta recommended (in Taxon 54: 536. 2005) that the two names should be treated as homonyms, and this was approved by the General Committee (later reported in Taxon 55: 799. 2006) and ratified by the XVII International Botanical Congress in Vienna in 2005 (see App. VII). The name *Gilmania* is therefore to be accepted as legitimate.

53.5. When two or more legitimate homonyms have equal priority (see Art. 53 Note 1), the first of them that is adopted in an effectively published text (Art. 29–31) by an author who simultaneously rejects the other(s) is treated as having priority. Likewise, if an author in an effectively published text replaces with other names all but one of these homonyms, the homonym for the taxon that is not renamed is treated as having priority (see also Rec. F.5A.2).

Ex. 22. Linnaeus simultaneously published "10." *Mimosa cinerea* (Sp. Pl.: 517. 1753) and "25." *M. cinerea* (Sp. Pl.: 520. 1753). In 1759 (Syst. Nat., ed. 10: 1311), he renamed species 10 as *M. cineraria* L. and retained the name *M. cinerea* for species 25, so that the latter is treated as having priority over its homonym.

Ex. 23. Rouy & Foucaud (Fl. France 2: 30. 1895) published the name *Erysimum hieraciifolium* var. *longisiliquum,* with two different types, for two different taxa under different subspecies. Only one of these names can be maintained.

ⓘ **Note 6.** A homonym renamed or rejected under Art. 53.5 remains legitimate and has priority over a later synonym at the same rank should it be transferred to another genus or species.

Ex. 24. Mimosa cineraria L. (Syst. Nat., ed. 10: 1311. 1759), based on *M. cinerea* L. (Sp. Pl.: 517 [non 520]. 1753; see Art. 53 Ex. 22), was transferred to *Prosopis* L. by Druce (in Rep. Bot. Exch. Club Soc. Brit. Isles 3: 422. 1914) as *P. cineraria* (L.) Druce. However, the correct name in *Prosopis* would have been a combination based on *M. cinerea* (l.c. 1753) had not that name been successfully proposed for rejection (see App. V).

ARTICLE 54
INTER-*CODE* HOMONYMY

54.1. Consideration of homonymy does not extend to the names of taxa not treated under this *Code,* except as stated below (see also Art. F.6.1):

(a) Later homonyms of the names of taxa once treated as algae, fungi, or plants are illegitimate, even when the taxa have been reassigned to a different group of organisms to which this *Code* does not apply.

(b) A name applied to an organism covered by this *Code* and validly published under it (Art. 32–45) but originally published for a taxon other than an alga, fungus, or plant, i.e. under another *Code,* is illegitimate if it:

(1) is unavailable for use under the provisions of the other *Code*[1], usually because of homonymy; or

(2) becomes a homonym of an algal, fungal, or plant name when the taxon to which it applies is first treated as an alga, fungus, or plant (see also Art. 45.1).

(c) A name of a genus is treated as an illegitimate later homonym if it is spelled identically with a previously published intergeneric graft hybrid "name" established[2] under the provisions of the *International Code of Nomenclature for Cultivated Plants.*

Ex. 1. (b)(1) *Cribrosphaerella* Deflandre ex Góka (in Acta Palaeontol. Polon. 2: 239, 260, 280. 5 Sep 1957) was published under the provisions of the *International Code of Zoological Nomenclature* for the Cretaceous coccolith algae previously known as *Cribrosphaera* Arkhang. (in Mater. Geol. Rossii 25: 411. 1912), an objectively invalid (equivalent to illegitimate) name under that *Code* because it is a later homonym of *Cribrosphaera* Popofsky (in Ergebn. Plankton-Exped. 3(L.f.β): 22, 32, 63. 1906), a radiolarian genus. Although *Cribrosphaera* Arkhang. is not a later homonym under this *Code,* it is illegitimate because it is not available for use according to the provisions of the *Code* under which it was published; consequently, *Cribrosphaerella* is the correct name for the coccolith genus under both *Codes.*

ⓘ **Note 1.** The *International Code of Nomenclature of Prokaryotes* provides that a prokaryotic name is illegitimate if it is a later homonym of a name of a taxon of prokaryotes, fungi, algae, protozoa, or viruses.

Recommendation 54A

54A.1. Authors naming new taxa under this *Code* should, as far as is practicable, avoid using such names as already exist for zoological and prokaryotic taxa (see also Art. F.6.1).

---

1 Such names are termed "objectively invalid" in the *International Code of Zoological Nomenclature* and "illegitimate" in the *International Code of Nomenclature of Prokaryotes.*

2 The term "established" is used by the *International Code of Nomenclature for Cultivated Plants* for the concept of validly published in the *International Code of Nomenclature for algae, fungi, and plants.*

ARTICLE 55
LIMITATION OF ILLEGITIMACY

55.1. A name of a species or subdivision of a genus may be legitimate even if its epithet was originally placed under an illegitimate generic name (see also Art. 22.5).

> *Ex. 1.* *Agathophyllum neesianum* Blume (Mus. Bot. 1: 339. 1851) is legitimate even though *Agathophyllum* Juss. (Gen. Pl.: 431. 1789) is illegitimate (it is a superfluous replacement name for *Ravensara* Sonn., Voy. Indes Orient. 3: 248. 1782). Because Meisner (in Candolle, Prodr. 15(1): 104. 1864) cited *A. neesianum* as a synonym of his new *Mespilodaphne mauritiana*, *M. mauritiana* Meisn. is illegitimate under Art. 52.

> *Ex. 2.* *Calycothrix* sect. *Brachychaetae* Nied. (in Engler & Prantl, Nat. Pflanzenfam. 3(7): 100. 1893) is legitimate even though it was published under *Calycothrix* Meisn. (Pl. Vasc. Gen.: 107. 1838), a superfluous replacement name for *Calytrix* Labill. (Nov. Holl. Pl. 2: 8. 1806).

55.2. An infraspecific name may be legitimate even if its final epithet was originally placed under an illegitimate species name (see also Art. 27.2).

> *Ex. 3.* *Agropyron japonicum* var. *hackelianum* Honda (in Bot. Mag. (Tokyo) 41: 385. 1927) is legitimate even though it was published under the illegitimate *A. japonicum* Honda (l.c.: 384. 1927), a later homonym of *A. japonicum* (Miq.) P. Candargy (in Arch. Biol. Vég. Pure Appl. 1: 42. 1901) (see also Art. 27 Ex. 1).

ⓘ **Note 1.** A name falling under the provisions of Art. 55.1 or 55.2 is unavailable for use, but may serve as a replaced synonym or, if not itself illegitimate, a basionym of another name or combination.

55.3. The names of species and of subdivisions of genera assigned to genera the names of which are conserved, protected, or sanctioned later homonyms, and that had earlier been assigned to the genera under the rejected homonyms, are legitimate under the conserved, protected, or sanctioned names without change of authorship or date if there is no other obstacle under the rules.

> *Ex. 4.* When published, *Alpinia galanga* (L.) Willd. (Sp. Pl. 1: 12. 1797) was assigned to *Alpinia* L. (Sp. Pl.: 2. 1753). When the name *Alpinia* was conserved from a later publication (Art. 14.9(b)), as *Alpinia* Roxb. (in Asiat. Res. 11: 350. 1810), this species was included in the newly named genus and its name *A. galanga* is to be accepted without any change in status under this *Code*.

55.4. The epithet of the name of a species or subdivision of a genus that was originally placed under a generic name that is a later homonym, or the final epithet of the name of an infraspecific taxon that was originally placed under a species name that is a later homonym, may be placed under the respective legitimate earlier homonym without change of authorship and date.

Ex. 5. The epithet of *Haplanthus hygrophiloides* T. Anderson (in J. Linn. Soc., Bot. 9: 503. 1867) was originally placed under the illegitimate generic name *Haplanthus* T. Anderson (l.c. 1867), a later homonym of *Haplanthus* Nees (in Wallich, Pl. Asiat. Rar. 3: 77, 115. 1832). When *H. hygrophiloides* is considered to belong instead to *Haplanthus* Nees, it is so accepted without change of authorship and date.

Ex. 6. When the homonyms *Acidosasa* B. M. Yang (in J. Hunan Teachers' Coll. (Nat. Sci. Ed.) 1981(2): 53, 54. 1981) and *Acidosasa* C. D. Chu & C. S. Chao (in J. Bamboo Res. 1: 165. 1982) are considered to apply to the same genus, *A. chinensis* C. D. Chu & C. S. Chao (l.c. 1982) is so accepted even though its epithet was originally placed under the illegitimate *Acidosasa* C. D. Chu & C. S. Chao.

ARTICLE 56
REJECTED NAMES

56.1. Any name that would cause a disadvantageous nomenclatural change (Art. 14.1) or that is derogatory to a group of people (Art. 51.2) may be proposed for rejection. A name thus rejected, or its basionym if it has one, is placed on a list of nomina utique rejicienda (see Rec. 50E.2; suppressed names, App. V). Along with each listed name, all names for which it is the basionym are similarly rejected, and none is to be used.

ⓘ *Note 1.* A name rejected under Art. 56.1 does not become illegitimate on account of its rejection and can continue to provide the type of a name at higher rank. Similarly, a combination under a rejected name, although unavailable for use because of the inclusion of the rejected name, may be legitimate, and may serve as basionym for another combination.

56.2. The list of nomina utique rejicienda (suppressed names) will remain permanently open for additions and changes. Any proposal for rejection of a name must be accompanied by a detailed statement of the cases both for and against its rejection, including considerations of typification. Such proposals must be submitted to the General

Committee, which will refer them for examination to the specialist committees for the various taxonomic groups (see Rec. 56A.1, Div. III Prov. 2.2, 7.10(b), 7.11, and 8.13(a); see also Art. 14.12 and 34.1).

56.3. When a proposal for the rejection of a name under Art. 56 or F.7 has been approved by the General Committee after study by the specialist committee for the taxonomic group concerned, rejection of that name is authorized subject to the decision of a later International Botanical Congress (see also Art. 14.15, 34.2, 38.5, and 53.4). Rejection takes effect on the date of effective publication (Art. 29–31) of the General Committee's approval.

ⓘ **Note 2.** The date of the General Committee decision on a particular rejection proposal can be determined by consulting the *International Code of Nomenclature for algae, fungi, and plants* Appendices database (https://naturalhistory.si.edu/research/botany/codes-proposals).

Recommendation 56A

56A.1. When a proposal for the rejection of a name under Art. 56 or F.7 has been referred to the appropriate specialist committee for study, authors should follow existing usage of names as far as possible pending the General Committee's recommendation on the proposal (see also Rec. 14A.1 and 34A.1).

ARTICLE 57
NAMES USED FOR TAXA NOT INCLUDING THEIR TYPES

57.1. A name that has been widely and persistently used for a taxon or taxa not including its type is not to be used in a sense that conflicts with current usage unless and until a proposal to deal with it under Art. 14.1 or 56.1 has been submitted and rejected.

Ex. 1. The name *Boletus erythropus* Pers. (in Ann. Bot. (Usteri) 15: 23. 1795), or combinations based on it (e.g. *Neoboletus erythropus* (Pers.) Hahn in Mycol. Bavar. 16: 33. 2015), has been and still is widely and persistently used in the sense of Fries (Syst. Mycol. 1: 391. 1821). This sense, however, does not include the neotype designated by Simonini & al. (in Boll. Assoc. Micol. Ecol. Romana 23: 81. 2017), which follows the original sense of the protologue but is referable to the taxon known as *B. queletii* Schulzer (in Hedwigia 24: 143. 1885) or *Suillellus queletii* (Schulzer) Vizzini & al. (in Index Fungorum 188: 1. 2014). The name *B. erythropus* (*N. erythropus*) is not to be used in the original sense unless and until a proposal to reject it or to conserve *B. queletii* against it has been submitted and rejected. The option of conserving the name *B. erythropus* with a conserved type is also available to protect the current usage that is not compatible with the neotype.

ARTICLE 58
REUSE OF ILLEGITIMATE NAMES

58.1. If there is no obstacle under the rules, the final epithet in an illegitimate name may be reused in a different name, at either the same or a different rank; or an illegitimate generic name may be reused as the epithet in the name of a subdivision of a genus. The resulting name is then treated either as a replacement name with the same type as the illegitimate name (Art. 7.4; see also Art. 7.5 and Art. 41 Note 4) or as the name of a new taxon with a different type. Its priority does not date back to the publication of the illegitimate name (see Art. 11.3 and 11.4).

Ex. 1. The name *Talinum polyandrum* Hook. (in Bot. Mag. 81: ad t. 4833. 1855) is illegitimate under Art. 53.1 because it is a later homonym of *T. polyandrum* Ruiz & Pav. (Fl. Peruv. Prodr.: 65. 1794). When Bentham (Fl. Austral. 1: 172. 1863) transferred *T. polyandrum* Hook. to *Calandrinia* Kunth, he called it *C. polyandra*. This name has priority from 1863, and is cited as *C. polyandra* Benth., not *C. polyandra* "(Hook.) Benth."

Ex. 2. *Cymbella subalpina* Hust. (in Int. Rev. Gesamten Hydrobiol. Hydrogr. 42: 98. 1942) is illegitimate under Art. 53.1 because it is a later homonym of *C. subalpina* F. Meister (Kieselalg. Schweiz: 182, 236. 1912). When Mann (in Round & al., Diatoms: 667. 1990) transferred *C. subalpina* Hust. to *Encyonema* Kütz., he called it *E. subalpinum* D. G. Mann. This name is a replacement name with priority from 1990 and as such is illegitimate under Art. 52.1 because *C. mendosa* VanLand. (Cat. Fossil Recent Gen. Sp. Diatoms Syn. 3: 1211, 1236. 1969) had already been published as a replacement name for *C. subalpina* Hust.

Ex. 3. *Hibiscus ricinifolius* E. Mey. ex Harv. (Fl. Cap. 1: 171. 1860) is illegitimate under Art. 52.1 because *H. ricinoides* Garcke (in Bot. Zeitung (Berlin) 7: 834. 1849) was cited in synonymy. When the epithet *ricinifolius* was combined at varietal rank under *H. vitifolius* by Hochreutiner (in Annuaire Conserv. Jard. Bot. Genève 4: 170. 1900) his name was legitimate and is treated as a replacement name, typified (Art. 7.4) by the type of *H. ricinoides*. The name is cited as *H. vitifolius* var. *ricinifolius* Hochr., not *H. vitifolius* var. *ricinifolius* "(E. Mey. ex Harv.) Hochr."

Ex. 4. *Geiseleria* Klotzsch (in Arch. Naturgesch. 7: 254. 1841) is illegitimate under Art. 52.1 because Klotzsch's circumscription included *Croton glandulosus* L., the original type of *Decarinium* Raf. (Neogenyton: 1. 1825). Later, Asa Gray (Manual, ed. 2: 391. 1856) published *Croton* subg. *Geiseleria,* which has priority from that date and is cited as *C.* subg. *Geiseleria* A. Gray, not *C.* subg. *Geiseleria* "(Klotzsch) A. Gray". Because the subgeneric name is a replacement name, its type is *C. glandulosus,* the type (Art. 7.4) of *Decarinium* and automatic type (Art. 7.5) of *Geiseleria.*

ⓘ Note 1. When the epithet of a name illegitimate under Art. 52.1 is reused at the same rank, the resulting name is illegitimate unless either the type of the name causing illegitimacy is explicitly excluded or its epithet is unavailable for use.

Ex. 5. *Menispermum villosum* Lam. (Encycl. 4: 97. 1797) is illegitimate under Art. 52.1 because *M. hirsutum* L. (Sp. Pl.: 341. 1753) was cited in synonymy. The name *Cocculus villosus* DC. (Syst. Nat. 1: 525. 1817), based on *M. villosum,* is also illegitimate because the type of *M. hirsutum* was not excluded and the epithet *hirsutus* was available for use in *Cocculus.*

Ex. 6. *Cenomyce ecmocyna* Ach. (Lichenogr. Universalis: 549. 1810) is an illegitimate renaming of *Lichen gracilis* L. (Sp. Pl.: 1152. 1753). *Scyphophorus ecmocynus* Gray (Nat. Arr. Brit. Pl. 1: 421. 1821), based on *C. ecmocyna,* is also illegitimate because the type of *L. gracilis* was not excluded and the epithet *gracilis* was available for use. When proposing the combination *Cladonia ecmocyna,* Leighton (in Ann. Mag. Nat. Hist., ser. 3, 18: 406. 1866) explicitly excluded *L. gracilis* and thereby published the legitimate name of a new species, *Cladonia ecmocyna* Leight.

Ex. 7. *Ferreola ellipticifolia* Stokes (Bot. Mat. Med. 4: 556. 1812) is illegitimate under Art. 52.1 because *Maba elliptica* J. R. Forst. & G. Forst. (Char. Gen. Pl., ed. 2: 122. 1776) was cited in synonymy. Bakhuizen van den Brink published *Diospyros ellipticifolia* Bakh. (in Gard. Bull. Straits Settlem. 7: 162. 1933) as a replacement name for *F. ellipticifolia* and did not exclude the type of *M. elliptica. Diospyros ellipticifolia* is nevertheless a legitimate name because in 1933 the epithet *elliptica* was not available for use in *Diospyros* due to the existence of *D. elliptica* Knowlt. (in Bull. U. S. Geol. Surv. 204: 83. 1902), of which *D. elliptica* (J. R. Forst. & G. Forst.) P. S. Green (in Kew Bull. 23: 340. 1969) is an illegitimate later homonym (Art. 53.1).

ARTICLE 59
NAMES OF FUNGI WITH A PLEOMORPHIC LIFE CYCLE
SEE ARTICLE F.8

CHAPTER VIII
ORTHOGRAPHY AND GENDER OF NAMES

ARTICLE 60
ORTHOGRAPHY OF NAMES

60.1. The original spelling of a name or epithet is to be retained, except for the correction of typographical or orthographical errors and the standardizations imposed by Art. 60.4 (letters and ligatures foreign to classical Latin), 60.5 and 60.6 (interchange between *u/v, i/j*, or *eu/ev*), 60.7 (diacritical signs and ligatures), 60.8 (terminations; see also Art. 32.2), 60.10 (intentional latinizations), 60.11 (compounding forms), 60.12 and 60.13 (hyphens), 60.14 (apostrophes and full stops), 60.15 (abbreviations), and F.9.1 (epithets of fungal names) (see also Art. 14.8, 14.11, and F.3.2).

> ***Ex. 1.*** Retention of original spelling: The generic names *Mesembryanthemum* L. (Sp. Pl.: 480. 1753) and *Amaranthus* L. (l.c.: 989. 1753) were deliberately so spelled by Linnaeus and the spelling is not to be altered to '*Mesembrianthemum*' and '*Amarantus*', respectively, although these latter forms are linguistically correct (see Bull. Misc. Inform. Kew 1928: 113, 287. 1928). *Phoradendron* Nutt. (in J. Acad. Nat. Sci. Philadelphia, ser. 2, 1: 185. 1848) is not to be altered to '*Phoradendrum*'. *Triaspis mozambica* A. Juss. (in Ann. Sci. Nat., Bot., ser. 2, 13: 268. 1840) is not to be altered to '*T. mossambica*', as in Engler (Pflanzenw. Ost-Afrikas C: 232. 1895). *Alyxia ceylanica* Wight (Icon. Pl. Ind. Orient. 4: t. 1293. 1848) is not to be altered to '*A. zeylanica*', as in Trimen (Handb. Fl. Ceylon 3: 127. 1895). *Fagus sylvatica* L. (l.c.: 998. 1753) is not to be altered to '*F. silvatica*'. Although the classical spelling is *silvatica,* the mediaeval spelling *sylvatica* is not an orthographical error (see also Rec. 60E). *Scirpus cespitosus* L. (l.c.: 48. 1753) is not to be altered to '*S. caespitosus*'.

> ***Ex. 2.*** The published epithet '*callunigera*' in the new combination *Scleroderris callunigena* (P. Karst.) Nannf. (in Nova Acta Regiae Soc. Sci. Upsal., ser. 4, 8(2): 287. 1932) is to be corrected because the basionym was spelled *Peziza callunigena* P. Karst. (in Not. Sällsk. Fauna Fl. Fenn. Förh. 10: 171. 1869) (see Art. 6.10).

> ****Ex. 3.*** The epithet of *Agaricus rhacodes* Vittad. (Descr. Fung. Mang.: 158. 1833) is to be so spelled, even though it was originally spelled '*rachodes*' (see Wilson in Taxon 66: 189. 2017).

> ****Ex. 4.*** Typographical errors: *Globba* '*brachycarpa*' Baker (in Hooker, Fl. Brit. India 6: 205. 1890) and *Hetaeria* '*alba*' Ridl. (in J. Linn. Soc., Bot. 32: 404. 1896) are typographical errors for *G. trachycarpa* Baker and *H. alta* Ridl., respectively (see Sprague in J. Bot. 59: 349. 1921).

Ex. 5. *'Torilis' taihasenzanensis* Masam. (in J. Soc. Trop. Agric. 6: 570. 1934) was a typographical error for *Trollius taihasenzanensis,* as noted on the errata slip inserted between pages 4 and 5 of the same volume.

Ex. 6. The misspelled *Indigofera 'longipednnculata'* Y. Y. Fang & C. Z. Zheng (in Acta Phytotax. Sin. 21: 331. 1983) is presumably a typographical error and is to be corrected to *I. longipedunculata.*

*Ex. 7. Orthographical error: *Gluta 'benghas'* L. (Mant. Pl.: 293. 1771), which is an orthographical error for *G. renghas,* is cited as *G. renghas* L. (see Engler in Candolle & Candolle, Monogr. Phan. 4: 225. 1883); the vernacular name used as a specific epithet by Linnaeus is "renghas", not "benghas".

Ex. 8. The original spelling of the generic name *'Nilsonia'* Brongn. (in Ann. Sci. Nat. (Paris) 4: 210. 1825) is an orthographical error correctable under Art. 60.1 to *Nilssonia,* the conservation of which is not therefore required. Brongniart named the genus after Sven Nilsson, whose name he consistently misspelled as "Nilson" in his 1825 publication.

ⓘ ***Note 1.*** Art. 14.11 provides for the conservation of a particular spelling of a name of a family, genus, or species (see Art. 14.8).

Ex. 9. *Bougainvillea* Comm. ex Juss. *('Buginvillaea'),* orth. cons. (see App. III).

Ex. 10. *Wisteria* Nutt., nom. cons., is not to be altered to *'Wistaria',* although the genus was named in honour of Caspar Wistar, because *Wisteria* is the spelling used in App. III (see Art. 14.8).

60.2. The words "original spelling" mean the spelling used when a name of a new taxon or a replacement name was validly published. They do not refer to the use of an initial capital or lower-case letter, which is a matter of typography (see Art. 20.1, 21.2, and Rec. 60G.1).

60.3. The liberty of correcting a name is to be used with reserve, especially if the change affects the first syllable and, above all, the first letter of the name (but see Art. 60 *Ex. 7).

*Ex. 11. The spelling of the generic name *Lespedeza* Michx. (Fl. Bor.-Amer. 2: 70. 1803) is not to be altered, although it commemorates Vicente Manuel de Céspedes (see Ricker in Rhodora 36: 130–132; Hochreutiner in Rhodora 36: 390–392. 1934). *Cereus jamacaru* DC. (Prodr. 3: 467. 1828) may not be altered to *C. 'mandacaru',* even if *jamacaru* is believed to be a corruption of the vernacular name "mandacaru".

60.4. The letters *w* and *y,* foreign to classical Latin, and *k,* rare in that language, are permissible in scientific names (see Art. 32.1(b)). Other letters and ligatures foreign to classical Latin that may appear in scientific names are to be transcribed, for example the German *ß* is to be replaced by *ss*. Ligatures are to be replaced by the separate letters comprising those ligatures, e.g. *æ (ae)* and *œ (oe)*.

60.5. When a name has been published in a work where the letters *u, v* or *i, j* are used interchangeably or in any other way incompatible with modern typographical practices (e.g. one letter of a pair not being used in capitals, or not at all), those letters are to be transcribed in conformity with modern nomenclatural usage.

> *Ex. 12. Curculigo* Gaertn. (Fruct. Sem. Pl. 1: 63. 1788), not *'Cvrcvligo'; Taraxacum* Zinn (Cat. Pl. Hort. Gott.: 425. 1757), not *'Taraxacvm'; Uffenbachia* Fabr. (Enum., ed. 2: 21. 1763), not *'Vffenbachia'*.

> *Ex. 13. 'Geastrvm hygrometricvm'* and *'Vredo pvstvlata'* of Persoon (in Syn. Meth. Fung.: 135, 219. 1801) are spelled, respectively, *Geastrum hygrometricum* Pers., nom. sanct. and *Uredo pustulata* Pers., nom. sanct.

60.6. When the original publication of a name adopted a use of the letters *u, v* or *i, j* in any way incompatible with modern nomenclatural practices, those letters are to be transcribed in conformity with modern nomenclatural usage. When names or epithets are derived from Greek words that include the diphthong *ey* (ευ), its transcription as *ev* is treated as an error correctable to *eu*. When names or epithets of Latin but not Greek origin include the letter *i* used as a semi-vowel (followed by another vowel), it is treated as an error correctable to *j*.

> *Ex. 14.* The generic name *'Mezonevron'* Desf. is correctable to *Mezoneuron* Desf., and the basionym of *Neuropteris* (Brongn.) Sternb. (nom. & orth. cons.), *Filicites* sect. *'Nevropteris'* Brongn., is correctable to *Filicites* sect. *Neuropteris* Brongn. Similarly, *'Evonymus'* L. is correctable to *Euonymus* L. (nom. & orth. cons.).

> *Ex. 15. Jatropha* L., *Jondraba* Medik., and *Clypeola jonthlaspi* L., because they are of Greek origin, are not to be altered to *'Iatropha', 'Iondraba'*, and *Clypeola 'ionthlaspi';* nor are *Ionopsidium* Rchb. and *Ionthlaspi* Adans. to be altered to *'Jonopsidium'* and *'Jonthlaspi',* respectively.

> *Ex. 16. Brachypodium 'iaponicum'* Miq. is correctable to *Brachypodium japonicum* because the epithet is Latin and, in Latin, an initial *i* followed by a vowel is a semi-vowel. *Meiandra 'maior'* Markgr. is correctable to *Meiandra major* because the epithet is Latin and, in Latin, an *i* between two vowels is a semi-vowel, but the generic name is of Greek origin, and so the spelling *"Meiandra"* is correct.

60.7. Diacritical signs are not used in scientific names. When such signs appear in the spelling of a name at valid publication, the signs are to be suppressed with the necessary transcription of the letters so modified; for example *ä, ö, ü* become, respectively, *ae, oe, ue* (not

167

æ or *œ*, see Art. 60.4); *é, è, ê* become *e; ñ* becomes *n; ø* becomes *oe* (not *œ*); *å* becomes *ao*. The diaeresis, indicating that a vowel is to be pronounced separately from the preceding vowel (as in *Cephaëlis, Isoëtes*), is a phonetic device that is not considered to alter the spelling; as such, its use is optional.

> **Ex. 17.** Transcription (e.g. umlaut): *'Lühea'*, dedicated to Carl Emil von der Lühe, is spelled *Luehea* Willd. (in Neue Schriften Ges. Naturf. Freunde Berlin 3: 410. 1801); suppression (e.g. tilde): *Vochysia 'kosñipatae'*, named after the valley of Kosñipata, is spelled *V. kosnipatae* Huamantupa (in Arnaldoa 12: 82. 2005).

60.8. The termination of specific or infraspecific epithets derived from personal names that are not already in Greek or Latin and do not possess a well-established latinized form (see Rec. 60C.1) is as follows (but see Art. 60.9 for epithets derived from abbreviation of personal names):

(a) If the personal name ends with a vowel (including *y*) or -*er*, substantival epithets are formed by adding the genitive inflection appropriate to the gender and number of the person(s) honoured (e.g. *scopoli-i* for Scopoli (masculine), *fedtschenko-i* for Fedtschenko (m), *fedtschenko-ae* for Fedtschenko (feminine), *glaziou-i* for Glaziou (m), *lace-ae* for Lace (f), *gray-i* for Gray (m), *hooker-i* for Hooker (m), *hooker-ae* for Hooker (f), *hooker-orum* for the Hookers (m)), except when the name ends with -*a*, in which case -*e* (singular) or -*rum* (plural) is added (e.g. *triana-e* for Triana (m), *pojarkova-e* for Pojarkova (f), *orlovskaja-e* for Orlovskaja (f), *espinosa-rum* for the Espinosas (m)).

(b) If the personal name ends with a consonant (but not in -*er*), substantival epithets are formed by latinizing them with -*ius,* then removing the -*us* and adding the genitive inflection appropriate to the gender and number of the person(s) honoured (e.g. *lecardi-i* for Lecard (masculine), *wilsoni-ae* for Wilson (feminine), *verloti-orum* for the Verlot brothers, *brauni-arum* for the Braun sisters, *masoni-orum* for Mason, father and daughter).

(c) If the personal name ends with a vowel (including *y*), adjectival epithets are formed by adding -*an*- plus the nominative singular inflection appropriate to the gender of the generic name (e.g. *Cyperus heyne-anus* for Heyne, *Vanda lindley-ana* for Lindley, *Aspidium bertero-anum* for Bertero), except when the personal name ends with -*a* in which case -*n*- plus the appropriate inflection

is added (e.g. *balansa-nus* (masculine), *balansa-na* (feminine), and *balansa-num* (neuter) for Balansa).

(d) If the personal name ends with a consonant, adjectival epithets are formed by latinizing the personal name with *-ius,* then removing the *-us* and adding *-an-* (stem of adjectival suffix) plus the nominative singular inflection appropriate to the gender of the generic name (e.g. *Rosa webbi-ana* for Webb, *Desmodium griffithi-anum* for Griffith, *Verbena hassleri-ana* for Hassler).

Terminations contrary to the above standards are treated as errors to be corrected to *-[i]i, -[i]ae, -[i]ana, -[i]anus, -[i]anum, -[i]arum,* or *-[i]orum,* as appropriate (see also Art. 32.2). However, epithets formed in accordance with Rec. 60C.1 are not correctable (see also Art. 60.10), nor are those with terminations conforming to other classical Latin adjectival usage, namely *-[i]a, -[i]us,* or *-[i]um,* or such epithets ending in an *-ea.*

Ex. 18. In *Rhododendron 'potanini'* Batalin (in Trudy Imp. S.-Peterburgsk. Bot. Sada 11: 489. 1892), commemorating G. N. Potanin, the epithet is to be spelled *potaninii* under Art. 60.8(b) because Potanin is first put in Latin form by adding *-ius* to create *potaninius;* then the genitive is formed by first removing the *-us* and then adding the masculine genitive singular ending *-i,* resulting in the epithet *potaninii*). However, in *Phoenix theophrasti* Greuter (in Bauhinia 3: 243. 1967), commemorating Theophrastus, it is not spelled *'theophrastii'* because Rec. 60C.1 applies.

Ex. 19. Rosa 'pissarti' Carrière (in Rev. Hort. (Paris) 1880: 314. 1880) is a typographical error for *R. 'pissardi'* (see Rev. Hort. (Paris) 1881: 190. 1881), which is to be spelled *R. pissardii* under Art. 60.8(b).

Ex. 20. In *Caulokaempferia 'dinabandhuensis'* Biseshwori & Bipin (in J. Jap. Bot. 92: 84. 2017), commemorating Dinabandhu Sahoo, the adjectival epithet was wrongly given the geographical termination *-ensis* (see Rec. 60D.1). Instead it is to be spelled *C. dinabandhuana* under Art. 60.8(c).

Ex. 21. In *Uladendron codesuri* Marc.-Berti (in Pittieria 3: 10. 1971) the epithet derives from an acronym (CODESUR, Comisión para el Desarrollo del Sur de Venezuela), not a personal name, and is not to be changed to *'codesurii'* (as in Brenan, Index Kew., Suppl. 16: 296. 1981).

Ex. 22. In *Asparagus tamaboki* Yatabe (in Bot. Mag. (Tokyo) 7: 61. 1893) and *Agropyron kamoji* Ohwi (in Acta Phytotax. Geobot. 11: 179. 1942) the epithets correspond, respectively, to a Japanese vernacular designation, "tamaboki", or to part of such a designation, "kamojigusa", and are not therefore spelled *'tamabokii'* and *'kamojii'.*

Ex. 23. Gladiolus watsonius Thunb. (Gladiolus: 14. 1784), *Syringa josikaea* J. Jacq. ex Rchb. (Iconogr. Bot. Pl. Crit. 8: 32. 1830), *Taxus harringtonia* Knight ex

J. Forbes (Pinet. Woburn.: 217. 1839), and *Cephalotaxus harringtonia* (Knight ex J. Forbes) K. Koch (Dendrologie 2(2): 102. 1873) are not to be changed to *G. 'watsonianus'*, *S. 'josikaeana'*, *T. 'harringtonii'*, and *C. 'harringtonii'*, respectively.

ⓘ *Note 2.* The hyphens in Art. 60.8 are given solely for explanatory reasons. For the use of hyphens in epithets see Art. 23.1 and 60.12.

ⓘ *Note 3.* Rec. 60C.1 does not preclude the formation of substantival epithets following the process outlined in Art. 60.8. Authors publishing new substantival epithets derived from personal names that have a well-established latinized form may choose whether or not to use that form.

Ex. 24. Substantival epithets derived from the personal name Martin may be correctly published either as *martini* (from the latinized form Martinus) or *martinii*. *Grevillea martini* F. Muell. (Fragm. 4: 129. 1864), named for James Martin, is not to be changed to *G. 'martinii'*. *Comatricha martinii* Alexop. & Beneke (in Mycologia 46: 245. 1954), named for George W. Martin, is not to be changed to *C. 'martini'*. *Nitella martinii* Casanova & Karol (in Austral. Syst. Bot. 36: 337. 2023), named for Martin O'Brien, is not to be changed to *N. 'martini'*.

ⓘ *Note 4.* Art. 60.8 does not preclude the use, as epithets, of names of genera commemorating persons, or feminine nouns formed by analogy (see Rec. 20A.1(h)), placed in apposition (Art. 23.1).

ⓘ *Note 5.* If the gender and/or number of a substantival epithet derived from a personal name is inappropriate for the gender and/or number of the person(s) whom the name commemorates, the termination is to be corrected in conformity with Art. 60.8.

Ex. 25. *Rosa* ×*'toddii'* Wolley-Dod (in J. Bot. 69, Suppl.: 106. 1931) was named for "Miss E. S. Todd"; the epithet is to be spelled *toddiae*.

Ex. 26. *Astragalus 'matthewsii'* Podlech & Kirchhoff (in Mitt. Bot. Staatssamml. München 11: 432. 1974) commemorates Victoria A. Matthews; the epithet is to be spelled *matthewsiae* and the name is not to be treated as a later homonym of *A. matthewsii* S. Watson (in Proc. Amer. Acad. Arts 18: 192. 1883) commemorating Washington Matthews (see App. VII).

Ex. 27. *Codium 'geppii'* (Schmidt in Biblioth. Bot. 91: 50. 1923), which commemorates Ant(h)ony Gepp and Ethel S. B. Gepp, is to be corrected to *C. geppiorum* O. C. Schmidt.

Ex. 28. *Acacia 'Bancrofti'* Maiden (in Proc. Roy. Soc. Queensland 30: 26. 1918) "commemorates the Bancrofts, father and son, the former the late Dr. Joseph Bancroft, and the latter Dr. Thomas Lane Bancroft"; the epithet is to be spelled *bancroftiorum*.

Ex. 29. *Chamaecrista leonardiae* Britton (N. Amer. Fl. 23: 281. 1930, *'Leonardae'*), *Scolosanthus leonardii* Alain (in Brittonia 20: 160. 1968), and *Frankenia leonardiorum* Alain (l.c.: 155. 1968, *'leonardorum'*) were all based on type material collected by Emery C. Leonard and Genevieve M. Leonard. Because there is no explicit contradicting statement, these names are to be accepted as dedicated to either or both, as indicated by the termination of the epithet.

60.9. An epithet, or in the case of a compound epithet its final portion, formed from abbreviation of one or more personal names is considered to have been composed arbitrarily (Art. 23.2) and is not subject to modification, e.g. under the provisions of Art. 60.8.

🛈 *Note 6.* If the epithet itself is indicated as being abbreviated, Art. 60.15 applies.

Ex. 30. *Silene karekirii* Bocquet (in Candollea 22: 10. 1967), published as a replacement name for *Lychnis sordida* Kar. & Kir. (in Bull. Soc. Imp. Naturalistes Moscou 15: 170. 1842) to avoid creating a later homonym in *Silene*, is an arbitrarily formed epithet (Art. 23.2) constructed by abbreviating the names of Karelin and Kirilov, authors of the replaced synonym. The epithet is not to be changed to '*karekiriorum*' or '*karelinkirilovii*'.

Ex. 31. *Lepanthes carvii* Archila (Lepanthes Guatemala: 99. 2001) was said to be "dedicated to the family of Carlos Villela [a two-word family name] especially LIC [Licenciado] Jorge A Carlos who directed the photography in this investigation". As an epithet apparently formed from the abbreviations "Car" from Carlos and "V" from Villela, it is considered to have been composed arbitrarily (Art. 23.2) and is not to be changed in any way.

Ex. 32. *Telipogon 'crisariasae'* Baquero & Iturralde (in Phytotaxa 564: 249. 2022), commemorating María Cristina Arias (female), in which the final portion of a compound epithet is not formed from an abbreviation, is correctable to *Telipogon crisariasiae* (see Art. 60.8(b)).

60.10. The original spelling (Art. 60.2) of a name or epithet is to be retained (Art. 60.1) if it resulted from the intentional latinization of a personal, geographical, or vernacular name. Excepted from this are epithets formed from personal names when the latinization involves:

(a) only a termination to which Art. 60.8 applies; or
(b) only *(1)* omission of the terminal vowel or terminal consonant; or *(2)* conversion of the terminal vowel to a different vowel, for which the omitted or converted letter is to be restored.

Ex. 33. *Clutia* L. (Sp. Pl.: 1042. 1753), *Gleditsia* J. Clayton (in Linnaeus, l.c.: 1056. 1753), and *Valantia* L. (l.c.: 1051. 1753), commemorating Cluyt, Gleditsch, and Vaillant, respectively, are not to be altered to '*Cluytia*', '*Gleditschia*', and '*Vaillantia*'; these personal names were deliberately latinized as Clutius, Gleditsius, and Valantius.

Ex. 34. *Abies alcoquiana* Veitch ex Lindl. (in Gard. Chron. 1861: 23. 1861), commemorating "Rutherford Alcock Esq.", implies an intentional latinization of his family name to Alcoquius. In transferring the epithet to *Picea,* Carrière (Traité Gén. Conif., ed. 2: 343. 1867) deliberately changed the spelling to *'alcockiana'.* The resulting combination is nevertheless correctly cited as *P. alcoquiana* (Veitch ex Lindl.) Carrière (see Art. 61.4).

***Ex. 35.** Abutilon glaziovii* K. Schum. (in Martius, Fl. Bras. 12(3): 408. 1891), *Desmodium bigelovii* A. Gray (in Smithsonian Contr. Knowl. 5(6): 47. 1853), and *Rhododendron bureavii* Franch. (in Bull. Soc. Bot. France 34: 281. 1887), commemorating A. F. M. Glaziou, J. Bigelow, and L. E. Bureau, respectively, are not to be changed to *A. 'glazioui', D. 'bigelowii',* or *R. 'bureaui'.* In these three cases, the implicit latinizations Glaziovius, Bigelovius, and Bureavius result from conversion of the terminal vowel or consonant to a consonant and do not affect merely the termination of the names.

***Ex. 36.** Arnica chamissonis* Less. (in Linnaea 6: 238. 1831) and *Tragus berteronianus* Schult. (Mant. 2: 205. 1824), commemorating L. K. A. von Chamisso and C. L. G. Bertero, are not to be changed to *A. 'chamissoi'* or *T. 'berteroanus'.* The derivation of these epithets from the third declension genitive (Rec. 60C Ex. 1(b)), a practice normally discouraged (see Rec. 60C.1), involves the addition of letters to the personal name and does not affect merely the termination.

***Ex. 37.** Asa Gray (in Boston J. Nat. Hist. 6: 209. 1850) published the name *Eryngium ravenellii* A. Gray to honour Henry W. Ravenel. In forming the epithet, Gray's implicit latinization of Ravenel is Ravenellius, which is to be preserved because it does not affect merely the termination, there is no omission of the terminal consonant "l" in Ravenel, and there is no terminal vowel to omit or convert. The epithet is not to be changed to *'ravenelii'.*

***Ex. 38.** Acacia 'brandegeana', Blandfordia 'backhousii', Cephalotaxus 'fortuni', Chenopodium 'loureirei', Convolvulus 'loureiri', Glochidion 'melvilliorum', Hypericum 'buckleii', Solanum 'rantonnei',* and *Zygophyllum 'billardierii'* were published to commemorate T. S. Brandegee, J. Backhouse, R. Fortune, J. de Loureiro, R. Melville and E. F. Melville, S. B. Buckley, V. Rantonnet, and J. J. H. de Labillardière (de la Billardière). The implicit latinizations are Brandegeus, Backhousius, Fortunus, Loureireus or Loureirus, Melvillius, Buckleius, Rantonneus, and Billardierius, but these are not acceptable under Art. 60.10. The names are correctly cited as *A. brandegeeana* I. M. Johnst. (in Contr. Gray Herb. 75: 27. 1925), *B. backhousei* Gunn & Lindl. (in Edwards's Bot. Reg. 31: t. 18. 1845), *Cephalotaxus fortunei* Hook. (in Bot. Mag. 76: ad t. 4499. 1850), *Chenopodium loureiroi* Steud. (Nomencl. Bot., ed. 2, 1: 348. 1840), *Convolvulus loureiroi* G. Don (Gen. Hist. 4: 290. 1837), *G. melvilleorum* Airy Shaw (in Kew Bull. 25: 487. 1971), *H. buckleyi* M. A. Curtis (in Amer. J. Sci. Arts 44: 80. 1843), *S. rantonnetii* Carrière (in Rev. Hort. (Paris) 1859: 135. 1859), and *Z. billardierei* DC. (Prodr. 1: 705. 1824).

***Ex. 39.** Mycena seynii* Quél. (in Bull. Soc. Bot. France 23: 351. 1877), commemorating Jules de Seynes, is not to be altered to *M. 'seynesii'.* The implicit latinization of that name to Seynius results from omission of more than the terminal letter.

ⓘ **Note 7.** The provisions of Art. 60.8, 60.10, and Rec. 60C deal with the latinization of names through their modification. Latinization is not the same as translation of a name (e.g. Tabernaemontanus, Latin for Bergzabern; Nobilis, Latin for Noble). Epithets resulting from or derived from Latin translations are not subject to standardization under Art. 60.8, although Rec. 60C.1 and 60C.2 may apply.

Ex. 40. In *Wollemia nobilis* W. G. Jones & al. (in Telopea 6: 174. 1995), *nobilis* is the translation into Latin of the family name of the discoverer David Noble. *Cladonia abbatiana* S. Stenroos (in Ann. Bot. Fenn. 28: 107. 1991) honours the French lichenologist H. des Abbayes, where Abbayes can be translated to Abbatiae (abbeys). Neither epithet may be altered.

Ex. 41. The epithet, in apposition, in *Crataegus spes-aestatum* J. B. Phipps (in Novon 16: 382. 2006) honours Bill Summers, one person, but the genitive plural *spes-aestatum* (hope of summers) is not to be altered to the singular; it is a translation to Latin.

60.11. Adjectival epithets (and substantival epithets that semantically serve as if they were adjectives) that combine elements derived from two or more Greek or Latin words are to be compounded as follows:

A noun or adjective in a non-final position appears as a compounding form generally obtained by:

(a) removing the case ending of the genitive singular (Latin *-ae, -i, -us, -is;* transcribed Greek *-ou, -os, -es, -as, -ous* and its equivalent *-eos*); and

(b) before a consonant, adding a connecting vowel (*-i-* for Latin elements, *-o-* for Greek elements).

Adjectival epithets not formed in accordance with this provision are to be corrected to conform with it, unless Rec. 60H.1(a) or (b) applies. In particular, the use of the genitive singular case ending of Latin first-declension nouns instead of a connecting vowel is treated as an error to be corrected unless it serves to make a semantic distinction.

Ex. 42. The epithet meaning "having leaves like those of *Quercus*" is *quercifolia (Querc-,* connecting vowel *-i-,* and ending *-folia).*

Ex. 43. The epithet *'aquilegifolia'*, derived from the name *Aquilegia* must be changed to *aquilegiifolia (Aquilegi-,* connecting vowel *-i-,* and ending *-folia).*

Ex. 44. The epithet of *Pereskia 'opuntiaeflora'* DC. (in Mém. Mus. Hist. Nat. 17: 76. 1828) is to be spelled *opuntiiflora,* and that of *Myrosma 'cannaefolia'* L. f. (Suppl. Pl. 80. 1782), *cannifolia.*

Ex. 45. The epithet of *Cacalia 'napeaefolia'* DC. (Prodr. 6: 328. 1838) and *Senecio 'napeaefolius'* (DC.) Sch. Bip. (in Flora 28: 498. 1845) is to be spelled *napaeifolia (-us);* it refers to the resemblance of the leaves to those found in *Napaea* L. (not *'Napea'*), and the connecting vowel *-i-* should have been used instead of the genitive singular inflection *-ae-*.

Ex. 46. In *Andromeda polifolia* L. (Sp. Pl.: 393. 1753), the epithet is taken from a pre-Linnaean generic designation (*"Polifolia"* of Buxbaum) and is a noun used in apposition, not an adjective; it is not to be altered to *'poliifolia'* (*Polium*-leaved).

Ex. 47. *Tetragonia tetragonoides* (Pall.) Kuntze (Revis. Gen. Pl. 1: 264. 1891) was based on *Demidovia tetragonoides* Pall. (Enum. Hort. Demidof: 150. 1781), the specific epithet of which was derived from the generic name *Tetragonia* and the suffix *-oides*. Because this is a compound epithet derived from a noun and a suffix, not two Greek or Latin words, it is not to be altered to *'tetragonioides'*.

60.12. The use of a hyphen in an epithet is treated as an error to be corrected by deletion of the hyphen. A hyphen is permitted only when at least one condition of (a) and at least one condition of (b) are met:

(a) *(1)* the hyphen was present at the valid publication of the name or its basionym (if the name has a basionym); or

   (2) the epithet is hyphenated according to Art. 23.1 or 23.3 (if the epithet consisted originally of two or more words, or a word and a symbol);

and

(b) *(1)* the epithet is formed of words that usually stand independently; or

   (2) the letters before and after the hyphen are the same.

Ex. 48. Hyphen deleted: *Acer pseudoplatanus* L. (Sp. Pl.: 1054. 1753, *'pseudo-platanus'*) (a1); *Eugenia costaricensis* O. Berg (in Linnaea 27: 213. 1856, *'costa-ricensis'*) (a1); *Ficus neoebudarum* Summerh. (in J. Arnold Arbor. 13: 97. 1932, *'neo-ebudarum'*) (a1); *Lycoperdon atropurpureum* Vittad. (Monogr. Lycoperd.: 42. 1842, *'atro-purpureum'*) (a1); *Mesospora vanbosseae* Børgesen (in Skottsberg, Nat. Hist. Juan Fernandez 2: 258. 1924, *'van-bosseae'*) (a1); *Peperomia lasierrana* Trel. & Yunck. (Piperac. N. South Amer.: 530. 1950, *'la-sierrana'*) (a1, see Rec. 60C.4(c)); *Scirpus* sect. *Pseudoeriophorum* Jurtzev (in Byull. Moskovsk. Obshch. Isp. Prir., Otd. Biol. 70(1): 132. 1965, *'Pseudo-eriophorum'*) (a1).

Ex. 49. Hyphen maintained: *Athyrium austro-occidentale* Ching (in Acta Bot. Boreal.-Occid. Sin. 6: 152. 1986) (a1, b2); *Piper pseudo-oblongum* McKown (in Bot. Gaz. 85: 57. 1928) (a1, b2); *Ribes non-scriptum* (Berger) Standl. (in Publ.

Field Mus. Nat. Hist., Bot. Ser. 8: 140. 1930) (a1, b1); *Vitis novae-angliae* Fernald (in Rhodora 19: 146. 1917) (a1, b1); *Pleurothyrium roberto-andinoi* C. Nelson (in Phytologia 72: 402. 1992) (a1, b1); *Kalanchoe adolphi-engleri* Raym.-Hamet (in Bull. Soc. Bot. France 102: 239. 1955) (a1, b1).

Ex. 50. Hyphen inserted: *Arctostaphylos uva-ursi* (L.) Spreng. (Syst. Veg. 2: 287. 1825, *'uva ursi'*) (a2, b1); *Aster novae-angliae* L. (Sp. Pl.: 875. 1753, *'novae angliae'*) (a2, b1); *Coix lacryma-jobi* L. (l.c.: 972. 1753, *'lacryma jobi'*) (a2, b1); *Marattia rolandi-principis* Rosenst. (in Repert. Spec. Nov. Regni Veg. 10: 162. 1911, *'rolandi principis'*) (a2, b1); *Veronica anagallis-aquatica* L. (l.c.: 12. 1753, *'anagallis ▽'*) (a2, b1); *Veronica argute-serrata* Regel & Schmalh. (in Trudy Imp. S.-Peterburgsk. Bot. Sada 5: 626. 1878, *'argute serrata'*) (a2, b1); *Sclerospora graminicola* var. *andropogonis-sorghi* Kulk. (in Mem. Dept. Agric. India, Bot. Ser. 5: 272. 1913, *'andropogonis sorghi'*) (a2, b1, b2).

Ex. 51. Hyphen not inserted: *Synsepalum letestui* Aubrév. & Pellegr. (in Notul. Syst. (Paris) 16: 263. 1961, *'Le Testui'*) (a2, see Rec. 60C.4(c)); *Allelochaeta neoorbicularis* Crous (in Fungal Syst. Evol. 2: 294. 2018) (b2); *Acacia circummarginata* Chiov. (in Ann. Bot. (Rome) 13: 394. 1915) (b2); *Sporisorium andrewmitchellii* R. G. Shivas & al. (in Persoonia 28: 155. 2012) (b1).

ⓘ **Note 8.** Art. 60.12 refers only to epithets (in combinations), not to names of genera (for names of fossil-genera see Art. 60.13) or taxa at higher ranks; a non-fossil generic name published with a hyphen can be changed only by conservation (Art. 14.11; see also Art. 20.3; but see Art. H.6.5).

Ex. 52. *Pseudo-fumaria* Medik. (Philos. Bot. 1: 110. 1789) may not be changed to *'Pseudofumaria'*, whereas by conservation *'Pseudo-elephantopus'* was changed to *Pseudelephantopus* Rohr (in Skr. Naturhist.-Selsk. 2: 214. 1792).

60.13. The use of a hyphen in the name of a fossil-genus is in all cases treated as an error to be corrected by deletion of the hyphen.

Ex. 53. *'Cicatricosi-sporites'* R. Potonié & Gelletich (in Sitzungsber. Ges. Naturf. Freunde Berlin 1932: 522. 1932) and *'Pseudo-Araucaria'* Fliche (in Bull. Soc. Sci. Nancy 14: 181. 1896) are names of fossil-genera. They are treated as errors to be corrected by deletion of the hyphen to *Cicatricosisporites* and *Pseudoaraucaria*, respectively.

60.14. The use of an apostrophe or quotation mark in an epithet is treated as an error to be corrected by deletion of the apostrophe or quotation mark unless it follows *m* to represent the patronymic prefix Mc (or Mᶜ), in which case it is replaced by the letter *c*. The use of a full stop (period) in an epithet that is derived from a personal or geographical name that contains this full stop is treated as an error to be corrected by expansion or, when nomenclatural tradition does not support expansion (Art. 60.15), deletion of the full stop.

Ex. 54. In *Cymbidium 'i'ansoni'* Rolfe (in Orchid Rev. 8: 191. 1900), *Lycium 'o'donellii'* F. A. Barkley (in Lilloa 26: 202. 1953), and *Solanum tuberosum* var. *'muru'kewillu'* Ochoa (in Phytologia 65: 112. 1988), the final epithet is to be spelled *iansonii, odonellii,* and *murukewillu,* respectively.

Ex. 55. In *Nesoluma 'St.-Johnianum'* H. J. Lam & B. Meeuse (in Occas. Pap. Bernice Pauahi Bishop Mus. 14: 153. 1938), derived from St. John, the family name of one of the collectors, the epithet is to be spelled *st-johnianum.*

Ex. 56. Harvey (Fl. Cap. 3: 494. 1865) published *Stobaea 'M'Kenii'*. The name commemorates one of the collectors of the type specimen, Mark Johnston McKen (1823–1872). The spelling has been changed to *S. 'mkenii'* but must be corrected to *S. mckenii.*

60.15. Names or epithets indicated as abbreviated are to be expanded in conformity with nomenclatural tradition (see also Art. 23 *Ex. 29 and Rec. 60C.4(d)).

Ex. 57. In *Allium 'a. bolosii'* P. Palau (in Anales Inst. Bot. Cavanilles 11: 485. 1954), dedicated to Antonio de Bolòs y Vayreda, the epithet is to be spelled *antonii-bolosii.*

Recommendation 60A

60A.1. When a name of a new taxon or a replacement name, or its epithet, is to be derived from Greek, the transcription to Latin should conform to classical usage.

Ex. 1. The Greek spiritus asper (an inverted apostrophe) in words transcribed to Latin should be replaced by the letter h, as in *Hyacinthus* (from ὑάκινθος) and *Rhododendron* (from ῥοδόδενδρον).

Recommendation 60B

60B.1. When a new generic name, or epithet in a new name of a subdivision of a genus, is taken from a personal name, it should be formed as follows (see also Rec. 20A.1(h); but see Rec. 21B.2):

(a) When the personal name ends with a vowel, the letter -*a* is added (e.g. *Ottoa* after Otto; *Sloanea* after Sloane), except when the name ends with -*a,* when -*ea* is added (e.g. *Collaea* after Colla), or with -*ea,* when nothing is added (e.g. *Correa* after Correa).

(b) When the personal name ends with a consonant, the letters -*ia* are added, but when the name ends with -*er,* either of the terminations -*ia* and -*a* is appropriate (e.g. *Sesleria* after Sesler and *Kernera* after Kerner).

(c) In latinized personal names ending with -*us* this termination is dropped before applying the procedure described under (a) and (b) (e.g. *Dillenia* after Dillenius).

ⓘ **Note 1.** The syllables not modified by these endings are unaffected unless they contain letters, ligatures, or diacritical signs that must be transcribed under Art. 60.4 and 60.7.

ⓘ **Note 2.** More than one generic name, or epithet of a subdivision of a genus, may be based on the same personal name, e.g. by adding a prefix or suffix to that personal name or by using an anagram or abbreviation of it (but see Art. 53.2 and 53.3).

Ex. 1. *Bouchea* Cham. (in Linnaea 7: 252. 1832) and *Ubochea* Baill. (Hist. Pl. 11: 103. 1891); *Engleria* O. Hoffm. (in Bot. Jahrb. Syst. 10: 273. 1888), *Englerella* Pierre (Not. Bot.: 46. 1891), and *Englerastrum* Briq. (in Bot. Jahrb. Syst. 19: 178. 1894); *Gerardia* L. (Sp. Pl.: 610. 1753) and *Graderia* Benth. (in Candolle, Prodr. 10: 521. 1846); *Lapeirousia* Pourr. (in Hist. & Mém. Acad. Roy. Sci. Toulouse 3: 79. 1788) and *Peyrousea* DC. (Prodr. 6: 76. 1838); *Martia* Spreng. (Anleit. Kenntn. Gew., ed. 2, 2: 788. 1818) and *Martiusia* Schult. (Mant. 1: 69, 226. 1822); *Orcuttia* Vasey (in Bull. Torrey Bot. Club 13: 219. 1886) and *Tuctoria* Reeder (in Amer. J. Bot. 69: 1090. 1982); *Urvillea* Kunth (in Humboldt & al., Nov. Gen. Sp. 5, ed. qu.: 105; ed. fol.: 81. 1821) and *Durvillaea* Bory (Dict. Class. Hist. Nat. 9: 192. 1826) (see Art. 53 *Ex. 15).

Recommendation 60C

60C.1. When forming specific and infraspecific epithets from personal names already in Greek or Latin, or that possess a well-established latinized form, the epithets, when substantival, should (despite Art. 60.8) be given the appropriate Latin genitive form (e.g. *alexandri* from Alexander or Alexandre, *alberti* from Albert, *arnoldi* from Arnold, *augusti* from Augustus or August or Auguste, *ferdinandi* from Ferdinand or Fernando or Fernand, *martini* from Martinus or Martin, *linnaei* from Linnaeus, *martii* from Martius, *wislizeni* from Wislizenus, *edithae* from Editha or Edith, *elisabethae* from Elisabetha or Elisabeth, *murielae* from Muriela or Muriel, *conceptionis* from Conceptio or Concepción, *beatricis* from Beatrix or Béatrice, *hectoris* from Hector; but not *'cami'* from Edmond Gustave Camus or Aimée Camus). Treating modern family names, i.e. ones that do not have a well-established latinized form, as if they were in third declension should be avoided (e.g. *munronis* from Munro, *richardsonis* from Richardson).

60C.2. New epithets based on personal names that have a well-established latinized form should maintain the traditional use of that latinized form.

Ex. 1. In addition to the epithets in Rec. 60C.1, the following epithets commemorate personal names already in Latin or possessing a well-established latinized form: *(a)* second declension: *afzelii* based on Afzelius; *allemanii* based on Allemanius (Freire Allemão); *bauhini* based on Bauhinus (Bauhin); *clusii* based on Clusius; *rumphii* based on Rumphius (Rumpf); *solandri* based on Solandrus (Solander); *(b)* third declension (otherwise discouraged, see Rec. 60C.1): *bellonis* based on Bello; *brunonis* based on Bruno (Robert Brown); *chamissonis* based on Chamisso; *(c)* adjectives (see Art. 23.5): *afzelianus, clusianus, linnaeanus, martianus, rumphianus, brunonianus,* and *chamissonianus.*

60C.3. In forming new epithets based on personal names the customary spelling of the personal name should not be modified unless it contains letters, ligatures, or diacritical signs that must be transcribed under Art. 60.4 and 60.7.

60C.4. In forming new epithets based on personal names prefixes and particles should be treated as follows:

(a) The Scottish and Irish patronymic prefix Mac, Mc, Mᶜ, or Mʼ, meaning "son of", should either all be spelled as *mac* or the latter three as *mc* and united with the rest of the name (e.g. *macfadyenii* after Macfadyen, *macgillivrayi* after MacGillivray, *macnabii* or *mcnabii* after McNab, *macclellandii* or *mcclellandii* after MʼClelland).

(b) The Irish patronymic prefix O should be united with the rest of the name (Art. 60.14) or omitted (e.g. *obrienii, brienianus* after OʼBrien, *okellyi* after OʼKelly).

(c) A prefix consisting of an article (e.g. le, la, lʼ, les, el, il, lo), or containing an article (e.g. du, de la, des, del, della), should be united to the name (e.g. *leclercii* after Le Clerc, *dubuyssonii* after Du Buysson, *lafarinae* after La Farina, *logatoi* after Lo Gato). See Art. 23.1 and Art. 60 Ex. 51 for cases where such epithets were originally spelled in two words.

(d) A prefix to a person's family name indicating ennoblement or canonization should be omitted (e.g. *candollei* after de Candolle, *jussieui* after de Jussieu, *hilairei* after Saint-Hilaire, *remyi* after St Rémy); in geographical epithets, however, "St" should be expanded as *sanctus* (masculine) or *sancta* (feminine) (e.g. *sancti-johannis,* of St John, *sanctae-helenae,* of St Helena).

(e) A German or Dutch prefix should be omitted (e.g. *iheringii* after von Ihering, *martii* after von Martius, *steenisii* after van Steenis, *strassenii* after zu Strassen, *vechtii* after van der Vecht), but when it is normally treated as part of the family name it should be included in the epithet (e.g. *vonhausenii* after Vonhausen, *vanderhoekii* after Vanderhoek, *vanbruntiae* after Van Brunt).

Recommendation 60D

60D.1. An epithet derived from a geographical name is preferably an adjective and usually takes one of the terminations *-ensis, -(a)nus, -inus,* or *-icus.*

Ex. 1. Rubus quebecensis L. H. Bailey (from Quebec), *Ostrya virginiana* (Mill.) K. Koch (from Virginia), *Eryngium amorginum* Rech. f. (from Amorgos), *Fraxinus pennsylvanica* Marshall (from Pennsylvania).

Recommendation 60E

60E.1. The epithet in a name of a new taxon or replacement name should be written in conformity with the customary spelling of the word or words from which it is derived and in accordance with the accepted usage of Latin and latinization (see also Art. 23.5).

Ex. 1. sinensis (not *chinensis*).

Recommendation 60F

60F.1. In forming names or epithets that are based on personal, geographical, or vernacular names or on other words, in which signs (such as diacritical signs or ligatures) or letters appear that do not belong to the twenty-six letters of the modern Latin alphabet (Art. 32.1(b)), authors should suppress or transcribe these signs or letters in conformity with modern nomenclatural usage (see also Art. 60.4 and 60.7).

Recommendation 60G

60G.1. All specific and infraspecific epithets should be written with an initial lower-case letter.

Recommendation 60H

60H.1. A name or epithet that combines elements derived from two or more Greek or Latin words should be formed, as far as practicable, in accordance with classical usage, subject to the provisions of Art. 60.11.

(a) Exceptions to the procedure outlined in Art. 60.11 are frequent, and one should review earlier usages of a particular compounding form. In forming apparently irregular compounds, classical usage is often followed.

Ex. 1. The compounding forms *hydro-* and *hydr- (Hydro-phyllum)* stem from water (hydor, hydatos); *calli- (Calli-stemon)* derives from the adjective beautiful (kalos); and *meli- (Meli-osma, Meli-lotus)* stems from honey (meli, melitos).

(b) In pseudocompounds, a noun or adjective in a non-final position appears as a word with a case ending, not as a modified stem. Examples are: *nidus-avis* (nest of bird, nominative), *Myos-otis* (mouse ear, genitive), *albo-marginatus* (white-margined, ablative), etc. In epithets where tingeing is expressed, the modifying colour is often in the ablative because the preposition e or ex is implicit, e.g. *atropurpureus* (blackish purple) from "ex atro purpureus" (purple tinged with black). Pseudocompounds, in particular those using the genitive singular of Latin first-declension nouns, are considered as correctable errors

under Art. 60.11, except when they serve to reveal semantic differences between identically spelled regular compounds formed from different elements.

Ex. 2. The Latin words for tube (tubus, tubi) and for trumpet (tuba, tubae) in regular compounds result in identical epithets (e.g. *tubiformis*), whereas the pseudocompound *tubaeformis* can only mean trumpet-formed, as in *Cantharellus tubaeformis* Fr. (Syst. Mycol. 1: 319. 1821), nom. cons.

Ex. 3. Regular compounds derived from papaya *(Carica, Caricae)* and sedge *(Carex, Caricis)* are identical, whereas the pseudocompound *caricaefolius* can only mean papaya-leaved, as in *Solanum caricaefolium* Rusby (in Bull. New York Bot. Gard. 8: 118. 1912).

ⓘ **Note 1.** The hyphens in the above examples (*nidus-avis* excepted) are given solely for explanatory reasons. For the use of hyphens in generic names and in epithets see Art. 20.3, 23.1, 60.12, and 60.13.

Recommendation 60I

60I.1. When naming new genera or lower-ranked taxa or providing replacement names, authors should explicitly state the etymology of the names and epithets, especially when their meaning is not obvious.

ARTICLE 61
ORTHOGRAPHICAL VARIANTS OF NAMES

61.1. Only one orthographical variant of any one name is treated as validly published: the form that appears in the original publication (but see Art. 6.10 and 61.6), except as provided in Art. 60 and F.9 (typographical or orthographical errors and standardizations), Art. 14.8 and 14.11 (spelling of conserved names), Art. F.3.2 (spelling of sanctioned names), and Art. 16.3, 18.4, 19.7, and 32.2 (improper Latin or transcribed Greek terminations).

61.2. For the purposes of this *Code,* orthographical variants are the various spelling, compounding, and inflectional forms of a name or its final epithet (including typographical errors) when only one nomenclatural type is involved.

Ex. 1. *Nelumbo* Adans. (Fam. Pl. 2: 76. 1763) and *'Nelumbium'* (Jussieu, Gen. Pl.: 68. 1789) are spelling forms of a generic name based on *Nymphaea nelumbo* L., and are treated as orthographical variants. Similarly, *'Musenium'* (Nuttall in Torrey & Gray, Fl. N. Amer. 1: 642. 1840), for which Pfeiffer (Nomencl. Bot. 2: 377. 1873) designated *Seseli divaricatum* Pursh as type, is an orthographical variant of *Musineon* Raf. (in J. Phys. Chim. Hist. Nat. Arts 91: 71. 1820), of which *S. divaricatum* is the original type.

Ex. 2. The epithet of *Selaginella apus* Spring (in Martius, Fl. Bras. 1(2): 119. 1840) is a noun in apposition, so that *apus* cannot be treated as an orthographical variant of the adjective *apodus,* used in *Lycopodium apodum* L. (Sp. Pl.: 1105. 1753). Spring cited *L. apodum* as a synonym of *S. apus,* but instead he should have adopted the former epithet and published *"S. apoda".* Consequently, *S. apus* was nomenclaturally superfluous when published and is illegitimate under Art. 52.1.

61.3. If orthographical variants of a name of a new taxon or replacement name appear in the original publication, the one that conforms to the rules and best suits the recommendations of Art. 60 is to be retained. If the variants conform and suit equally well, the first author who, in an effectively published text (Art. 29–31), explicitly adopts one of the variants and rejects the other(s) must be followed (see also Rec. F.5A.2).

61.4. The orthographical variants of a name are to be corrected to the validly published form of that name. Whenever such a variant appears in a publication, it is to be treated as if it appeared in its corrected form.

🛈 ***Note 1.*** In full citations it is desirable that the original form of a corrected orthographical variant of a name be added (Rec. 50F).

61.5. Confusingly similar names based on the same type are treated as orthographical variants. (For confusingly similar names based on different types, see Art. 53.2–53.4.)

Ex. 3. *'Geaster'* (Fries, Syst. Mycol. 3: 8. 1829) and *Geastrum* Pers. (in Neues Mag. Bot. 1: 85. 1794), nom. sanct., are similar names with the same type (see Taxon 33: 498. 1984); they are treated as orthographical variants even though they are derived from two different nouns, aster and astrum, that both mean star.

61.6. Epithets with the root *caf[f][e]r-,* such as *cafra, caffra, cafrorum,* and *cafrum,* are not permitted in the nomenclature of organisms covered by this *Code.* Where these epithets were used in validly published names, they are to be treated as orthographical variants that are to be replaced by epithets with the root *af[e]r-,* such as *afra, afrorum,* and *afrum,* respectively. If this results in a later homonym, the correct name is determined by Art. 11.4.

Ex. 4. *Portulaca 'caffra'* Thunb. (Prodr. Pl. Cap.: [85]. 1800) is to be treated as having been published as *P. afra* (with one *f*) Thunb. (l.c. 1800), i.e. with retention of author attribution and date and place of publication. The new combination in *Talinum* Adans. is to be treated as having been published as *T. afrum* (with one *f*) (Thunb.) Eckl. & Zeyh. (Enum. Pl. Afric. Austral.: 282. 1836).

Ex. 5. When the epithet *'cafra'* in *Plantago 'cafra'* Decne. (in Candolle, Prodr. 13(1): 719. 1852) is replaced by *afra, P. afra* Decne. (l.c. 1852), i.e. with retention of author attribution and date and place of publication, is a later homonym of *P. afra* L. (Sp. Pl., ed. 2: 168. 1762) and therefore illegitimate. Under Art. 11.4, the name that has to be adopted is *P. capillaris* E. Mey. ex Decne. (in Candolle, Prodr. 13(1): 719. 1852), which is widely treated as a heterotypic synonym of *P. afra* Decne.

ARTICLE 62
GENDER OF NAMES

62.1. A generic name retains the gender assigned by nomenclatural tradition, regardless of classical usage or the author's original usage (but see Art. 62.2–62.4). A generic name without a nomenclatural tradition retains the gender assigned by its author (but see Art. 62.4).

ⓘ **Note 1.** Tradition for generic names usually maintains the classical gender of the corresponding Greek or Latin word, if such exists, but may differ.

*Ex. 1. In accordance with tradition, *Adonis* L., *Atriplex* L., *Diospyros* L., *Eucalyptus* L'Hér., *Hemerocallis* L., *Orchis* L., *Stachys* L., and *Strychnos* L. must be treated as feminine while *Lotus* L. and *Melilotus* (L.) Mill. must be treated as masculine. Although their ending suggests masculine gender, *Cedrus* Trew and *Fagus* L., like most other classical tree names, were traditionally treated as feminine and therefore retain that gender; similarly, *Rhamnus* L. is feminine, even though Linnaeus assigned it masculine gender. *Erigeron* L. (masculine, not neuter), *Phyteuma* L. (neuter, not feminine), and *Sicyos* L. (masculine, not feminine) are other names for which tradition has re-established the classical gender despite another choice by Linnaeus.

Ex. 2. *Glomus* Tul. & C. Tul. (in Giorn. Bot. Ital. 1(2): 63. 1845), despite having been introduced with a masculine gender, must be treated as neuter, because it has generally been accepted as such since Trappe (in Phytopathology 72: 1102–1108. 1982) proposed to re-establish its classical gender; he was followed from then on, thus establishing a significant tradition.

62.2. Compound generic names take the gender of the last word in the nominative case in the compound (but see Art. 14.11). If the termination is altered, however, the gender is altered accordingly. An exception is made for compounds, with endings other than those listed in clauses (a)–(c) of this Article, that were classical Latin words and in which tradition has adopted the classical Latin gender of that word even though the gender of the last word differs in the original language (usually Greek). In such cases the classical Latin gender is adopted.

Ex. 3. Even though the name *Parasitaxus* de Laub. (Fl. Nouv.-Calédonie & Dépend. 4: 44. 1972) was treated as masculine when published, its gender is feminine: it is a compound of which the last part coincides with the generic name *Taxus* L., which is feminine by tradition (Art. 62.1).

Ex. 4. Compound generic names in which the termination of the last word is altered: *Dipterocarpus* C. F. Gaertn., *Stenocarpus* R. Br., and all other compounds ending in the Greek masculine *-carpos* (or *-carpus*), e.g. *Hymenocarpos* Savi, are masculine; those in *-carpa* or *-carpaea,* however, are feminine, e.g. *Callicarpa* L. and *Polycarpaea* Lam.; and those in *-carpon, -carpum,* or *-carpium* are neuter, e.g. *Polycarpon* L., *Ormocarpum* P. Beauv., and *Pisocarpium* Link.

(a) Compounds ending in *-botrys, -codon, -dens, -myces, -odon, -panax, -pogon, -stemon,* and other masculine words, are masculine.

Ex. 5. Even though the generic names *Andropogon* L. and *Oplopanax* (Torr. & A. Gray) Miq. were originally treated as neuter by their authors, they are masculine.

(b) Compounds ending in *-achne, -anthes, -chlamys, -daphne, -glochin, -mecon, -osma* (the modern transcription of the feminine Greek word οσμή, osmē), and other feminine words, are feminine. An exception is made in the case of names ending in *-gaster,* which strictly speaking should be feminine but are treated as masculine in accordance with tradition.

Ex. 6. Even though *Tetraglochin* Poepp., *Triglochin* L., *Dendromecon* Benth., and *Hesperomecon* Greene were originally treated as neuter, they are feminine.

(c) Compounds ending in *-ceras, -dendron, -derma, -doma, -nema, -sperma, -stigma, -stoma,* and other neuter words, are neuter. An exception is made for names ending in *-anthos* (or *-anthus*), *-chilos (-chilus* or *-cheilos),* and *-phykos (-phycos* or *-phycus),* which should be neuter, because that is the gender of the Greek words άνθος, anthos, χείλος, cheilos, and φύκος, phykos, but are treated as masculine in accordance with tradition.

Ex. 7. Even though *Aceras* R. Br. and *Xanthoceras* Bunge were treated as feminine when first published, they are neuter.

Ex. 8. The classical Latin feminine noun *polygala,* which applies to milkworts, was derived from the Greek word with the same meaning, πολύγᾰλον (polygalon), itself a compound of πολύ- (poly-), many, much, and γάλα (gala), milk, a neuter noun. Linnaeus (Sp. Pl.: 701–706. 1753) adopted the classical Latin feminine gender for *Polygala,* which is to be maintained.

Ex. 9. The classical gender both of the Latin *onosma* and the original Greek ὄνοσμα (onosma) is neuter. Linnaeus (Sp. Pl., ed. 2: 196. 1762), in taking up the name for a new genus, treated *Onosma* as feminine. Because the ending *-osma* is listed in Art. 62.2(b) as feminine, *Onosma* maintains its feminine gender.

ⓘ **Note 2.** Art. 14.11 provides for the conservation of a generic name in order to preserve a particular gender.

> **Ex. 10.** As an exception to Art. 62.2, the generic name *Bidens* L., formed from the Latin masculine noun dens (tooth), has been assigned feminine gender by conservation (see App. III).

62.3. Arbitrarily formed generic names or vernacular names or adjectives used as generic names, of which the gender is not apparent, take the gender assigned to them by their authors. If the original author did not indicate the gender, a subsequent author may choose a gender, and the first such choice, if effectively published (Art. 29–31), is to be accepted (see also Rec. F.5A.2).

> **Ex. 11.** *Taonabo* Aubl. (Hist. Pl. Guiane 1: 569. 1775) is feminine because Aublet's two species were *T. dentata* and *T. punctata.*

> **Ex. 12.** *Agati* Adans. (Fam. Pl. 2: 326. 1763) was published without indication of gender; feminine gender was assigned to it by Desvaux (in J. Bot. Agric. 1: 120. 1813), who was the first subsequent author to adopt the name in an effectively published text, and his choice is to be accepted.

> **Ex. 13.** The original gender of *Manihot* Mill. (Gard. Dict. Abr., ed. 4: *Manihot.* 1754), as apparent from some of Miller's phrase names (e.g. *"Manihot spinossimima, folio vitigineo"*), was feminine, and *Manihot* is therefore to be treated as feminine.

> **Ex. 14.** *Ailanthus* Desf. (in Hist. Acad. Roy. Sci. Mém. Math. Phys. (Paris, 4to) 1786 (Mém.): 265. 1788), nom. cons., could appear to be of Greek origin because of the *-anthus* ending and, if so, would be treated as masculine under Art. 62.2(c). The generic name was, however, derived from a vernacular name originating in a language from the Maluku Islands of Indonesia. The gender of *Ailanthus* is therefore feminine as originally assigned by Desfontaines.

62.4. Generic names ending in *-oides,* or *-odes* are treated as feminine and those ending in *-ites* as masculine, regardless of the gender assigned to them by the original author.

Recommendation 62A

62A.1. When a genus is divided into two or more genera, the gender of the new generic name or names should, if there is no obstacle under the rules, be that of the generic name that is retained (see also Rec. 20A.1(h) and 60B).

> **Ex. 1.** When *Boletus* L., nom. cons. (masculine) was divided, the segregated new genera were usually given masculine names, e.g. *Xerocomus* Quél. (in Mougeot & Ferry, Fl. Vosges, Champ.: 477. 1887) and *Boletellus* Murrill (in Mycologia 1: 9. 1909).

CHAPTER F
NAMES OF ORGANISMS TREATED AS FUNGI
(MAASTRICHT VERSION)

This Chapter brings together the provisions of this *Code* that deal solely with names of organisms treated as fungi.

Content in this Chapter may be modified by action of the Fungal Nomenclature Session of an International Mycological Congress (IMC) (see Div. III Prov. 8). The current version of this Chapter, the *Maastricht Chapter F,* embodies the decisions accepted by the 12th IMC in Maastricht, The Netherlands, on 15 August 2024.

Always consult the online version of this *Code* in case of changes resulting from an IMC that takes place before the next International Botanical Congress.

The following changes are introduced in the *Maastricht Chapter F:*

Art. F.2.1 and F.7.1. The procedures for lists of protected and/or rejected names were revised to more clearly set out the steps involved.
Art. F.2 Note 1. When preparing lists of names for protection, included names may be proposed with or without the listing of synonyms.
Art. F.3.5. An earlier homonym of a sanctioned name remains unavailable if the sanctioned name is rejected.
Art. F.5 Note 3. An addition was made to clarify that an identifier is not required when proposing a conserved type.
Rec. F.11A. A new section within Chapter F (Section 7) was added, concerning types that are living cultures, including two Recommendations: Rec. F.11A.1 about depositing ex-type cultures in public collections and Rec. F.11A.2 about utilizing ex-type cultures when selecting neotypes.

Mycologists should note that the content of this *Code* outside of Chapter F pertains to all organisms covered by this *Code,* including fungi, unless expressly limited. This content includes rules about effective publication, valid publication, typification, legitimacy, and priority of names; citation and orthography; and names of hybrids.

Some provisions in the Preamble, Principles, Articles, and Recommendations elsewhere in this *Code,* such as those listed below, while not restricted to fungi, are of particular relevance to mycologists.

The full wording of these and all other relevant provisions of this *Code* should be consulted in all cases.

Pre. 8. The provisions of this *Code* apply to all organisms traditionally treated as fungi, whether fossil or non-fossil, including chytrids, oomycetes, and slime moulds (but excluding *Microsporidia*).

Principle I. This *Code* applies to names of taxonomic groups treated as fungi, whether or not these groups were originally treated as such.

Art. 4 Note 4. In classifying parasites, especially fungi, authors may distinguish within the species special forms (formae speciales) characterized by their adaptation to different hosts, but the nomenclature of special forms is not governed by the provisions of this *Code*.

Art. 8.4 (see also Art. 8 Ex. 12, Rec. 8B, Art. 40 Note 4, and Art. 40.7). Cultures of fungi are acceptable as types if preserved in a metabolically inactive state, and on or after 1 January 2019 this must be stated in the protologue.

Art. 14.15 and Art. 14 Note 4(c)(2). Before 1 January 1954, decisions on conservation of names made by the Special Committee for Fungi, became effective on 20 July 1950 at the VII International Botanical Congress in Stockholm.

Art. 16.3. Automatically typified suprafamilial names of fungi end as follows: *(a)* division or phylum in *-mycota; (b)* subdivision or subphylum in *-mycotina; (c)* class in *-mycetes* and subclass in *-mycetidae*. Automatically typified names not in accordance with these terminations are to be corrected.

Rec. 38E.1. The hosts should be indicated in descriptions or diagnoses of new taxa of parasitic organisms, especially fungi.

Art. 40.6. The type of a name of a new species or infraspecific taxon of non-fossil microfungi may be an effectively published illustration if there are technical difficulties of specimen preservation or if it is impossible to preserve a specimen that would show the features attributed to the taxon by the author of the name (but see Art. 40 Ex. 10, which treats representations of DNA sequences as falling outside of the definition of illustrations in Art. 6.1 footnote).

Art. 41.8(b) (see also Art. 41 Ex. 27). Failure to cite the place of valid publication of a basionym or replaced synonym, when explained by the backward shift of the starting date for some fungi, is a correctable error.

Art. 45.1 (see also Art. 45 Ex. 6 and 7 and Note 1). If a taxon originally assigned to a group not covered by this *Code* is treated as belonging to the algae or fungi, any of its names need satisfy only the requirements of the relevant other *Code* that the author was using for status equivalent to valid publication under this *Code*. Note especially that names of *Microsporidia* are not covered by this *Code* even when *Microsporidia* are considered as fungi.

SECTION 1
LIMITATION OF PRIORITY

ARTICLE F.1
NOMENCLATURAL STARTING-POINT

F.1.1. Valid publication of names for non-fossil fungi (Pre. 8) is treated as beginning at 1 May 1753 (Linnaeus, *Species plantarum*, ed. 1, treated as having been published on that date; see Art. 13.1). For nomenclatural purposes, names given to lichens apply to their fungal component. Names of *Microsporidia* are governed by the *International Code of Zoological Nomenclature* (see Pre. 8).

ⓘ **Note 1.** For fossil fungi, see Art. 13.1(f).

ARTICLE F.2
PROTECTED NAMES

F.2.1. In the interest of nomenclatural stability, for organisms treated as fungi, subcommittees may be established by the Nomenclature Committee for Fungi (see Div. III Prov. 7.2 and 8.13(f)) in consultation with the General Committee and appropriate international bodies for the purpose of preparing lists of names proposed for protection and/or rejection (see Art. F.7.1) for submission to the General Committee (see Div. III Prov. 2.2, 7.10(j), 7.11, and 8.13(a)). Protected names on these lists, which become part of the Appendices of the *Code* (see App. IIA, III, and IV) once reviewed and approved by the Nomenclature Committee for Fungi and the General Committee (see Art. 14.15 and Rec. 14A.1), are to be listed with their types and are treated as conserved against any competing synonyms or homonyms (including sanctioned names), although conservation under Art. 14 overrides this protection. The lists of protected names remain open for revision through the procedures described in this Article.

ⓘ **Note 1.** Names in lists of names proposed for protection may be proposed with or without the listing of synonyms.

ARTICLE F.3
SANCTIONED NAMES

F.3.1. Names in *Uredinales, Ustilaginales,* and *Gasteromycetes* (s. l.) adopted by Persoon (*Synopsis methodica fungorum,* 1801) and names of other fungi (excluding slime moulds) adopted by Fries (*Systema mycologicum,* vol. 1–3. 1821–1832, with additional *Index,* 1832; and *Elenchus fungorum,* vol. 1–2. 1828), are sanctioned.

F.3.2. Names sanctioned are treated as if conserved against earlier homonyms and competing synonyms. Such names, once sanctioned, remain sanctioned even if elsewhere in the sanctioning works the sanctioning author does not recognize them. The spelling used when the name was sanctioned is treated as conserved, except for changes required by Art. 60 and F.9.

> **Ex. 1.** The name *Strigula smaragdula* Fr. (in Linnaea 5: 550. 1830) was accepted by Fries (Syst. Mycol., Index: 184. 1832) and therefore sanctioned. It is treated as if conserved against the competing earlier synonym *Phyllochoris elegans* Fée (Essai Crypt. Écorc.: xciv. 1825), which is the basionym of *Strigula elegans* (Fée) Müll. Arg. (in Linnaea 43: 41. 1880).

> **Ex. 2.** *Agaricus ericetorum* Pers. (Observ. Mycol. 1: 50. 1796) was accepted by Fries (Syst. Mycol. 1: 165. 1821), but later (Elench. Fung. 1: 22. 1828) regarded by him as a synonym of *A. umbelliferus* L. (Sp. Pl.: 1175. 1753), nom. sanct., and not included in his *Index* (Syst. Mycol., Index: 18. 1832) as an accepted name. Nevertheless *A. ericetorum* Pers. is a sanctioned name.

> **Ex. 3.** The spelling used when the name *Merulius lacrimans* (Wulfen) Schumach. was sanctioned (Fries, Syst. Mycol. 1: 328. 1821) is to be maintained, even though the epithet was spelled *'lacrymans'* by Schumacher (Enum. Pl. 2: 371. 1803) and the basionym was originally published as *Boletus 'lacrymans'* Wulfen (in Jacquin, Misc. Austriac. 2: 111. 1781).

F.3.3. A sanctioned name is illegitimate if it is a later homonym of another sanctioned name (see also Art. 53).

F.3.4. An earlier homonym of a sanctioned name is not made illegitimate by that sanctioning but is unavailable for use; if not otherwise illegitimate, it may serve as a basionym of another name or combination based on the same type (see also Art. 55.3).

Ex. 4. *Patellaria* Hoffm. (Descr. Pl. Cl. Crypt. 1: 55. 1789) is an earlier homonym of the sanctioned generic name *Patellaria* Fr. (Syst. Mycol. 2: 158. 1822). Hoffmann's name is legitimate but unavailable for use. *Lecanidion* Endl. (Fl. Poson.: 46. 1830), based on the same type as *Patellaria* Fr., nom. sanct., is illegitimate under Art. 52.1.

Ex. 5. *Antennaria* Gaertn. (Fruct. Sem. Pl. 2: 410. 1791), in order to become available for use, required conservation against the later homonym *Antennaria* Link (in Neues J. Bot. 3(1, 2): 16. 1809), nom. sanct. (Fries, Syst. Mycol. 1: xlvii. 1821).

Ex. 6. *Agaricus cervinus* Schaeff. (Fung. Bavar. Palat. Nasc. 4: 6. 1774) is an earlier homonym of the sanctioned name *A. cervinus* Hoffm. (Nomencl. Fung. 1: t. 2, fig. 2. 1789), nom. sanct. (Fries, Syst. Mycol. 1: 82. 1821); Schaeffer's name is unavailable for use, but it is legitimate and may serve as basionym for combinations in other genera. In *Pluteus* Fr. the combination is cited as *P. cervinus* (Schaeff.) P. Kumm. and has priority over the heterotypic (taxonomic) synonym *P. atricapillus* (Batsch) Fayod, based on *A. atricapillus* Batsch (Elench. Fung. Cont. Prima: 77. 1786).

F.3.5. An earlier homonym of a sanctioned name remains unavailable if the sanctioned name is rejected (under Art. 56 or F.7).

F.3.6. When, for a taxon at a rank from family to genus, inclusive, two or more sanctioned names compete, Art. 11.3 governs the choice of the correct name (see also Art. F.3.8).

F.3.7. When, for a taxon at a rank lower than genus, two or more sanctioned names and/or two or more names with the same final epithet and type as a sanctioned name compete, Art. 11.4 governs the choice of the correct name.

ⓘ *Note 1.* The date of sanctioning does not affect the date of valid publication, and therefore priority (Art. 11), of a sanctioned name. In particular, when two or more homonyms are sanctioned, only the earliest of them may be used because the later one(s) are illegitimate under Art. F.3.3.

Ex. 7. Fries (Syst. Mycol. 1: 41. 1821) accepted and thus sanctioned *Agaricus flavovirens* Pers. (in Hoffmann, Abbild. Schwämme: t. [24]. 1793) and treated *A. equestris* L. (Sp. Pl.: 1173. 1753) as a synonym. He later (Elench. Fung. 1: 6. 1828) accepted *A. equestris,* stating "Nomen prius et aptius certe restituendum [The prior and more apt name is certainly to be restored]". Both names are sanctioned, but, when they are treated as synonyms, *A. equestris* L., nom. sanct., is to be used because it has priority.

F.3.8. A name that neither is sanctioned nor has the same type and final epithet as a sanctioned name at the same rank may not be used for a taxon that includes the type of a sanctioned name at that rank unless the final epithet of the sanctioned name is not available for the required combination (see Art. 11.4(c)).

Ex. 8. The name *Agaricus involutus* Batsch (Elench. Fung. Cont. Prima: 39. 1786) was sanctioned by Fries (Syst. Mycol. 1: 271. 1821) and therefore, when treated in *Paxillus* Fr. with the earlier but non-sanctioned name *A. contiguus* Bull. (Herb. Fr. 5: t. 240. 1785) as a synonym, the correct name is *P. involutus* (Batsch) Fr.

Ex. 9. The name *Polyporus brumalis* (Pers.) Fr. (Observ. Mycol. 2: 255. 1818), nom. sanct. (Fries, Syst. Mycol. 1: 348. 1821), based on *Boletus brumalis* Pers. (in Neues Mag. Bot. 1: 107. 1794), was treated by Zmitrovich & Kovalenko (in Int. J. Med. Mushr. 18: 23–38, suppl. 2: [2]. 2015) as synonymous with *B. hypocrateriformis* Schrank (Baier. Fl. 2: 621. 1789) and placed in *Lentinus* Fr., nom. sanct., in which the correct name is *L. brumalis* (Pers.) Zmitr. (in Int. J. Med. Mushr. 12: 88. 2010).

F.3.9. Conservation (Art. 14), protection (Art. F.2), and explicit rejection (Art. 56 and F.7) override sanctioning.

F.3.10. The type of a name of a species or infraspecific taxon adopted in one of the works specified in Art. F.3.1, and thereby sanctioned, may be selected from among the elements associated with the name in the protologue and/or the sanctioning treatment.

ⓘ **Note 2.** For names falling under Art. F.3.10, elements from the context of the protologue are original material and those from the context of the sanctioning work are considered as equivalent to original material.

Ex. 10. When Stadler & al. (in IMA Fungus 5: 61. 2014) designated the lectotype of *Clavaria hypoxylon* L. (Sp. Pl.: 1182. 1753), sanctioned by Fries (Syst. Mycol. 2: 327. 1823) as *Sphaeria hypoxylon* (L.) Pers. (Observ. Mycol. 1: 20. 1796), they selected a specimen in K distributed by Fries (Scler. Suec. No. 181) and cited by him in the sanctioning treatment rather than any of the elements associated with the protologue.

Ex. 11. In the absence of any specimens or illustrations from the context of the protologue that are original material, Peterson (in Amer. J. Bot. 63: 313. 1976) designated a specimen in L as the neotype of *Clavaria formosa* Pers. (Comm. Fung. Clav.: 41. 1797), nom. sanct. However, when sanctioning *C. formosa,* Fries (Syst. Mycol. 1: 466. 1821) cited several illustrations, which are therefore considered as equivalent to original material. Peterson's neotypification was not therefore designated in conformity with Art. 9.13 and is not to be followed (Art. 9.19). Instead, Franchi & Marchetti (in Riv. Micol. 59: 323. 2017) designated as the lectotype of *C. formosa* one of the illustrations (Persoon, Icon. Desc. Fung. Min. Cognit. 1: t. III, fig. 6. 1798) that was cited by Fries (l.c., as "f. 5").

F.3.11. When a sanctioning author accepted an earlier name but did not include, even implicitly, any element associated with its protologue, or when the protologue did not include the subsequently designated type of the sanctioned name, the sanctioning author is considered to have created a later homonym, which is treated as if conserved (see also Art. 48).

ⓘ **Note 3.** For typification of sanctioned generic names, see Art. 10.2. Note that automatic typification under Art. 7.5 does not apply to sanctioned names. For legitimacy of sanctioned names (or names based on them), see also Art. 6.4, 52.1, 53.1, and 55.3.

Recommendation F.3A

F.3A.1. When it is considered useful to indicate the nomenclatural status of a sanctioned name (Art. F.3.1), the abbreviation "nom. sanct." (nomen sanctionatum) should be added in a formal citation; the place of sanctioning should also be added in full nomenclatural citations.

> **Ex. 1.** *Boletus piperatus* Bull. (Herb. France: t. 451, fig. 2. 1790) was adopted in Fries (Syst. Mycol. 1: 388. 1821) and was thereby sanctioned. Depending on the level of nomenclatural information being presented, it should be cited as *B. piperatus* Bull., nom. sanct.; or *B. piperatus* Bull. 1790, nom. sanct.; or *B. piperatus* Bull., Herb. France: t. 451, fig. 2. 1790, nom. sanct.; or *B. piperatus* Bull., Herb. France: t. 451, fig. 2. 1790, nom. sanct. (Fries, Syst. Mycol. 1: 388. 1821).

> **Ex. 2.** *Agaricus compactus* [unranked] *sarcocephalus* (Fr.) Fr. was sanctioned when adopted by Fries (Syst. Mycol. 1: 290. 1821). That status should be indicated by citing it as *A. compactus* [unranked] *sarcocephalus* (Fr.) Fr., nom. sanct. The abbreviation "nom. sanct." should not be added when citing its basionym *A. sarcocephalus* Fr. (Observ. Mycol. 1: 51. 1815) or when citing subsequent combinations such as *Psathyrella sarcocephala* (Fr.) Singer (in Lilloa 22: 468. 1949).

SECTION 2
VALID PUBLICATION AND TYPIFICATION OF NAMES

ARTICLE F.4
MISPLACED RANK-DENOTING TERMS

F.4.1. A name is not validly published if it is given to a taxon of which the rank is at the same time, contrary to Art. 5, denoted by a misplaced term (Art. 37.7), but an exception is made for names of the subdivisions of genera termed tribes (tribus) in Fries's *Systema* *mycologicum,* which are treated as validly published names of un-ranked subdivisions of genera.

Ex. 1. *Agaricus* "tribus" [unranked] *Pholiota* Fr. (Syst. Mycol. 1: 240. 1821), sanctioned in the same work, is the validly published basionym of the generic name *Pholiota* (Fr.) P. Kumm. (Führer Pilzk.: 22. 1871) (see Art. 41 Ex. 9).

ARTICLE F.5
REGISTRATION OF NAMES AND NOMENCLATURAL ACTS

F.5.1. The Nomenclature Committee for Fungi (see Div. III Prov. 7) has the power to:

(a) Appoint one or more localized or decentralized, open and accessible, electronic repositories (recognized repositories) to accession the information required by Art. F.5.3 and F.5.5 and issue the identifiers required by Art. F.5.2 and F.5.4.
(b) Cancel such appointment at its discretion.
(c) Set aside the requirements of Art. F.5.2, F.5.3, F.5.4, and F.5.5, should the repository mechanism, or essential parts thereof, cease to function.

Decisions made by this Committee under these powers are subject to ratification by a later International Mycological Congress.

F.5.2. To be validly published, nomenclatural novelties (Art. 6 Note 4) applied to organisms treated as fungi under this *Code* (Pre. 8; including fossil fungi and lichen-forming fungi) and published on or after 1 January 2013 must, in the protologue, include citation of the identifier issued for the name by a recognized repository (Art. F.5.1).

Ex. 1. The protologue of *Albugo arenosa* Mirzaee & Thines (in Mycol. Prog. 12: 50. 2013) complies with Art. F.5.2 because it includes citation of "MB 564515", an identifier issued by MycoBank, one of three recognized repositories. The decision by the Nomenclature Committee for Fungi to appoint (Art. F.5.1) Fungal Names, Index Fungorum, and MycoBank as repositories (Redhead & Norvell in Taxon 62: 173–174. 2013) was ratified (Art. F.5.1) by the 10th International Mycological Congress (May in Taxon 66: 484. 2017).

Ex. 2. The designation *"Austropleospora archidendri"* (Ariyawansa & al. in Fungal Diversity 75: 64. 2015) is not a validly published new combination based on *Paraconiothyrium archidendri* Verkley & al. (in Persoonia 32: 37. 2014) because it was published without citing an identifier issued by a recognized repository, even though the recognized repository Index Fungorum had previously issued the identifier "IF 551419" for the intended new combination.

Ex. 3. The designation *"Priceomyces fermenticarens"* (Gouliamova & al. in Persoonia 36: 429. 2016), intended as a new combination, was published with the identifier "MB 310255", which refers to the identifier "IF 310255" that had been assigned to the intended basionym, *Candida fermenticarens* Van der Walt & P. B. Baker (in Bothalia 12: 561. 1978) by Index Fungorum before registration became mandatory. The recognized repository MycoBank assigned the identifier "MB 818676" for the intended new combination after its publication, but because no identifier was issued before its publication the intended combination was not validly published. *Priceomyces fermenticarens* (Van der Walt & P. B. Baker) Gouliam. & al. (in Persoonia 39: 289. 2017) was subsequently validly published with citation of the identifier "MB 818692", newly issued by MycoBank.

F.5.3. For an identifier to be issued by a recognized repository as required by Art. F.5.2, the minimum elements of information that must be accessioned by author(s) of scientific names are the proposed name itself and those elements required for valid publication under Art. 38.1(a) and 39.2 (validating description or diagnosis) and Art. 40.1 and 40.5 (type) or Art. 41.5 (reference to the basionym or replaced synonym). When the accessioned and subsequently published information for a name with a given identifier differ, the published information is considered as definitive.[1]

ⓘ **Note 1.** Issuance of an identifier by a recognized repository presumes subsequent fulfilment of the requirements for valid publication of the name (Art. 32–45, F.5.2, and F.5.3) but does not by itself constitute or guarantee valid publication.

ⓘ **Note 2.** The words "name" and "names" are used in Art. F.5.2 and F.5.3 for names that may not yet be validly published, in which case the definition in Art. 6.3 does not apply.

F.5.4. For purposes of priority (Art. 9.19, 9.20, and 10.5), designation of a type, on or after 1 January 2019, of the name of an organism treated as a fungus under this *Code* (Pre. 8), is achieved only if an identifier issued for the type designation by a recognized repository (Art. F.5.1) is cited.

ⓘ **Note 3.** Art. F.5.4 applies only to the designation of lectotypes and neotypes (and their equivalents under Art. 10.2) and epitypes; it does not apply to the designation of a holotype when publishing the name of a new taxon, for which see Art. F.5.3, nor does it apply to proposing a conserved type when publishing a proposal to conserve a name (Art. 14.9).

---

1   It is the practice of repositories to assign a new identifier when an orthographical correction is made to a name subsequent to the protologue.

F.5.5. For an identifier to be issued by a recognized repository as required by Art. F.5.4, the minimum elements of information that must be accessioned by author(s) of type designations are the name being typified, the author designating the type, and those elements required by Art. 9.21, 9.22, and 9.23.

Note 4. Issuance of an identifier by a recognized repository presumes subsequent fulfilment of the requirements for effective type designation (Art. 7.8–7.11 and F.5.4) but does not by itself constitute or guarantee a type designation.

F.5.6. When the identifier issued for a name by a recognized repository is cited incorrectly in the protologue, this is treated as a correctable error not preventing valid publication of the name, provided that the identifier was issued prior to the protologue.

> *Ex. 4.* The identifier "MB 564220" was issued by MycoBank for *Cortinarius peristeris* Soop (in Bresadoliana 1: 22. 2013) before publication of the name. Even though the identifier was incorrectly cited as "MB 564" in the protologue, the name is validly published.

> *Ex. 5.* The identifier "MB 805372" was issued by MycoBank on 22 August 2013 for *Leucoagaricus vindobonensis* (Tratt.) L. A. Parra (Fungi Eur. 1A: 721. 30 Oct 2013) before publication of the name. Even though the identifier was incorrectly cited as "MB 807352" in the protologue, the name is validly published.

F.5.7. An identifier remains associated with the name or designation for which it was issued. If, when published, a designation for which an identifier has been issued does not meet other requirements for valid publication, in order for that designation to become a validly published name, a new identifier must be obtained.

> *Ex. 6.* The designation *"Nigelia"* (Luangsa-ard & al. in Mycol. Progr. 16: 378. 2017) was published without citation of an identifier. MycoBank assigned the identifier "MB 823565" for this designation after publication. The designation was later validated as *Nigelia* Luangsa-ard & al. (in Index Fungorum 345: 1. 2017) with citation of the identifier "IF 553229" newly issued by Index Fungorum.

F.5.8. When the identifier issued for a type designation by a recognized repository is cited incorrectly in the typifying publication, this is treated as a correctable error not preventing designation of the type, provided that the identifier was issued prior to the typifying publication.

Recommendation F.5A

F.5A.1. Authors of names of organisms treated as fungi are encouraged to:

(a) Deposit the required elements of information for any nomenclatural novelty in a recognized repository as soon as possible after a work is accepted for publication, so as to obtain identifiers for each nomenclatural novelty.

(b) Inform the recognized repository that issued the identifier of the complete bibliographic details upon publication of the name, including volume and part number, page number, date of publication, and (for books) the publisher and place of publication.

(c) Upon publication of a name, supply an electronic version of the publication to the recognized repository that issued the identifier associated with the name.

F.5A.2. In addition to meeting the requirements for effective publication of choices of name (Art. 11.5 and 53.5), orthography (Art. 61.3), or gender (Art. 62.3), those publishing such choices for names of organisms treated as fungi are encouraged to record the choice in a recognized repository (Art. F.5.1) and cite the identifier in the place of publication.

SECTION 3
REJECTION OF NAMES

ARTICLE F.6
HOMONYMS

F.6.1. The name of a taxon treated as a fungus published on or after 1 January 2019 is illegitimate if it is a later homonym of a prokaryotic or protozoan name (see also Art. 54 and Rec. 54A).

ARTICLE F.7
REJECTED NAMES

F.7.1. Rejected names on lists prepared by the subcommittees defined in Art. F.2.1, which become part of the Appendices of the *Code* (see App. V) once reviewed and approved by the Nomenclature Committee for Fungi and the General Committee (see Art. 56.3 and Rec. 56A.1), are to be treated as rejected under Art. 56.1, except that they may become eligible for use by conservation under Art. 14.

SECTION 4
NAMES OF FUNGI WITH A PLEOMORPHIC LIFE CYCLE

ARTICLE F.8

F.8.1. A name published before 1 January 2013 for a taxon of non-lichen-forming *Ascomycota* and *Basidiomycota,* with the intent or implied intent of applying to or being typified by one particular morph (e.g. anamorph or teleomorph; see Art. F.8 Note 2), may be legitimate even if it otherwise would be illegitimate under Art. 52 on account of the protologue including a type (as defined in Art. 52.2) referable to a different morph. If the name is otherwise legitimate, it competes for priority (Art. 11.3 and 11.4).

> *Ex. 1.* *Penicillium brefeldianum* B. O. Dodge (in Mycologia 25: 92. 1933) was described and based on a type with both the anamorph and teleomorph (and therefore necessarily typified by the teleomorph element alone under editions of the *Code* prior to the *Melbourne Code* of 2012). The combination *Eupenicillium brefeldianum* (B. O. Dodge) Stolk & D. B. Scott (in Persoonia 4: 400. 1967) for the teleomorph is legitimate. *Penicillium dodgei* Pitt (Gen. Penicillium: 117. 1980), typified by the anamorph in a dried culture "derived from Dodge's type", did not include the teleomorphic type of *P. brefeldianum* and therefore it too is legitimate. However, when considered to be a species of *Penicillium*, the correct name for all its states is *P. brefeldianum.*

ⓘ *Note 1.* Except as provided in Art. F.8.1, names of fungi with mitotic asexual morphs (anamorphs) as well as a meiotic sexual morph (teleomorph) must conform to the same provisions of this *Code* as all other fungi.

ⓘ *Note 2.* Editions of the *Code* prior to the *Melbourne Code* of 2012 provided for separate names for mitotic asexual morphs (anamorphs) of certain pleomorphic fungi and required that the name applicable to the whole fungus be typified by a meiotic sexual morph (teleomorph). Under the current *Code,* however, all legitimate fungal names are treated equally for the purpose of establishing priority, regardless of the morph of the type (see also Art. F.2.1).

> *Ex. 2.* *Mycosphaerella aleuritis* (I. Miyake) S. H. Ou (in Sinensia 11: 183. 1940, *'Aleuritidis'*), when published as a new combination, was accompanied by a Latin diagnosis of the newly discovered teleomorph corresponding to the anamorph on which the basionym *Cercospora aleuritis* I. Miyake (in Bot. Mag. (Tokyo) 26: 66. 1912, *'Aleuritidis'*) was typified. Under editions of the *Code* before the *Melbourne Code* of 2012, *M. aleuritis* was considered to be the name of a new species with a teleomorph type, dating from 1940, with authorship attributed solely to Ou. Under the current *Code,* the name is cited as originally published, *M. aleuritis* (I. Miyake) S. H. Ou, and is typified by the type of the basionym.

Ex. 3. In the protologue of the teleomorph-typified *Venturia acerina* Plakidas ex M. E. Barr (in Canad. J. Bot. 46: 814. 1968) the anamorph-typified *Cladosporium humile* Davis (in Trans. Wisconsin Acad. Sci. 19: 702. 1919) was included as a synonym. Because it was published before 1 January 2013, the name *V. acerina* is not illegitimate, but *C. humile* is the earliest legitimate name at the rank of species.

🛈 **Note 3.** Names proposed simultaneously for separate morphs (e.g. anamorph and teleomorph) of a taxon of non-lichen-forming *Ascomycota* and *Basidiomycota* are necessarily heterotypic and are not therefore alternative names as defined by Art. 36.3.

Ex. 4. *Hypocrea dorotheae* Samuels & Dodd and *Trichoderma dorotheae* Samuels & Dodd were simultaneously validly published (in Stud. Mycol. 56: 112. 2006) for what the authors considered to be a single species with *Samuels & Dodd 8657* (PDD 83839) as the holotype. Because these names were published before 1 January 2013 (see Art. F.8.1 and Art. F.8 Note 2), and because the authors explicitly indicated that the name *T. dorotheae* was typified by the anamorphic element of PDD 83839, both names are validly published and legitimate. They are not alternative names as defined in Art. 36.3.

SECTION 5
ORTHOGRAPHY OF NAMES

ARTICLE F.9

F.9.1. Epithets of fungal names derived from the generic name of an associated organism are to be spelled in accordance with the accepted spelling of the name of that organism; other spellings are regarded as orthographical variants to be corrected (see Art. 61).

Ex. 1. *Phyllachora 'anonicola'* Chardón (in Mycologia 32: 190. 1940) is to be corrected to *P. annonicola* in accordance with the accepted spelling of *Annona* L.; *Meliola 'albizziae'* Hansf. & Deighton (in Mycol. Pap. 23: 26. 1948) is to be corrected to *M. albiziae* in accordance with the accepted spelling of *Albizia* Durazz.

Ex. 2. *Dimeromyces 'corynitis'* Thaxter (in Proc. Amer. Acad. Arts 48: 157. 1912) was stated to occur "On the elytra of *Corynites ruficollis* Fabr.", but the name of the host, a species of beetle, is correctly spelled *Corynetes ruficollis*. The fungal name is therefore to be spelled *D. corynetis*.

Ex. 3. *Tricholomopsis 'pteridicola'* Olariaga & al. (in Mycol. Progr. 14(4/21): 6. 2015) was stated to occur in association with *Pteridium aquilinum.* Therefore, the name is to be corrected to *T. pteridiicola* because the genitive singular of *Pteridium* is *Pteridii* (in contrast to that of *Pteris,* which is *Pteridis;* see Art. 60.11).

SECTION 6
AUTHOR CITATIONS

ARTICLE F.10

F.10.1. For names of organisms treated as fungi, the identifier issued for the name by a recognized repository (Art. F.5.2) may be used subsequent to the protologue in place of an author citation for the name but not to replace the name itself (see also Art. 22.1 and 26.1).

Recommendation F.10A

F.10A.1. An identifier used in place of an author citation as permitted by Art. F.10.1 should be presented with the symbol # preceding the numerical part of the identifier, and the resulting string should be enclosed in square brackets. In electronic publications, this string should be provided with a direct and stable link to the corresponding record in one of the recognized repositories.

> **Ex. 1.** *Astrothelium meristosporoides* [#816706]. The direct and stable link to a record in a recognized repository would be either https://www.mycobank.org/MB/816706 or https://www.indexfungorum.org/Names/NamesRecord.asp?RecordID=816706.

SECTION 7
RECOMMENDATIONS ON TYPES THAT ARE LIVING CULTURES

Recommendation F.11A

F.11A.1. When applying Rec. 8B.1 to the name of a fungus, the collections should be public (see also Rec. 7A.1).

F.11A.2. When the type of a name of a fungus has been lost or destroyed and was a living culture (preserved in a metabolically inactive state), a neotype (if permissible, see Art. 9.8) selected to replace it should be the oldest progeny (preserved in a metabolically inactive state) of an ex-type culture (Rec. 8B.2).

CHAPTER H
NAMES OF HYBRIDS

ARTICLE H.1
INDICATION OF HYBRIDS

H.1.1. Hybridity is indicated by use of the multiplication sign × or by addition of the prefix "notho-"[1] to the term denoting the rank of the taxon.

ARTICLE H.2
HYBRID FORMULAE

H.2.1. A hybrid between named taxa may be indicated by placing the multiplication sign × between the names of the taxa; the whole expression is then called a hybrid formula.

Ex. 1. Agrostis L. × *Polypogon* Desf.; *Agrostis stolonifera* L. × *Polypogon monspeliensis* (L.) Desf.; *Melampsora medusae* Thüm. × *M. occidentalis* H. S. Jacks.; *Mentha aquatica* L. × *M. arvensis* L. × *M. spicata* L.; *Polypodium vulgare* subsp. *prionodes* (Asch.) Rothm. × *P. vulgare* L. subsp. *vulgare; Salix aurita* L. × *S. caprea* L.; *Tilletia caries* (DC.) Tul. & C. Tul. × *T. foetida* (Wallr.) Liro.

Ex. 2. Kunzea linearis (Kirk) de Lange × *Kunzea robusta* de Lange & Toelken or *Kunzea linearis* (Kirk) de Lange × *K. robusta* de Lange & Toelken, but not "*Kunzea linearis* (Kirk) de Lange × *robusta* de Lange & Toelken", which omits the generic name or its abbreviation from the second species name contrary to Art. 23.1.

Recommendation H.2A

H.2A.1. It is usually preferable to place the names or epithets in a formula in alphabetical order. The direction of a cross may be indicated by including the gender-denoting symbols (♀: female; ♂: male) in the formula, or by placing the female parent first. If a non-alphabetical sequence is used, its basis should be clearly indicated.

---

1   From the Greek νόθος, nothos, meaning hybrid.

ARTICLE H.3
NAMES OF NOTHOTAXA

H.3.1. Hybrids between representatives of two or more taxa may receive a name. For nomenclatural purposes, the hybrid nature of a taxon is indicated by placing the multiplication sign × before the name of an intergeneric hybrid or before the epithet in the name of an interspecific hybrid, or by prefixing the term "notho-" (optionally abbreviated "n-") to the term denoting the rank of the taxon (see Art. 3.2 and 4.4). All such taxa are designated nothotaxa.

> *Ex. 1.* ×*Agropogon* P. Fourn. (Quatre Fl. France: 50. 1934); ×*Agropogon littoralis* (Sm.) C. E. Hubb. (in J. Ecol. 33: 333. 1946); *Melampsora* ×*columbiana* G. Newc. (in Mycol. Res. 104: 271. 2000); *Mentha* ×*smithiana* R. A. Graham (in Watsonia 1: 89. 1949); *Polypodium vulgare* nothosubsp. (or nsubsp.) *mantoniae* (Rothm.) Schidlay (in Futák, Fl. Slov. 2: 225. 1966); *Salix* ×*capreola* Andersson (in Kongl. Svenska Vetensk. Acad. Handl., n.s., 6(1): 71. 1867). (The putative or known parentage of these nothotaxa is found in Art. H.2 Ex. 1.)

H.3.2. A nothotaxon cannot be designated unless at least one parental taxon is known or can be postulated.

H.3.3. For purposes of homonymy and synonymy the multiplication sign × and the prefix "notho-" are disregarded.

> *Ex. 2.* ×*Hordelymus* Bachteev & Darevsk. (in Bot. Zhurn. (Moscow & Leningrad) 35: 191. 1950) (*Elymus* L. × *Hordeum* L.) is a later homonym of *Hordelymus* (Jess.) Harz (Landw. Samenk.: 1147. 1885).

ⓘ **Note 1.** Taxa that are believed to be of hybrid origin need not be designated as nothotaxa.

> *Ex. 3.* The true-breeding tetraploid raised from the artificial cross *Digitalis grandiflora* L. × *D. purpurea* L. may, if desired, be referred to as *D. mertonensis* B. H. Buxton & C. D. Darl. (in Nature 77: 94. 1931); *Triticum aestivum* L. (Sp. Pl.: 85. 1753), which provides the type of *Triticum* L., is treated as a species although it is not found in nature and its genome has been shown to be composed of those of several wild species; the taxon known as *Phlox divaricata* subsp. *laphamii* (A. W. Wood) Wherry (in Morris Arbor. Monogr. 3: 41. 1955) was believed by Levin (in Evolution 21: 92–108. 1967) to be a stabilized product of hybridization between *P. divaricata* L. subsp. *divaricata* and *P. pilosa* subsp. *ozarkana* Wherry; *Rosa canina* L. (l.c.: 492. 1753), a polyploid believed to be of ancient hybrid origin, is treated as a species.

Recommendation H.3A

H.3A.1. In named hybrids, the multiplication sign × belongs with the name or epithet but is not actually part of it, and its placement should reflect that relation. The exact amount of space, if any, between the multiplication sign and the initial letter of the name or epithet should depend on what best serves readability.

ⓘ ***Note 1.*** The multiplication sign × in a hybrid formula is always placed between, and separate from, the names of the parents.

H.3A.2. If the multiplication sign × is not available it should be approximated by the lower-case letter "x" (not italicized).

ARTICLE H.4
CIRCUMSCRIPTION OF NOTHOTAXA

H.4.1. When all the parental taxa can be postulated or are known, a nothotaxon is circumscribed to include all individuals recognizably derived from the crossing of representatives of the stated parental taxa (i.e. not only the F_1 but subsequent filial generations and also back-crosses and combinations of these). There can thus be only one correct name corresponding to a particular hybrid formula; this is the earliest legitimate name (Art. 6.5) at the appropriate rank (Art. H.5), and other names corresponding to the same hybrid formula are synonyms of it (but see Art. 52 Note 4).

> ***Ex. 1.*** The names *Oenothera* ×*drawertii* Renner ex Rostański (in Acta Bot. Acad. Sci. Hung. 12: 341. 1966) and *O.* ×*wienii* Renner ex Rostański (in Fragm. Florist. Geobot. 23: 289. 1977) are both considered to apply to the hybrid *O. biennis* L. × *O. villosa* Thunb.; the types of the two nothospecific names are known to differ by a whole gene complex; nevertheless, the earlier name is the correct name, and the later name is treated as a synonym of it.

ⓘ ***Note 1.*** Variation within nothospecies and infraspecific nothotaxa may be treated according to Art. H.12 or, if appropriate, according to the *International Code of Nomenclature for Cultivated Plants*.

ARTICLE H.5
RANKS OF NOTHOTAXA

H.5.1. The appropriate rank of a nothotaxon is that of the postulated or known parental taxa.

H.5.2. If the postulated or known parental taxa are at unequal ranks, the appropriate rank of the nothotaxon is the lowest of these ranks, unless the nothotaxon is the only one known for hybrids between the species to which the parental taxa of the nothotaxon belong.

i **Note 1.** When a nothotaxon is designated by a name at a rank inappropriate to its hybrid formula, the name is incorrect in relation to that hybrid formula but may nevertheless be correct or may become correct later (see also Art. 52 Note 4).

> ***Ex. 1.*** The combination *Elymus* ×*laxus* (Fr.) Melderis & D. C. McClint. (in Watsonia 14: 394. 1983), based on *Triticum laxum* Fr. (Novit. Fl. Suec. Mant. 3: 13. 1842), was published for hybrids with the formula *E. farctus* subsp. *boreoatlanticus* (Simonet & Guin.) Melderis × *E. repens* (L.) Gould, so that the combination is at a rank inappropriate to the hybrid formula. It is, however, the correct name applicable to all hybrids between *E. farctus* (Viv.) Melderis and *E. repens*.

> ***Ex. 2.*** Radcliffe-Smith published the nothospecific name *Euphorbia* ×*cornubiensis* Radcl.-Sm. (in Kew Bull. 40: 445. 1985) for *E. amygdaloides* L. × *E. characias* subsp. *wulfenii* (W. D. J. Koch) Radcl.-Sm., but the correct nothospecific name for all hybrids between *E. amygdaloides* and *E. characias* L. is *E.* ×*martini* Rouy (Ill. Pl. Eur. Rar.: 107. 1900); later, Radcliffe-Smith published the appropriate combination *E.* ×*martini* nothosubsp. *cornubiensis* (Radcl.-Sm.) Radcl.-Sm. (in Taxon 35: 349. 1986). However, the name *E.* ×*cornubiensis* is potentially correct for hybrids with the formula *E. amygdaloides* × *E. wulfenii* W. D. J. Koch.

> ***Ex. 3.*** *Aloe* ×*engelbrechtii* Gideon F. Sm. & Figueiredo (in Phytotaxa 464: 253. 2020) was published for the nothospecies with *A. arborescens* Mill. and *A. hardyi* Glen as parents. The rank of nothospecies is appropriate because both parents are species. The appropriate rank for the name of a hybrid between *A. arborescens* subsp. *mzimnyati* van Jaarsv. & A. E. van Wyk and *A. hardyi* would be nothosubspecies. Valid publication of the name of this nothosubspecies under *A.* ×*engelbrechtii* would establish an autonym that would apply to *A. arborescens* subsp. *arborescens* × *A. hardyi*.

Recommendation H.5A

H.5A.1. When publishing a name of a new nothotaxon at the rank of species or below, authors should provide any available information on the taxonomic identity, at lower ranks, of the known or postulated parents of the type of the name.

ARTICLE H.6
NOTHOGENERIC NAMES AND CONDENSED FORMULAE

H.6.1. A nothogeneric name (i.e. the name at generic rank for a hybrid between representatives of two or more genera) is a condensed formula or is equivalent to a condensed formula (but see Art. 11.9 and 54.1(c)).

ℹ️ **Note 1.** The provisions for nothogeneric names do not apply to graft hybrids, which are dealt with in the *International Code of Nomenclature for Cultivated Plants* (but see Art. 54.1(c)).

H.6.2. The nothogeneric name of a bigeneric hybrid is a condensed formula in which the names adopted for the parental genera are combined into a single word, using the first part or the whole of one, the last part or the whole of the other (but not the whole of both) and, optionally, a connecting vowel.

Ex. 1. ×*Agropogon* P. Fourn. (Quatre Fl. France: 50. 1934) (*Agrostis* L. × *Polypogon* Desf.); ×*Gymnanacamptis* Asch. & Graebn. (Syn. Mitteleur. Fl. 3: 854. 1907) (*Anacamptis* Rich. × *Gymnadenia* R. Br.); ×*Cupressocyparis* Dallim. (Hand-List Conif., Roy. Bot. Gard., Kew, ed. 4: 37. 1938) (*Chamaecyparis* Spach × *Cupressus* L.); ×*Seleniphyllum* G. D. Rowley (in Backeberg, Cactaceae 6: 3557. 1962) (*Epiphyllum* Haw. × *Selenicereus* (A. Berger) Britton & Rose).

Ex. 2. ×*Amarcrinum* Coutts (in Gard. Chron., ser. 3, 78: 411. 1925) is correct for *Amaryllis* L. × *Crinum* L., not "×*Crindonna*". The latter formula was proposed by Ragionieri (in Gard. Chron., ser. 3, 69: 32. 1921) for the same nothogenus, but was formed from the generic name adopted for one parent *(Crinum)* and a synonym *(Belladonna* Sweet) of the generic name adopted for the other *(Amaryllis)*. Because it is contrary to Art. H.6, it is not validly published under Art. 32.1(c).

Ex. 3. The name ×*Leucadenia* Schltr. (in Repert. Spec. Nov. Regni Veg. 16: 290. 1919) is correct for *Leucorchis* E. Mey. × *Gymnadenia* R. Br., but if the generic name *Pseudorchis* Ség. is adopted instead of *Leucorchis,* ×*Pseudadenia* P. F. Hunt (in Orchid Rev. 79: 141. 1971) is correct.

Ex. 4. Boivin (in Naturaliste Canad. 94: 526. 1967) published ×*Maltea* for what he considered to be the intergeneric hybrid *Phippsia* (Trin.) R. Br. × *Puccinellia* Parl. Because this is not a condensed formula, the name cannot be used for that intergeneric hybrid, for which the correct name is ×*Pucciphippsia* Tzvelev (in Novosti Sist. Vyssh. Rast. 8: 76. 1971). *Maltea* B. Boivin is nevertheless a validly published generic name, because Boivin provided a Latin description and designated a type, and may be correct if its type is not treated as belonging to a nothogenus.

H.6.3. The nothogeneric name of an intergeneric hybrid derived from four or more genera is formed from a personal name to which is added the termination *-ara,* except when the personal name already ends with *-a* in which case the termination *-ra* is added; no such name may exceed eight syllables. Such a name is equivalent to a condensed formula.

Ex. 5. ×*Beallara* Moir (in Orchid Rev. 78(929): New Orch. Hybr. [1, 3]. 1970) commemorating J. Ferguson Beall (*Brassia* R. Br. × *Cochlioda* Lindl. × *Miltonia* Lindl. × *Odontoglossum* Kunth); ×*Cogniauxara* Garay & H. R. Sweet (see Art. H.8 Ex. 3) commemorating Célestin A. Cogniaux (*Arachnis* Blume × *Euanthe* Schltr. × *Renanthera* Lour. × *Vanda* W. Jones ex R. Br.); ×*Derosara* Hort. (in Orchid. Rev. 104(1209): 166. 1996, *'Derosaara'*) commemorating Victor De Rosa (*Aspasia* Lindl. × *Odontoglossum* Kunth × *Miltonia* Lindl. × *Brassia* R. Br.); ×*Hayatara* J. M. H. Shaw (in Sander's List Orchid Hybrids Addendum 2002–2004: xxxv. 2005, *'Hayataara'*) commemorating Bunzô Hayata (*Brassavola* R. Br. × *Cattleya* Lindl. × *Laelia* Lindl. × *Myrmecophila* Rolfe × *Pseudolaelia* Porto & Brade).

H.6.4. The nothogeneric name of a trigeneric hybrid is either:

(a) a condensed formula in which the three names adopted for the parental genera are combined into a single word not exceeding eight syllables, using the whole or first part of one, followed by the whole or any part of another, followed by the whole or last part of the third (but not the whole of all three) and, optionally, one or two connecting vowels; or

(b) a name formed like that of a nothogenus derived from four or more genera, i.e. from a personal name to which is added the termination *-ara,* except when the personal name already ends with *-a* in which case the termination *-ra* is added.

Ex. 6. (a) ×*Sophrolaeliocattleya* Hurst (in J. Roy. Hort. Soc. 21: 468. 1898) (*Cattleya* Lindl. × *Laelia* Lindl. × *Sophronitis* Lindl.); ×*Rodrettiopsis* Moir (in Orchid Rev. 84: ix. 1976) (*Comparettia* Poepp. & Endl. × *Ionopsis* Kunth × *Rodriguezia* Ruiz & Pav.).

Ex. 7. (b) ×*Holttumara* Holttum (see Art. H.8 Ex. 3) commemorating Richard E. Holttum (*Arachnis* Blume × *Renanthera* Lour. × *Vanda* W. Jones ex R. Br.); ×*Kagawara* Kagawa & Wreford (Orchid Rev. 76: New Orch. Hybr. [2, 4]. 1968) commemorating Hiroshi Kagawa (*Ascocentrum* J. J. Sm. × *Renanthera* × *Vanda*).

ⓘ *Note 2.* The termination *-ara* does not necessarily indicate the name of a nothogenus derived from three or more genera.

Ex. 8. Kumara Medik. (in Theodora 69: t. 4. 1786) is not a nothogeneric name. ×*Gonimara* Gideon F. Sm. & Molteno (in Bradleya 36: 54. 2018) was published for bigeneric hybrids between *Gonialoe* (Baker) Boatwr. & J. C. Manning and *Kumara* Medik.

H.6.5. The use of a hyphen instead of or in addition to a connecting vowel in a nothogeneric name that is a condensed formula is treated as an error to be corrected by deletion of the hyphen(s) (but see Art. 20.3 for non-hybrid generic names; see also Art. 60.13 for names of fossil-genera).

Ex. 9. The nothogeneric name ×*Anthematricaria* Asch. (in Ber. Deutsch. Bot. Ges. 9: (99). 1892), proposed for bigeneric hybrids with the parentage *Anthemis* L. × *Matricaria* L., was originally published as *'Anthe-Matricaria'*; ×*Brassocattleya* Rolfe (in Gard. Chron., ser. 3, 5: 438. 1889), proposed for bigeneric hybrids with the parentage *Brassavola* R. Br. × *Cattleya* Lindl., was originally published as *'Brasso-Cattleya'*; ×*Brassolaeliacattleya* J. G. Fowler (in Gard. Chron., ser. 3, 41: 290. 1907), proposed for trigeneric hybrids with the parentage *Brassavola* × *Cattleya* × *Laelia* Lindl., was originally published as *'Brasso-Laelia-Cattleya'*; ×*Sophrolaeliocattleya* Hurst (in J. Roy. Hort. Soc. 21: 468. 1898), proposed for trigeneric hybrids with the parentage *Cattleya* × *Laelia* × *Sophronitis* Lindl., was originally published as *'Sophro-Laelio-Cattleya'*.

Recommendation H.6A

H.6A.1. When a nothogeneric name is formed from a personal name by adding the termination -*ara,* that person should preferably be a collector, grower, or student of the group.

ARTICLE H.7
HYBRIDS BETWEEN SUBDIVISIONS OF GENERA

H.7.1. The name of a nothotaxon that is a hybrid between subdivisions of a genus is a combination of an epithet, which is a condensed formula formed in the same way as a nothogeneric name (Art. H.6.2–H.6.4), with the name of the genus.

Ex. 1. Ptilostemon nothosect. *Platon* Greuter (in Boissiera 22: 159. 1973) comprises hybrids between *P.* sect. *Platyrhaphium* Greuter and *P.* Cass. sect. *Ptilostemon. Ptilostemon* nothosect. *Plinia* Greuter (l.c.: 158. 1973) comprises hybrids between *P.* sect. *Cassinia* Greuter and *P.* sect. *Platyrhaphium.*

ARTICLE H.8
PARENTAGE, NOTHOGENERIC NAMES,
AND CONDENSED FORMULAE

H.8.1. When the name or the epithet in the name of a nothotaxon is a condensed formula (Art. H.6 and H.7), the parental names used in its formation must be those that are correct for the particular circumscription, position, and rank accepted for the parental taxa.

Ex. 1. If the genus *Triticum* L. is interpreted on taxonomic grounds as including *Triticum* (s. str.) and *Agropyron* Gaertn., and the genus *Hordeum* L. as including *Hordeum* (s. str.) and *Elymus* L., then hybrids between *Agropyron* and *Elymus* as well as between *Triticum* (s. str.) and *Hordeum* (s. str.) are placed in the same nothogenus, ×*Tritordeum* Asch. & Graebn. (Syn. Mitteleur. Fl. 2(1): 748. 1902). If, however, *Agropyron* is treated as a genus separate from *Triticum,* hybrids between *Agropyron* and *Hordeum* (s. str. or s. l.) are placed in the nothogenus ×*Agrohordeum* E. G. Camus ex A. Camus (in Bull. Mus. Natl. Hist. Nat. 33: 537. 1927). Similarly, if *Elymus* is treated as a genus separate from *Hordeum,* hybrids between *Elymus* and *Triticum* (s. str. or s. l.) are placed in the nothogenus ×*Elymotriticum* P. Fourn. (Quatre Fl. France: 88. 1935). If both *Agropyron* and *Elymus* are given generic rank, hybrids between them are placed in the nothogenus ×*Agroelymus* E. G. Camus ex A. Camus (l.c.: 538. 1927); ×*Tritordeum* is then restricted to hybrids between *Hordeum* (s. str.) and *Triticum* (s. str.), and hybrids between *Elymus* and *Hordeum* are placed in ×*Elyhordeum* Mansf. ex Tsitsin & Petrova (in Züchter 25: 164. 1955), replacing ×*Hordelymus* Bachteev & Darevsk. (in Bot. Zhurn. (Moscow & Leningrad) 35: 191. 1950) non *Hordelymus* (Jess.) Harz (Landw. Samenk.: 1147. 1885).

Ex. 2. When *Orchis fuchsii* Druce was renamed *Dactylorhiza fuchsii* (Druce) Soó, the name for its hybrid with *Coeloglossum viride* (L.) Hartm., ×*Orchicoeloglossum mixtum* Asch. & Graebn. (Syn. Mitteleur. Fl. 3: 847. 1907), had to be changed to ×*Dactyloglossum mixtum* (Asch. & Graebn.) Rauschert (in Feddes Repert. 79: 413. 1969).

H.8.2. Names ending in -*ara* for nothogenera, which are equivalent to condensed formulae (Art. H.6.3 and H.6.4(b)), are applicable only to hybrids that are accepted taxonomically as derived from the parents named.

Ex. 3. If *Euanthe* Schltr. is recognized as a distinct genus, hybrids simultaneously involving its only species, *E. sanderiana* (Rchb.) Schltr., and the three genera *Arachnis* Blume, *Renanthera* Lour., and *Vanda* W. Jones ex R. Br. must be placed in ×*Cogniauxara* Garay & H. R. Sweet (in Bot. Mus. Leafl. 21: 156. 1966); if, on the other hand, *E. sanderiana* is included in *Vanda,* the same hybrids are placed in ×*Holttumara* Holttum (in Malayan Orchid Rev. 5: 75. 1958) *(Arachnis × Renanthera × Vanda).*

ARTICLE H.9
VALID PUBLICATION OF NAMES OF NOTHOGENERA
AND THEIR SUBDIVISIONS

H.9.1. To be validly published, the name of a nothogenus or a notho-taxon at the rank of a subdivision of a genus (Art. H.6 and H.7) must be effectively published (Art. 29–31) with a statement of the names of the parental genera or subdivisions of genera, but no description or diagnosis is necessary, whether in Latin, English, or any other language.

Ex. 1. Validly published names: ×*Philageria* Mast. (in Gard. Chron. 1872: 358. 1872), published with a statement of parentage, *Lapageria* Ruiz & Pav. × *Philesia* Comm. ex Juss.; *Eryngium* nothosect. *Alpestria* Burdet & Miège (pro sect.) (in Candollea 23: 116. 1968), published with a statement of parentage, *E.* sect. *Alpina* H. Wolff × *E.* sect. *Campestria* H. Wolff; ×*Agrohordeum* E. G. Camus ex A. Camus (in Bull. Mus. Natl. Hist. Nat. 33: 537. 1927), published with a statement of parentage, *Agropyron* Gaertn. × *Hordeum* L.; and its later synonym ×*Hordeopyron* Simonet (in Compt. Rend. Hebd. Séances Acad. Sci. 201: 1212. 1935, '*Hordeopyrum*'; see Art. 32.2), published with an identical statement of parentage.

ⓘ **Note 1.** Because the names of nothogenera and nothotaxa at the rank of a subdivision of a genus are condensed formulae or equivalent to such, they do not have types.

Ex. 2. The name ×*Ericalluna* Krüssm. (in Deutsche Baumschule 12: 154. 1960) was published for plants that were thought to be the product of the cross *Calluna vulgaris* (L.) Hull × *Erica cinerea* L. If these plants are considered not to be hybrids but variants of *E. cinerea,* the name ×*Ericalluna* Krüssm. remains available for use should known or postulated hybrids of *Calluna* Salisb. × *Erica* L. be produced.

Ex. 3. ×*Arabidobrassica* Gleba & Fr. Hoffm. (in Naturwissenschaften 66: 548. 1979), a nothogeneric name that was validly published with a statement of parentage for the result of somatic hybridization by protoplast fusion of *Arabidopsis thaliana* (L.) Heynh. with *Brassica campestris* L., is also available for intergeneric hybrids resulting from normal crosses between *Arabidopsis* Heynh. and *Brassica* L., should any be produced.

ⓘ **Note 2.** A statement of the names of the parental species of a nothogenus, or of the names of the parental species of any of its included taxa, is sufficient to validly publish the name of a nothogenus, if the full names of all parental genera appear among the species names, when there is no separate statement of the names of the parental genera.

Ex. 4. When *Kalanchoe* Adans. (Fam. Pl. 2: 248. 1763) and *Bryophyllum* Salisb. (Parad. Lond.: ad t. 3. 1805) are treated as two separate genera, the nothogeneric name ×*Bryokalanchoe* Resende (in Bol. Soc. Portug. Ci. Nat., ser. 2, 6: 242. 1956) applies to hybrids between representatives of the two genera. Resende (l.c.: 241. 1956) provided the names of the species he used when making the intergeneric crosses.

ⓘ **Note 3.** Names published merely in anticipation of the existence of a hybrid are not validly published under Art. 36.1(a).

ARTICLE H.10
VALID PUBLICATION OF NAMES OF NOTHOSPECIES AND INFRASPECIFIC NOTHOTAXA

H.10.1. Names of nothotaxa at the rank of species or below must conform with:

(a) the provisions of the *Code* outside of Chapter H applicable to names at the same ranks (see Art. 32.4); and

(b) the provisions in Art. H.3.

Infringements of Art. H.3.1 are treated as errors to be corrected (see also Art. 11.9).

> **Ex. 1.** The nothospecific name *Melampsora ×columbiana* G. Newc. (in Mycol. Res. 104: 271. 2000) was validly published, with a Latin description and designation of a holotype, for the hybrid between *M. medusae* Thüm. and *M. occidentalis* H. S. Jacks.

ⓘ **Note 1.** Taxa previously published as species or infraspecific taxa that are later considered to be nothotaxa may be indicated as such, without change of rank, in conformity with Art. 3 and 4 and by the application of Art. 50 (which also operates in the reverse direction).

H.10.2. The following are considered to be formulae and not true epithets: designations consisting of the epithets of the names of the parents combined in unaltered form by a hyphen, or with only the termination of one epithet changed, or consisting of the specific epithet of the name of one parent combined with the generic name of the other (with or without change of termination).

> **Ex. 2.** The designation *"Potentilla atrosanguinea-pedata"* published by Maund (in Bot. Gard. 5: No. 385, t. 97. 1833) is considered to be a formula meaning *P. atrosanguinea* Lodd. ex D. Don × *P. pedata* Nestl.

> **Ex. 3.** *"Verbascum nigro-lychnitis"* (Schiede, Pl. Hybr.: 40. 1825) is considered to be a formula meaning *V. lychnitis* L. × *V. nigrum* L.; the correct binary name for this hybrid is *V. ×schiedeanum* W. D. J. Koch (Syn. Fl. Germ. Helv., ed. 2: 592. 1844).

Ex. 4. In *Acaena* ×*anserovina* Orchard (in Trans. Roy. Soc. South Australia 93: 104. 1969) (*A. anserinifolia* (J. R. Forst. & G. Forst.) J. Armstr. × *A. ovina* A. Cunn.) the epithet (contrary to Rec. H.10A) combines the first part of the first and the whole of the second epithet in the names of the parental species; because more than the termination of the first epithet is omitted, *anserovina* is a true epithet.

Ex. 5. In *Micromeria* ×*benthamineolens* Svent. (Index Seminum Hortus Acclim. Pl. Arautap.: 48. 1969) (*M. benthamii* Webb & Berthel. × *M. pineolens* Svent.) the epithet (contrary to Rec. H.10A) combines the first part of the first and the second part of the second epithet in the names of the parental species; because neither epithet is unaltered, *benthamineolens* is a true epithet.

🛈 ***Note 2.*** Because the name of a nothotaxon at the rank of species or below has a type, statements of parentage play a secondary part in determining the application of the name.

Ex. 6. *Quercus* ×*deamii* Trel. (in Mem. Natl. Acad. Sci. 20: 14. 1924) when described was considered to be the cross *Q. alba* L. × *Q. muehlenbergii* Engelm. However, progeny grown from acorns of the tree from which the type originated led Bartlett to conclude that the parents were in fact *Q. macrocarpa* Michx. and *Q. muehlenbergii*. If this conclusion is accepted, the name *Q.* ×*deamii* applies to *Q. macrocarpa* × *Q. muehlenbergii,* and not to *Q. alba* × *Q. muehlenbergii.*

Recommendation H.10A

H.10A.1. In forming epithets for names of nothotaxa at the rank of species and below, authors should avoid combining parts of the epithets of the names of the parents.

Recommendation H.10B

H.10B.1. When contemplating the publication of names for hybrids between named infraspecific taxa, authors should carefully consider whether these names are really needed, taking into account that formulae, though more cumbersome, are more informative.

ARTICLE H.11
NOTHOTAXA WITH PARENTS BELONGING TO
DIFFERENT HIGHER-RANKED TAXA

H.11.1. The name of a nothospecies of which the postulated or known parental species belong to different genera is a combination of a nothogeneric name with a nothospecific epithet.

Ex. 1. ×*Heucherella tiarelloides* (Lemoine & É. Lemoine) H. R. Wehrh. is considered to have originated from the cross between a garden hybrid of *Heuchera* L. and *Tiarella cordifolia* L. (see Stearn in Bot. Mag. 165: ad t. 31. 1948). Its basionym, *Heuchera* ×*tiarelloides* Lemoine & É. Lemoine (in Catalogue (Lemoine) 182: 3. 1912), is therefore incorrect.

H.11.2. The final epithet in the name of an infraspecific nothotaxon of which the postulated or known parental taxa are assigned to different species may be placed under the correct name of the corresponding nothospecies (but see Rec. H.10B).

Ex. 2. *Mentha* ×*piperita* L. nothosubsp. *piperita (M. aquatica* L. × *M. spicata* L. subsp. *spicata); M.* ×*piperita* nothosubsp. *pyramidalis* (Ten.) Harley (in Kew Bull. 37: 604. 1983) (*M. aquatica* L. × *M. spicata* subsp. *tomentosa* (Briq.) Harley).

ARTICLE H.12
SUBORDINATE TAXA WITHIN NOTHOSPECIES

H.12.1. Subordinate taxa within nothospecies may be recognized without an obligation to specify parental taxa at the subordinate rank. In this case non-hybrid infraspecific categories at the appropriate rank are used.

Ex. 1. *Mentha* ×*piperita* f. *hirsuta* Sole; *Populus* ×*canadensis* var. *serotina* (R. Hartig) Rehder and *P.* ×*canadensis* var. *marilandica* (Poir.) Rehder (see also Art. H.4 Note 1).

Note 1. Art. H.4 and H.5, governing the circumscription and appropriate rank of hybrid taxa, do not apply when there is no statement of parentage.

Note 2. Art. H.11.2 and H.12.1 cannot both be applied simultaneously at the same infraspecific rank.

H.12.2. Names published at the rank of nothomorph[1] are treated as having been published as names of varieties (see Art. 50).

---

1   Editions of the *Code* prior to the Sydney *Code* of 1983 permitted only one rank of infraspecific nothotaxa under provisions equivalent to Art. H.12. That rank was equivalent to variety and the category was termed "nothomorph".

DIVISION III
PROVISIONS FOR GOVERNANCE OF THE *CODE*

PROVISION 1
GENERAL PROVISIONS FOR GOVERNANCE OF THE *CODE*

1.1. The *International Code of Nomenclature for algae, fungi, and plants* is governed by its users, who are represented by members of a Nomenclature Section of an International Botanical Congress acting under the authority of that Congress and, between such Congresses, by the Permanent Nomenclature Committees and any Special-purpose Committees.

1.2. The *Code* may be modified only by action of a plenary session of an International Botanical Congress on a resolution moved by the Nomenclature Section of that Congress.

1.3. In the event that there should not be another International Botanical Congress, authority for the *International Code of Nomenclature for algae, fungi, and plants* shall be transferred to the International Union of Biological Sciences or to an organization at that time corresponding to it. The General Committee is empowered to define the mechanism to achieve this.

1.4. The *Code* is provided with logistical and financial support by the International Association for Plant Taxonomy, which liaises with the Permanent Nomenclature Committees and the Bureau of Nomenclature. The nomenclatural publications[1] required by Div. III are published as specified by the General Committee (currently in the journal *Taxon,* except for proposals to amend the *Code* relating solely to names of organisms treated as fungi and proposals to protect or reject names under Art. F.2 or F.7, submitted as lists, which are published in the journal *IMA Fungus*).

---

1 The nomenclatural publications required by Div. III include proposals to conserve, protect, or reject names or suppress works, requests for decisions, reports of Permanent Nomenclature Committees and Special-purpose Committees, proposals to amend the *Code* and a synopsis of these proposals, notices of institutional votes, and the results of the preliminary guiding vote and Congress-approved decisions and elections of the Nomenclature Section or Fungal Nomenclature Session.

1.5. The Provisions for governance of the *Code* (Div. III) apply to the edition of the *Code* of which they form a part. These Provisions are not retroactive (see Pre. 7).

PROVISION 2
PROPOSALS TO AMEND THE *CODE*

2.1. Proposals concerning the Preamble, Divisions I–III, and the Glossary are submitted by publication (see Prov. 1.4) to the Nomenclature Section of an International Botanical Congress.

2.2. Proposals concerning Appendices I–VII, i.e. proposals to conserve, protect, or reject names (Art. 14.12, 56.2, F.2.1, and F.7.1), proposals to suppress works (Art. 34.1), and requests for decisions (Art. 38.5 and 53.4), are submitted by publication (see Prov. 1.4) to the General Committee.

2.3. At least three years before an International Botanical Congress, the Rapporteur-général publishes an announcement that proposals to amend the *Code* may be published between specified dates.

2.4. Approximately six months before an International Botanical Congress, a synopsis of proposals to amend the *Code* is published. It is compiled by the Rapporteur-général and Vice-rapporteur, includes their comments on the proposals, and may include opinions of the Permanent Nomenclature Committees on certain proposals.

2.5. A guiding vote on proposals to amend the *Code* is organized by the Bureau of Nomenclature in conjunction with the International Association for Plant Taxonomy (IAPT) to coincide with the publication of the synopsis of proposals. No accumulation or transfer of votes is permissible in this vote. The following persons are entitled to vote:

(a) individual members of the IAPT; and
(b) authors of proposals to amend the *Code;* and
(c) members of the Permanent Nomenclature Committees.

2.6. The purpose of the guiding vote is to advise the Nomenclature Section of the International Botanical Congress of the level of support for proposals to amend the *Code*. The results of the vote and any Permanent Nomenclature Committee opinions are provided at the Nomenclature Section (see also Prov. 5.5).

PROVISION 3
INSTITUTIONAL VOTES

3.1. Before an International Botanical Congress, the Committee on Institutional Votes updates the list of institutions from the previous Congress and allocates one vote to each institution (see Prov. 5.9(b)). The list must be approved by the General Committee and published (see Prov. 1.4) before the Congress. No single institution, even in the broad sense of the term (e.g. mycological and botanical divisions together), is entitled to more than one vote.

3.2. Before an International Botanical Congress, any institution desiring to vote in the Nomenclature Section and not listed as having been allocated a vote in the previous Nomenclature Section should notify the Rapporteur-général of its wish to be allocated a vote and provide relevant information regarding its current level of taxonomic activity (e.g. active staff, collections, publications) and show that it is registered in an online, open-access international or regional index of herbaria, collections, or institutions.

3.3. An institution wishing to exercise its vote, as allocated in the published list (Prov. 3.1), must provide its official written authorization to be presented at the Nomenclature Section by its delegate (Prov. 5.9(b)).

3.4. A delegate who is a member of an institution that has not previously applied for, or been allocated, a vote may apply in person for one institutional vote at the Nomenclature Section.

PROVISION 4
NOMENCLATURE SECTION

4.1. The Nomenclature Section is part of an International Botanical Congress and meets prior to a plenary session of the Congress.

4.2. Registration for the Nomenclature Section is through the International Botanical Congress. Only registered members of the Nomenclature Section are entitled to vote at the Nomenclature Section.

4.3. The Nomenclature Section has the following functions:

(a) Approves the previous *Code* as published as a basis for discussion by the Section.
(b) Decides on proposals to amend the *Code*.
(c) Appoints ad hoc committees to consider specific questions and report back to the Section.
(d) Authorizes Special-purpose Committees, with a specific mandate, to be appointed by the General Committee and report back to the Nomenclature Section of the next Congress.
(e) Elects the ordinary members of the Permanent Nomenclature Committees (but see Prov. 8.5(e) and (f)).
(f) Elects the Rapporteur-général for the next Congress.
(g) Receives the reports of the Permanent Nomenclature Committees and Special-purpose Committees.
(h) Decides on the recommendations of the General Committee.

4.4. The decisions and appointments of the Nomenclature Section become binding upon their acceptance by a subsequent plenary session of the same International Botanical Congress acting on a resolution moved by the Nomenclature Section (see Prov. 1.2).

4.5. The Bureau of Nomenclature of the International Botanical Congress comprises the following officers: President of the Nomenclature Section; up to five Vice-presidents; the Rapporteur-général; the Vice-rapporteur; one or more Recorders. The Bureau of Nomenclature defines the sequence and timing of debates; appoints Tellers to collect and count voting cards in the event of a card vote (see Prov. 5.10); and advises the President on procedural matters.

4.6. The President of the Nomenclature Section is elected by the General Committee prior to the Congress. The President chairs the debates and is responsible for their harmony and timely conclusion; recognizes and silences speakers; may end a debate; decides on procedural matters not covered in Div. III; and is authorized to move a resolution on behalf of the Nomenclature Section at a plenary session of the same International Botanical Congress that the decisions and appointments of the Nomenclature Section be approved.

4.7. The Vice-presidents are appointed by the Bureau of Nomenclature, either in advance of the International Botanical Congress or from those present at the Nomenclature Section. A Vice-president serves in place of the President, if and when requested.

4.8. The Rapporteur-général is elected by the previous International Botanical Congress. The Rapporteur-général is responsible for: presentation of nomenclature proposals to the subsequent Congress; general duties in connection with the editing of the *Code* resulting from that Congress; and deposition of unpublished relevant material in the nomenclature archives of the International Association for Plant Taxonomy.

4.9. The Vice-rapporteur is appointed by the Rapporteur-général and approved by the General Committee no later than three years before the Congress. The Vice-rapporteur assists and, if necessary, serves in place of the Rapporteur-général.

4.10. Recorders are appointed by the Organizing Committee of the International Botanical Congress in consultation with the Rapporteur-général. Recorders are responsible for all local facilities needed by the Nomenclature Section, such as the venue and its equipment, and in particular for the detailed recording of the proceedings of the Section and for facilitating the voting.

4.11. The Nominating Committee comprises members who should preferably be unavailable to serve on the Permanent Nomenclature Committees or as Rapporteur-général. They are proposed by the President of the Nomenclature Section in consultation with the other members of the Bureau of Nomenclature and are elected by the Nomenclature Section.

4.12. The Nominating Committee is charged with preparing lists of candidates to serve on the Permanent Nomenclature Committees (except the Nomenclature Committee for Fungi and the Editorial Committee for Fungi; see Prov. 4.13), in consultation with the current secretaries of those committees, and proposing the Rapporteur-général for the next International Botanical Congress. The nominations of the Nominating Committee are subject to approval by the Nomenclature Section.

4.13. The Nominating Committee of the Fungal Nomenclature Session (Prov. 8.1) is charged with preparing lists of candidates to serve on the Nomenclature Committee for Fungi, in consultation with the current Secretary of that Committee, and on the Editorial Committee for Fungi, and proposing the Secretary of the Fungal Nomenclature Bureau for the next International Mycological Congress. The nominations of the Nominating Committee of the Fungal Nomenclature Session are subject to approval by the Fungal Nomenclature Session.

Recommendation 1. The Nominating Committee of the Nomenclature Section should represent the different taxonomic groups covered by the *Code* and both Nominating Committees, so far as is practicable, should be geographically balanced.

Recommendation 2. Individuals or groups should be able to observe the Nomenclature Section of an International Botanical Congress online on the World Wide Web. The Organizing Committee of the International Botanical Congress in consultation with the Bureau of Nomenclature should be responsible for ensuring that this is implemented.

PROVISION 5
PROCEDURES AND VOTING AT THE NOMENCLATURE SECTION

5.1. A qualified majority (at least 60%) of votes cast is required for the following decisions:

(a) Accepting a proposal to amend the *Code.*
(b) Referring items to the Editorial Committee.
(c) Accepting a motion to end discussion and proceed to a vote (to "call the question").
(d) Accepting a motion to set a time limit for a debate.

5.2. A simple majority (more than 50%) of votes cast is required for all other decisions, including the following:

(a) Electing the Nominating Committee for the Nomenclature Section.

(b) Accepting the *Code* that arose from the previous International Botanical Congress as the basis for discussion at the Nomenclature Section.

(c) Choosing between two alternative proposals.

(d) Accepting an amendment to a proposal.

(e) Establishing an ad hoc committee.

(f) Establishing and referring items to a Special-purpose Committee.

(g) Accepting recommendations of the General Committee.

(h) Approving the nominations made by the Nominating Committee.

5.3. When a report of the General Committee contains more than one recommendation, the Nomenclature Section may vote separately on an individual recommendation if such a procedure is proposed by a member of the Section, supported (seconded) by five other members (see Prov. 5.7), and approved by a simple majority (more than 50%) of the Section.

5.4. When a vote to accept a General Committee recommendation does not achieve the required majority (Prov. 5.2(g)), that recommendation is cancelled, and the matter is referred back to the General Committee. Retention or rejection of a name, suppression of a work, or a binding decision on valid publication or homonymy is no longer authorized (Art. 14.15, 34.2, 38.5, 53.4, and 56.3).

5.5. Any proposal to amend the *Code* that receives 75% or more "no" votes in the preliminary guiding vote is automatically rejected at the Nomenclature Section unless a proposal to discuss it is moved by a member of the Section and supported (seconded) by five other members.

5.6. Any proposal to amend the *Code* that concerns only Examples (excluding voted Examples) or the Glossary is automatically referred to the Editorial Committee unless a proposal to discuss it is moved by a member of the Section and supported (seconded) by five other members (but see Prov. 5.5).

5.7. A new proposal to amend the *Code* (i.e. one not previously published) or an amendment to a proposal to amend the *Code* may be introduced at the Nomenclature Section by a member of the Section only when supported (seconded) by five other members.

5.8. A member of the Nomenclature Section may propose a friendly amendment to a proposal to amend the *Code;* if accepted by the original proposer(s), such an amendment does not require the support of other members (seconders).

5.9. There are two kinds of votes at the Nomenclature Section:

(a) Personal votes. Each member of the Section has one personal vote. No accumulation or transfer of personal votes is permissible.
(b) Institutional votes (see Prov. 3). An institution may authorize in writing any member of the Section as a delegate to carry its one institutional vote.

A member of the Section may carry the institutional votes of more than one institution. No single person is allowed more than 15 votes, including personal vote and institutional votes.

5.10. A card vote requires members of the Nomenclature Section to deposit anonymous cards printed to indicate the kind of votes (personal or institutional), which are counted by the Tellers (see Prov. 4.5). A card vote may be conducted when the required majority cannot be detected by other means or may be requested in advance of the vote by at least five members.

PROVISION 6
AFTER AN INTERNATIONAL BOTANICAL CONGRESS

6.1. Certain publications, which may be electronic or printed or both, appear as soon as feasible after an International Botanical Congress, not necessarily in this sequence:

(a) The Congress-approved decisions and elections of the Nomenclature Section including the results (if not published before the Congress) of the preliminary guiding vote.
(b) The announcement of Special-purpose Committees and their membership.
(c) The new edition of the *Code,* including the Glossary.
(d) The Appendices of the *Code* (App. I–VII).
(e) A transcript of the Nomenclature Section.

PROVISION 7
PERMANENT NOMENCLATURE COMMITTEES

7.1. There are ten Permanent Nomenclature Committees, including five specialist committees (clauses (f)–(j)):

(a) General Committee
(b) Editorial Committee
(c) Editorial Committee for Fungi
(d) Committee on Institutional Votes
(e) Registration Committee
(f) Nomenclature Committee for Vascular Plants
(g) Nomenclature Committee for Bryophytes
(h) Nomenclature Committee for Fungi
(i) Nomenclature Committee for Algae
(j) Nomenclature Committee for Fossils.

Membership

7.2. Members of the Permanent Nomenclature Committees are elected by an International Botanical Congress (except where indicated otherwise). The committees have power to elect officers as desired, to fill vacancies, and to establish temporary subcommittees in consultation with the General Committee.

7.3. The General Committee has, in addition to its ordinary (elected) members, the following ex-officio members: the secretaries of the five specialist committees (Prov. 7.1(f)–(j)), the Rapporteur-général, the Vice-rapporteur, and the President and Secretary-general of the International Association for Plant Taxonomy.

7.4. The Editorial Committee comprises individuals who should preferably have been present at the Nomenclature Section of the relevant International Botanical Congress and includes at least one specialist in each of vascular plants, bryophytes, fungi, algae, and fossils and at least one individual nominated by the Nomenclature Committee for Fungi who attended the Fungal Nomenclature Session of the relevant International Mycological Congress; the Rapporteur-général and Vice-rapporteur of the relevant International Botanical Congress serve as Chair and Secretary, respectively, of the Editorial Committee.

7.5. The Editorial Committee for Fungi is elected by an International Mycological Congress and comprises individuals who should preferably have been present at the Fungal Nomenclature Session of the relevant International Mycological Congress and includes the Chair and Secretary of the Editorial Committee for this *Code*. The Secretary and Deputy Secretary of the Fungal Nomenclature Session of the relevant International Mycological Congress serve as Chair and Secretary, respectively, of the Editorial Committee for Fungi.

7.6. The Committee on Institutional Votes comprises six members, each to represent a different continent, plus the Rapporteur-général, who serves as its chair, plus any additional members that the Committee considers are required.

7.7. The Registration Committee includes at least five members appointed by the Nomenclature Section selected, in part, to ensure geographical balance, and representatives nominated by:

(a) the other Permanent Nomenclature Committees;
(b) prospective or functioning nomenclatural repositories;
(c) the International Association for Plant Taxonomy;
(d) the International Association of Bryologists;
(e) the International Federation of Palynological Societies;
(f) the International Mycological Association;
(g) the International Organisation of Palaeobotany;
(h) the International Phycological Society.

7.8. Each specialist committee includes the Rapporteur-général, the Vice-rapporteur, and the Secretary of the General Committee as non-voting ex-officio members.

7.9. The Nomenclature Committee for Fungi is elected by an International Mycological Congress and includes the Secretary and the Deputy Secretary of the Fungal Nomenclature Bureau (Prov. 8.1) as non-voting ex-officio members if they are not already members of the Nomenclature Committee for Fungi.

Recommendation 1. Each committee should, so far as is practicable, be geographically and gender balanced.

Functions

7.10. The General Committee has the following functions:

(a) Specifies where the nomenclatural publications required by Div. III are to be published (Prov. 1.4).

(b) Receives proposals to conserve, protect, or reject names, proposals to suppress works, and requests for decisions (Art. 14.12, 34.1, 38.5, 53.4, 56.2, F.2.1, and F.7.1) and refers these proposals or requests to the specialist committee(s) concerned (receipt and referral of proposals and requests are automatic upon their publication).

(c) Considers recommendations of the specialist committees and either approves or overturns those recommendations or refers them back to the specialist committees for further consideration.

(d) Ratifies a list of institutional votes drawn up by the Committee on Institutional Votes (see Prov. 3.1).

(e) Receives applications for recognition as nomenclatural repositories for organisms other than those treated as fungi, refers the applications to the Registration Committee, and acts upon its recommendation (Art. 42.2).

(f) Recognizes or appoints nomenclatural repositories for organisms other than those treated as fungi; may suspend, revoke, or cancel such recognition or appointment; and may set aside the requirements for identifiers to be issued if the repository mechanism, or essential parts of it, cease to function (Art. 42.2, 42.3, and 42.7).

(g) Appoints Special-purpose Committees:
  (1) authorized by a Nomenclature Section at an International Botanical Congress (see Prov. 4.3(d); or
  (2) proposed between International Botanical Congresses following an open call for expressions of interest in serving on such a committee based on *(i)* the initiative of the General Committee; or *(ii)* requests that have been submitted to the General Committee by at least five people.

(h) Decides which proposals relate solely to names of organisms treated as fungi, in consultation with the Nomenclature Committee for Fungi (Prov. 8.2).

(i) Is consulted by the Nomenclature Committee for Fungi on the appointment of Special-purpose Committees set up according to Prov. 8.5(d).

(j) Is consulted by the Nomenclature Committee for Fungi when subcommittees are established to prepare lists of protected or rejected names; and reviews and approves such lists (Art. F.2.1 and F.7.1).

(k) Is consulted by the Nomenclature Committee for Fungi about the election of the Chair of the Fungal Nomenclature Session (Prov. 8.6) and the approval of the appointment of the Deputy Secretary of the Fungal Nomenclature Session (Prov. 8.7).

(l) May communicate an international standard format in addition to, or as a successor to, Portable Document Format (PDF) for effective publication of electronic material (Art. 29.3).

7.11. Each of the five specialist committees examines proposals to conserve or reject names, proposals to suppress works, and requests for decisions (Art. 14.12, 34.1, 38.5, 53.4, and 56.2) referred to them by the General Committee, to which they then submit their recommendations. They may also submit opinions on proposals to amend the *Code* to the Bureau of Nomenclature. The Nomenclature Committee for Fungi has a mandate under Art. F.2.1 and F.7.1 with respect to lists of protected or rejected names proposed for approval and under Art. F.5.1 with respect to repositories for fungal names.

7.12. The Editorial Committee is charged with the preparation and publication of the *Code* in conformity with the decisions approved by the relevant International Botanical Congress. It is empowered to make any editorial modification not affecting the meaning of the provisions concerned, e.g. to change the wording of any Article, Note, or Recommendation and to avoid duplication, to add or remove non-voted Examples, and to place Articles, Notes, Recommendations, and Chapters of the *Code* in the most convenient place, while retaining the previous numbering insofar as possible.

7.13. The Editorial Committee for Fungi is charged with the preparation and publication of Chapter F in conformity with the decisions approved by the relevant International Mycological Congress. It is empowered to make the editorial modifications specified in Prov. 7.12.

7.14. The Committee on Institutional Votes maintains a list of institutions entitled to vote at the upcoming International Botanical Congress (see Prov. 3.1).

7.15. The Registration Committee is charged with considering applications for recognition as nomenclatural repositories for organisms other than those treated as fungi, assisting the design and implementation of repositories for nomenclatural novelties and/or any nomenclatural act, monitoring the functioning of existing repositories, and advising the General Committee on relevant matters.

Procedural rules

7.16. A specialist committee, provided that a qualified majority (at least 60%) of its members supports or opposes a proposal, may make any of the following recommendations to the General Committee: conserve or not conserve a name; reject or not reject a name; suppress or not suppress a publication; and for names of organisms treated as fungi, protect or not protect names on a list. In the case of binding decisions on valid publication (Art. 38.5) and homonymy (Art. 53.4), the qualified majority decides whether or not a binding decision should be recommended, then a simple majority (more than 50%) decides between the two alternatives: i.e. treat a name as validly published or not validly published; treat names as homonyms or not homonyms. If a specialist committee is unable to make a recommendation after voting at least twice, the proposal is referred to the General Committee without a recommendation from the specialist committee.

7.17. The General Committee may approve or overturn a recommendation of a specialist committee provided that a qualified majority (at least 60%) of the General Committee members supports or opposes the recommendation. In either case, the General Committee makes its own recommendation, which is subject to the decision of a later International Botanical Congress (see also Art. 14.15, 34.2, 38.5, 53.4, and 56.3). If the required majority is not achieved after voting at least twice, the General Committee is considered to have recommended against the proposal or against making a binding decision. The General Committee may also decide to refer the matter back to the specialist committee for further consideration.

Recommendation 2. The General Committee and the specialist committees should publish their recommendations at least annually.

PROVISION 8

PROVISIONS FOR GOVERNANCE OF THE *CODE* RELATING SOLELY TO
NAMES OF ORGANISMS TREATED AS FUNGI

8.1. For proposals to amend the *Code* relating to the content of
Chapter F, which brings together the provisions of this *Code* that deal
solely with names of organisms treated as fungi (but excluding any
other content), exactly the same procedures outlined in Prov. 1–7 are to
be followed except that in Prov. 1, 2, 4, and 5 mentions of International
Botanical Congress, Nomenclature Section (of that Congress), Bureau
of Nomenclature, Nominating Committee, and Editorial Committee
are to be replaced by International Mycological Congress, Fungal
Nomenclature Session (of that Congress), Fungal Nomenclature
Bureau, Nominating Committee of the Fungal Nomenclature Session,
and Editorial Committee for Fungi, respectively; and officers such
as President, Rapporteur-général, and Vice-rapporteur (these specifically renamed Chair, Secretary, and Deputy Secretary, respectively)
are to be understood as members of the Fungal Nomenclature Bureau
rather than the Bureau of Nomenclature (specifically in Prov. 1.1, 1.2,
1.4 footnote, 2.1, 2.3, 2.4, 2.6, 4.2, 4.4, 4.5, 4.7, 4.8, 4.10, 4.11, 5.1, 5.2,
5.5, 5.6, 5.7, and 5.8; but not in Prov. 5.3 and 5.4; and the following
clause does not apply: Prov. 5.2(g)). See also Prov. 4.12, 4.13, 7.1, 7.5,
and 7.13.

8.2. The General Committee in consultation with the Nomenclature
Committee for Fungi is responsible for deciding which proposals relate solely to names of organisms treated as fungi.

8.3. A guiding vote on proposals to amend the *Code* relating
solely to names of organisms treated as fungi is organized by the
Fungal Nomenclature Bureau in conjunction with the International
Mycological Association (IMA) to coincide with the publication of the
synopsis of proposals. No accumulation or transfer of votes is permissible in this vote. The following persons are entitled to vote:

(a) individual members of the IMA; and
(b) individual members of organizations affiliated with the IMA; and
(c) individual members of other organizations approved by the
      Fungal Nomenclature Bureau; and

(d) authors of proposals to amend the *Code* relating solely to names of organisms treated as fungi; and

(e) members of the Nomenclature Committee for Fungi.

8.4. The Fungal Nomenclature Session is part of an International Mycological Congress and meets prior to a plenary session of the Congress at a time and with a duration to be determined by consultation between the International Mycological Association and the Fungal Nomenclature Bureau.

8.5. The Fungal Nomenclature Session has the following functions:

(a) Approves the previous *Code* if amended at the last International Mycological Congress (in the circumstance where there has not been an International Botanical Congress since the last International Mycological Congress) as a basis for discussion by the Session, and otherwise utilizes the most recent published *Code*.

(b) Decides on proposals to amend the *Code* relating solely to organisms treated as fungi.

(c) Appoints ad hoc committees to consider specific questions and report back to the Session.

(d) Authorizes Special-purpose Committees, with a specific mandate, to deal with matters relating solely to names of organisms treated as fungi, to be appointed by the Nomenclature Committee for Fungi in consultation with the General Committee and report back to the Fungal Nomenclature Session of the next International Mycological Congress.

(e) Elects the ordinary members of the Nomenclature Committee for Fungi.

(f) Elects the ordinary members of the Editorial Committee for Fungi.

(g) Elects the Secretary of the Fungal Nomenclature Bureau for the next International Mycological Congress.

(h) Receives reports of Special-purpose Committees dealing with matters relating solely to names of organisms treated as fungi.

8.6. The Chair of the Fungal Nomenclature Session is elected by the Nomenclature Committee for Fungi in consultation with the General Committee prior to the International Mycological Congress. The Chair chairs the debates and is responsible for their harmony and timely conclusion; recognizes and silences speakers; may end a debate; decides on procedural matters not covered in Div. III; and is authorized to move a resolution on behalf of the Fungal Nomenclature Session at a plenary session of the same International Mycological Congress that the decisions and appointments of the Fungal Nomenclature Session with respect to matters relating solely to names of organisms treated as fungi be approved.

8.7. In the Fungal Nomenclature Bureau, the Deputy Secretary is appointed by the Secretary and approved by the Nomenclature Committee for Fungi in consultation with the General Committee no later than two years before the International Mycological Congress. The Deputy Secretary assists and, if necessary, serves in place of the Secretary.

8.8. The Rapporteur-général elected for the International Botanical Congress that follows the International Mycological Congress, or an alternate appointed by that Rapporteur-général, is invited to attend the Fungal Nomenclature Session as a non-voting advisor to the Session.

8.9. When proposals relating solely to names of organisms treated as fungi are dealt with in a Fungal Nomenclature Session, there are no institutional votes, and therefore Prov. 3, 7.6, and 7.14 do not apply. Each member of the Session has one personal vote. No accumulation or transfer of personal votes is permissible.

8.10. The decisions taken at the Fungal Nomenclature Session of an International Mycological Congress relating solely to names of organisms treated as fungi, once accepted by a subsequent plenary session of the same Congress, are binding on the Nomenclature Section convened at the subsequent International Botanical Congress. Such decisions will, however, be open for any editorial adjustments deemed necessary by the Editorial Committee for Fungi after consultation with the Editorial Committee for this *Code*.

8.11. Certain publications, which may be electronic or printed or both, appear as soon as feasible after an International Mycological Congress, not necessarily in this sequence:

(a) The Congress-approved decisions and elections of the Fungal Nomenclature Session including the results of the preliminary guiding vote.

(b) The announcement of Special-purpose Committees and their membership.

(c) The new edition of Chapter F of this *Code*.

(d) A transcript of the Fungal Nomenclature Session.

8.12. Where modifications to the *Code* have been authorized by a plenary session of an International Mycological Congress on a resolution moved by the Fungal Nomenclature Session of that Congress, such modifications are inserted into any online version of the *Code* in such a manner that it is clear that the modifications originated from that International Mycological Congress.

8.13. The Nomenclature Committee for Fungi has the following functions:

(a) Examines proposals to conserve, protect, or reject names, proposals to suppress works, and requests for decisions (Art. 14.12, 34.1, 38.5, 53.4, 56.2, F.2.1, and F.7.1) that relate to names of organisms treated as fungi, as referred by the General Committee (Prov. 7.10(b) and (c) and 7.11).

(b) Is consulted by the General Committee as to which proposals relate solely to names of organisms treated as fungi (Prov. 8.2).

(c) Elects the Chair of the Fungal Nomenclature Session, in consultation with the General Committee (Prov. 8.6).

(d) Approves the appointment of the Deputy Secretary of Fungal Nomenclature Bureau, in consultation with the General Committee (Prov. 8.7).

(e) Appoints Special-purpose Committees, in consultation with the General Committee:

    (1) authorized by a Nomenclature Session at an International Mycological Congress (Prov. 8.5(d)); or

    (2) proposed between International Mycological Congresses following an open call for expressions of interest in serving on such a committee based on: *(i)* the initiative of the Nomenclature Committee for Fungi; or *(ii)* requests that have been submitted to the Nomenclature Committee for Fungi by at least five people.

(f) Establishes subcommitees, in consultation with the General Committee, that prepare lists of protected or rejected names; and reviews and approves such lists (Art. F.2.1 and F.7.1).

(g) Appoints repositories that accession information on nomenclatural novelties and type designations (see Art. F.5.1).

GLOSSARY
DEFINITIONS OF TERMS USED IN THIS *CODE*

The Glossary includes two types of entries: (1) terms defined in the provisions of the *Code* given in bold; (2) other words, not defined in the provisions of the *Code,* given in italics with "[Not defined]" and an explanation of their use. Note that cross-references to relevant provisions of the *Code* and to other Glossary entries are not exhaustive.

admixture. [Not defined] – something mixed in; used for components of a gathering that represent a taxon or taxa other than that intended by the collector and, because the admixture is disregarded, do not prevent the gathering, or part thereof, from being a type specimen (Art. 8.2).

adopted name. [Not defined] – the name that is accepted for, and applied to, a taxon (see Art. 16.2, 19 Note 2, 22 Note 1, and 26 Note 1) (see also ***correct name***).

affirmation. The adoption in a publication that did not use a largely mechanical method of selection of a choice of type of a name of a genus or subdivision of a genus that had been made using such a method and that had not in the interval been superseded (Art. 10.5). Choices of type that have been so affirmed can no longer be superseded (see also *superseded*).

alternative names. Two or more different names based on the same type accepted simultaneously for the same taxon by at least one author in common in the same publication (Art. 36.3) (see also ***nomen alternativum***).

analysis. A figure or group of figures, often presented separately from the main illustration of the organism (though usually on the same page or plate), showing details aiding identification, with or without a separate caption (Art. 38.10; see also Art. 38.11).

anamorph. A mitotic asexual morph in pleomorphic fungi (Art. F.8 Notes 1 and 2).

ascription. The direct association of the name of a person or persons with a new name or description or diagnosis of a taxon (Art. 46.3).

attributed. [Not defined] – regarded as belonging to or produced by a person or a taxon, e.g. a name attributed to its author(s) as determined by Art. 46, a feature attributed to a taxon (Art. 40.6), or a specimen attributed to a taxon (Art. 26 Ex. 3 and 6).

author citation. A statement of the name(s) of the author(s) responsible for the establishment or introduction of a name; when used, it is appended to that name (Art. 46–50).

automatic typification. (1) Typification of a nomenclaturally superfluous and illegitimate name by the type of the name (the replaced synonym) that itself or the epithet of which ought to have been adopted under the rules (Art. 7.5), but not applying to names sanctioned under Art. F.3. (2) Typification of a new combination or a name at new rank by the type of the basionym (Art. 7.3) or of a replacement name by the type of the replaced synonym (Art. 7.4) (see also Art. 7 Note 4). (3) Typification of an autonym by the type of the name from which it is derived (Art. 7.7; see also Art. 7 Note 4). (4) Typification of a generic name by the type of a species name under Art. 10.9. (5) Typification of the name of a taxon above the rank of genus by the type of the generic name from which it is formed (Art. 10.10 and 10.11).

autonym. The automatically established name of a subdivision of a genus or of an infraspecific taxon that includes the type of the adopted, legitimate name of the genus or species, respectively. Its final epithet repeats unaltered the generic name or specific epithet and is not followed by an author citation (Art. 22.1 and 26.1). Autonyms need not be effectively published nor comply with the provisions for valid publication (Art. 32.1); they are automatically established, at any given rank, by the first instance of valid publication at that rank of a name of a subdivision of a genus under a legitimate generic name or of a name of an infraspecific taxon under a legitimate species name (Art. 22.3 and 26.3). Autonyms are not established under illegitimate names of genera or species (Art. 22.5 and 27.2); nor do they exist above the rank of genus.

available. [Not defined] – applied to an epithet in a name (Art. 11.4, 11.5, and F.3.8), the type of which falls within the circumscription of the taxon under consideration and where the use of the epithet would not be contrary to the rules (see also ***available name***).

available name. A name published under the *International Code of Zoological Nomenclature* with a status equivalent to that of a validly published name under the *International Code of Nomenclature for algae, fungi, and plants* (Art. 45 Ex. 1 footnote).

avowed substitute. See ***replacement name***.

basionym. The legitimate, previously published name on which a new combination or name at new rank is based. The basionym does not itself have a basionym; it provides the final epithet, name, or stem of the new combination or name at new rank (Art. 6.10) (see also ***name at new rank, new combination***).

binary combination. See *binomial*.

binary designation. [Not defined] – an apparent binomial that has not been validly published (see Art. 6.3) (see also *binomial, designation*).

binding decision. A recommendation made by the General Committee and ratified by an International Botanical Congress on (1) whether or not a name is validly published (Art. 38.5) or (2) whether or not names are to be treated as homonyms (Art. 53.4) (see also Art. 20 Note 1). Binding decisions are listed in (1) App. VI or (2) App. VII.

binomial (binary combination). A generic name combined with a specific epithet to form a species name (Art. 23.1) (see also *combination*).

circumscription. [Not defined] – a taxonomic rather than nomenclatural term for the limits of a taxon.

combinatio nova (comb. nov.). See *new combination*.

combination. A name of a taxon below the rank of genus, consisting of the name of a genus combined with one or two epithets (Art. 6.7).

compound. A name or epithet that combines elements derived from two or more words (usually nouns, adjectives, or adverbs), as well as prefixes or suffixes, in Greek, Latin, or other languages. A regular compound is one in which a noun or adjective in a non-final position appears as a modified stem (Art. 60.11) (see also *pseudocompound*).

confusingly similar names. Orthographically similar names at the rank of genus or below that are likely to be confused and are to be treated as homonyms if heterotypic (Art. 53.2 and 53.3) or as orthographical variants if homotypic (Art. 61.5). Binding decisions may be made on whether or not the former are to be treated as homonyms (Art. 53.4 and App. VII) (see also *binding decision, homonym*).

conserved name (nomen conservandum). (1) A name of a family, genus, or species, or in certain cases a name of a subdivision of a genus or of an infraspecific taxon, declared legitimate, even though it may have been illegitimate when published, and taking precedence over other specified names even if it lacks priority (Art. 14.1–14.7, 14.10, App. II, III, and IV). (2) A name for which the type, orthography, or gender has been determined by the conservation process (Art. 14.8, 14.9, 14.11, App. III, and IV) (see also *protected name*).

conserved type (typus conservandus). A type attached to a name by the conservation process (Art. 14.8 and 14.9) to serve as the nomenclatural type when it would otherwise not be the type under Art. 7–10, 40, and F.5.4 (see also *nomenclatural type*).

correct name. The name that must be adopted in accordance with the rules for a taxon with a particular circumscription, position, and rank (Art. 6.6, 11.1, 11.3, and 11.4).

cultivar. The basic independent category used for organisms in agriculture, forestry, and horticulture and defined and regulated in the *International Code of Nomenclature for Cultivated Plants* (Art. 28 Notes 2, 4, and 5).

date of name. The date of valid publication of a name (Art. 33.1).

descriptio generico-specifica. A single description simultaneously validating the names of a genus and its single species (Art. 38.6) (see ***description***).

description. A statement explicitly describing one or more features of an individual taxon. A description need not be diagnostic (Art. 38.4). It may include morphological, anatomical, biochemical, karyological, molecular, or similar features of the taxon but cannot serve as a validating description if it describes only properties such as those listed in Art. 38.3. Note that a description or a diagnosis is required for valid publication of a name of a new taxon (Art. 38.1(a)). A name, by itself, may convey some descriptive information about the taxon to which it is applied but is not sufficient to serve as a description or a diagnosis (see also ***diagnosis***).

descriptive name. A name of a taxon above the rank of family not formed from a generic name (Art. 16.1(b)).

designation. [Not defined] – the term used for what appears to be a name but that (1) has not been validly published and hence is not a name in the sense of the *Code* (Art. 6.3) or (2) is not to be regarded as a name (Art. 20.4 and 23.7) (see also *binary designation, type designation*).

diagnosis. A statement of properties that in the opinion of its author distinguishes a taxon from other taxa (Art. 38.2). Note that a diagnosis or a description is required for valid publication of a name of a new taxon (Art. 38.1(a)) (see also ***description***).

dual nomenclature. A nomenclatural relationship that allows both the name of a fossil-taxon (diatoms excepted) and that of a non-fossil taxon at the same rank to be considered as correct when these taxa are linked by a morphologically similar or identical part or life-history stage and when the names of these taxa are not considered to be synonyms (Art. 11.8) (see also ***taxonomic equivalence***).

duplicate. Part of a single gathering of a single species or infraspecific taxon (Art. 8.3 footnote). Note that for most fossils there are no duplicates (see also ***gathering***).

effective publication. Publication in accordance with Art. 29–31 (Art. 6.1).

element (as applied to typification). [Not defined] – applied to a specimen or illustration eligible as a type; also applied to a species name considered as the full equivalent of its type for the purpose of designation or citation of the type of a name of a genus or subdivision of a genus (Art. 10.1).

epithet. [Not defined] – used for the words in a combination other than the generic name and any rank-denoting term; hyphenated words are equivalent to a single word (Art. 6.7, 11.4, 21.1, 23.1, and 24.1; see also Art. H.10.2) (see also *final epithet*).

epitype. An interpretative type (specimen or illustration) designated to provide essential diagnostic characters, necessary for the precise application of the name of a taxon, that are lacking from the holotype, lectotype, or previously designated neotype that the epitype supports (Art. 9.9; see also Art. 9 Note 11).

ex-type (ex typo), ex-holotype (ex holotypo), ex-isotype (ex isotypo), etc. A living isolate obtained from the type of a name when this is a culture permanently preserved in a metabolically inactive state (Rec. 8B.2).

figure. See *analysis*.

final epithet. The last epithet in sequence in any particular combination, whether at the rank of a subdivision of a genus, or of a species, or of an infraspecific taxon (Art. 6.10 footnote).

forma specialis. See *special form*.

fossil-taxon. A taxon (diatom taxa excepted) the name of which is based on a fossil type (Art. 1.2 and 13.3).

gathering. Material collected by the same collector(s) at the same time from a single locality and presumed to be of a single taxon (Art. 8.2 footnote; see also Art. 8 Note 1) (see also *duplicate*).

heterotypic synonym (taxonomic synonym). A name based on a type different from that of another name referring to the same taxon (Art. 14.4); indicated by the symbol "=" in the Appendices of the *Code;* termed a "subjective synonym" in the *International Code of Zoological Nomenclature* (Art. 14.4 footnote).

holotype. The one specimen or illustration designated in the protologue as the nomenclatural type of a name of a new species or infraspecific taxon or, when no type was designated, used by the author(s) when preparing the protologue (Art. 9.1, 9.2, and Note 1) (see also *nomenclatural type*).

homonym. A name spelled exactly like another name published for a taxon at the same rank based on a different type (Art. 53.1). Note that names of subdivisions of the same genus or of infraspecific taxa within the same species that are based on different types and have the same final epithet are homonyms, even if they differ in rank (Art. 53.3), because the rank-denoting term is not part of the name (Art. 21 Note 1 and Art. 24 Note 2) (see also *confusingly similar names*).

homotypic synonym (nomenclatural synonym). A name based on the same type as that of another name (Art. 14.4); indicated by the symbol "≡" in the Appendices of the *Code;* termed an "objective synonym" in the *International Code of Zoological Nomenclature* (Art. 14.4 footnote).

hybrid formula. An expression consisting of the names of the parental taxa of a hybrid with a multiplication sign × placed between them (Art. H.2.1).

identifier. [Not defined] – (1) a unique number or string of characters issued by a recognized nomenclatural repository for the purpose of registering nomenclatural novelties and certain nomenclatural acts; required for names of fungi by Art. F.5.2 and F.5.4 and voluntary for names of algae and plants under Art. 42.5 and 42.6. (2) A unique number or string of characters applied to a specimen, e.g. an accession number or a barcode (see also *nomenclatural repository, registration*).

illegitimate name. A validly published name that is not in accordance with specified rules (Art. 6.4), principally those on superfluity (Art. 52) and homonymy (Art. 53 and 54) (see also *homonym, superfluous name*).

illustration. A work of art or a photograph depicting a feature or features of a species or infraspecific taxon, e.g. a drawing, a picture of a herbarium specimen, or a scanning or transmission electron micrograph, but not a photograph of habitat (Art. 6.1 footnote).

improper Latin termination. A termination of a name or epithet not in accordance with the termination required by the *Code* (Art. 16.3, 18.4, 19.7, and 32.2).

indelible autograph. Handwritten material reproduced by some mechanical or graphic process (such as lithography, offset, or metallic etching) (Art. 30.6).

indirect reference. A clear (if cryptic) indication, by an author citation or in some other way, that a previously and effectively published description or diagnosis applies (Art. 38.15) or that a basionym or replaced synonym exists (Art. 41.3).

informal usage. Usage of the same or equivalent rank-denoting term at more than one non-consecutive position in the taxonomic sequence. Note that names involved in such usage are validly published but unranked (Art. 37.9).

infraspecific. [Not defined] – below the rank of species.

isoepitype. A duplicate specimen of the epitype, regardless of whether or not it was cited or seen by the typifying author(s) (Art. 9.4 footnote).

isolectotype. A duplicate specimen of the lectotype, regardless of whether or not it was cited or seen by the typifying author(s) (Art. 9.4 footnote).

isoneotype. A duplicate specimen of the neotype, regardless of whether or not it was cited or seen by the typifying author(s) (Art. 9.4 footnote).

isonym. The same name based on the same type, published independently at different times by the same or different authors. Note that only the earliest isonym has nomenclatural status (Art. 6 Note 2; but see Art. 14.14).

isosyntype. A duplicate specimen of a syntype (Art. 9.4 footnote); it is not cited in the protologue. An isosyntype does not have to be seen by the author(s) of the name.

isotype. A duplicate specimen of the holotype regardless of whether or not it is cited in the protologue or seen by the author(s) of the name (Art. 9.5). Note that because there are no duplicates for fossils, names of fossil-taxa have no isotypes.

lectotype. One specimen or illustration designated from the original material as the nomenclatural type, in conformity with Art. 9.11 and 9.12, if the name was published without a holotype, or if the holotype is lost or destroyed, or if a type is found to belong to more than one taxon (Art. 9.3) (see also *nomenclatural type*).

legitimate name. A validly published name that is in accordance with the rules, i.e. one that is not illegitimate (Art. 6.5) (see also *illegitimate name*).

misplaced term. A rank-denoting term used contrary to the relative order specified in the *Code* (Art. 18.2, 19.2, 37.7, and 37 Note 1; but see Art. F.4.1).

monotypic genus. A genus for which a single binomial is validly published (Art. 38.7).

name. A name that has been validly published, whether it is legitimate or illegitimate (Art. 6.3) (see also *designation*).

name at new rank (status novus). A new name based on a legitimate, previously published name at a different rank, which is its basionym and which provides the final epithet, name, or stem of the name at new rank (Art. 6.10 and 7.3) (see also *basionym, new combination*).

name of a new taxon. A name validly published in its own right, i.e. one not based on a previously validly published name; it is not a new combination (combinatio nova), a name at new rank (status novus), or a replacement name (nomen novum) (Art. 6.9).

neotype. A specimen or illustration selected to serve as the nomenclatural type if no original material exists or as long as it is missing (Art. 9.8 and 9.13; see also Art. 9.16 and 9.19) (see also *nomenclatural type*).

new combination (combinatio nova). A new name for a taxon below the rank of genus based on a legitimate, previously published name, which is its basionym and which provides the final epithet of the new combination (Art. 6.10 and 7.3) (see also *basionym, name at new rank*).

new name. [Not defined] – a name as it appears in the place of its valid publication (see also *nomenclatural novelty*).

nomen alternativum (nom. alt.). One of eight family names, each regularly formed from a generic name in accordance with Art. 18.1, allowed as an alternative (Art. 18.6) to one of the family names of long usage treated as validly published under Art. 18.5. In addition, one subfamily name of long usage, *Papilionoideae,* may be used as an alternative to *Faboideae* (Art. 19.8) (see also *alternative names*).

nomen conservandum (nom. cons.). See *conserved name.*

nomen novum (nom. nov.). See *replacement name.*

nomen nudum (nom. nud.). A designation of a new taxon published without a description or diagnosis or reference to a description or diagnosis (Art. 38 Ex. 1, Rec. 50B) (see *description, designation, diagnosis*).

nomen rejiciendum (nom. rej.). A name rejected in favour of a name conserved under Art. 14 or a name ruled as rejected under Art. 56 (App. IIA, III, IV, and V) (see also *nomen utique rejiciendum, rejected name*).

nomen sanctionatum (nom. sanct.). See *sanctioned name.*

nomen utique rejiciendum (suppressed name). A name ruled as rejected under Art. 56 (see also Art. F.7). Note that, along with the rejected name, all names for which it is the basionym are similarly rejected, and none is to be used (see App. V).

nomenclatural act. An act requiring effective publication that results in a nomenclatural novelty or affects aspects of names such as typification

(Art. 7.10, 7.11, and F.5.4), priority (Art. 11.5 and 53.5), grammatical category (Art. 23.6), orthography (Art. 61.3), or gender (Art. 62.3) (Art. 34.1 footnote) (see also ***nomenclatural novelty***).

nomenclatural novelty. Any or all of the following categories: name of a new taxon, new combination, name at new rank, and replacement name (Art. 6 Note 4; see also Art. 6 Note 5) (see also *new name*).

nomenclatural repository. A repository that takes charge, for specified categories of organisms, of registering nomenclatural novelties and/or any nomenclatural act (Art. 42.1). A nomenclatural repository can be localized or decentralized, open and accessible, and electronic (Art. F.5.1), and it must be recognized by the General Committee (Art. 42.2, 42.3, and 42.7) or the Nomenclature Committee for Fungi (Art. F.5.1), which also have the power to suspend or revoke such recognition (see ***nomenclatural act, nomenclatural novelty, registration***).

nomenclatural synonym. See ***homotypic synonym***.

nomenclatural type. A specimen, or in some cases an illustration (Art. 8.1), to which the name of a taxon is permanently attached (Art. 7.2). The nomenclatural type of the name of a species or infraspecific taxon is a holotype (Art. 9.1), lectotype (Art. 9.3), neotype (Art. 9.8), or conserved type (Art. 14.9), any of which may be supported by an epitype (Art. 9.9).

non-fossil taxon. A taxon the name of which is based on a non-fossil type (Art. 13.3).

nothogenus. A hybrid genus (Art. 3.2).

nothomorph. A term formerly denoting the only rank of infraspecific nothotaxa, equivalent to variety, that was permitted in editions of the *Code* prior to the Sydney *Code* of 1983. Names published as nothomorphs are now treated as having been published as names of varieties (Art. H.12.2 and footnote).

nothospecies. A hybrid species (Art. 3.2).

nothotaxon. A hybrid taxon (Art. 3.2 and H.3.1).

objective synonym. See ***homotypic synonym***.

opera utique oppressa. See ***suppressed works***.

organism. As used in this *Code,* the term is applied only to organisms traditionally studied by botanists, mycologists, and phycologists (Pre. 2 footnote, Pre. 8).

original author(s). [Not defined] – the author(s) of a name of a new taxon or of a replacement name in its protologue (see Art. 9.2, 9.24, 10.8, 18.1, 23.5, Rec. 50D.1, Art. 62.3, and 62.4).

original material. The set of specimens and illustrations from which a lectotype may be chosen (see Art. 9.4, Art. 9 Notes 2–5, Art. F.3.10, and Art. F.3 Note 2 for details), or the holotype (see Art. 9.1).

original spelling. The spelling used when a name of a new taxon or a re-placement name was validly published (Art. 60.2) (see also ***orthographical variants***).

original type. [Not defined] – the type (1) that was designated or indicated in the protologue (see also *type designation, type indication*) or (2) of the single species name that was included in the protologue of the name of a monotypic genus or subdivision of a genus (Art. 10.2 and 10.3).

orthographical variants. Various spelling, compounding, and inflectional forms of a name or its final epithet when only one nomenclatural type is involved (Art. 61.2) (see also ***original spelling***).

page reference. Citation of the page or pages on which the basionym or replaced synonym was validly published or on which the protologue appears (Art. 41 Note 1; see also Art. 41 Note 2 and Rec. 41A.2).

paratype. Any specimen cited in the protologue that is neither the holotype nor an isotype, nor one of the syntypes if in the protologue two or more specimens were simultaneously designated as types (Art. 9.7).

personal name. [Not defined] – any or all of the individual component(s) or traditionally combined components of a person's name. For prefixes and particles of personal names see Rec. 60C.4.

phrase name. An apparent epithet consisting of one or more nouns and associated adjectives (see Art. 23.7(b)) (see also ***binomial***).

position. [Not defined] – used to denote the placement of a taxon relative to other taxa in a classification, regardless of rank (Principle IV, Art. 6.6, and 11.1).

priority. A right to precedence established by the date of valid publication of a legitimate name (Art. 11) or of an earlier homonym (Art. 53 Note 4), or by the date of designation of a type (Art. 7.10, 7.11, and F.5.4).

pro synonymo (pro syn., as synonym). A citation indicating that a name or a designation is merely cited as a synonym (Art. 36.1(b) and Rec. 50A) (see *designation*).

protected name. The name of an organism treated as a fungus listed (in App. IIA, III, or IV) with its type and treated as conserved against any competing listed or unlisted synonyms or homonyms (including sanctioned names), although conservation under Art. 14 overrides this protection (Art. F.2.1) (see also *conserved name*).

protologue. Everything associated with a name at its valid publication, e.g. description, diagnosis, illustrations, photographs of habitat, references, synonymy, geographical data, citation of specimens, discussion, and comments (Art. 6.13 footnote).

provisional name. A designation proposed in anticipation of the future acceptance of the taxon concerned, or of a particular circumscription, position, or rank of the taxon (Art. 36.1(a)) (see *designation*).

pseudocompound. A name or epithet that combines elements derived from two or more Greek or Latin words and in which a noun or adjective in a non-final position appears as a word with a case ending, not as a modified stem (Rec. 60H.1(b)) (see also *compound*).

rank. [Not defined] – used for the relative position of a taxon in the taxonomic hierarchy (Art. 2.1). For suprageneric names published on or after 1 January 1887, the rank is indicated by the termination of the name (see Art. 37.2 and footnote). For names published on or after 1 January 1953, a clear indication of the rank is required for valid publication (Art. 37.1).

recognized repository. See *nomenclatural repository*.

registration. The deposition of a name, type designation, or other nomenclatural act in a recognized nomenclatural repository (Art. 42 and F.5) (see *nomenclatural repository*).

rejected name. A name ruled as not to be used, either by formal action under Art. 14, 56, or F.7 overriding other provisions of the *Code* (see *nomen rejiciendum, nomen utique rejiciendum*) or because it was nomenclaturally superfluous when published (Art. 52) or a later homonym (Art. 53 and 54). A name treated as rejected under Art. F.7 may become eligible for use by conservation under Art. 14.

replaced synonym. The legitimate or illegitimate, previously published name on which a replacement name (nomen novum) is based. The replaced synonym, when legitimate, does not provide the final epithet, name, or stem of the replacement name (Art. 6.11) (see *replacement name*).

replacement name (nomen novum). A new name published as an explicit substitute (avowed substitute) for a legitimate or illegitimate, previously published name, which is its replaced synonym and which, when legitimate, does not provide the final epithet, name, or stem of the replacement name (Art. 6.11 and 7.4; for names not explicitly proposed as sub-stitutes see Art. 6.12 and 6.13) (see *replaced synonym*).

sanctioned name (nomen sanctionatum). The name of a fungus treated as if conserved against earlier homonyms and competing synonyms, through acceptance in a sanctioning work (Art. F.3.1).

special form (forma specialis). A taxon of parasites, especially fungi, char-acterized from a physiological standpoint but scarcely or not at all from a morphological standpoint, the nomenclature of which is not governed by this *Code* (Art. 4 Note 4).

specimen. A gathering, or part of a gathering, of a single species or infra-specific taxon, disregarding admixtures, mounted either as a single preparation or as more than one preparation with the parts clearly la-belled as being part of the same specimen or having a single, original label in common (Art. 8.2 and 8.3; for fossil-taxa see Art. 8.6). A speci-men may not be a living organism or an active culture (Art. 8.4) (see *gathering*).

status. (1) Nomenclatural standing with regard to effective publication, valid publication, legitimacy, and correctness (Art. 6 and 12.1). (2) Rank of a taxon within the taxonomic hierarchy (see *name at new rank*). (3) Category of nomenclatural novelty (Art. 6.14).

status novus (stat. nov.). See *name at new rank*.

subdivision of a family. Any taxon at a rank between and not including family and genus (Art. 4 Note 2).

subdivision of a genus. Any taxon at a rank between and not including genus and species (Art. 4 Note 2).

subjective synonym. See *heterotypic synonym*.

superfluous name. A name that, when published, was applied to a taxon that, as circumscribed by its author, definitely included the type of a name that ought to have been adopted, or of which the epithet ought to have been adopted, under the rules (Art. 52.1). A superfluous name is illegitimate except as provided by Art. 52.4 or unless conserved (Art. 14), protected (Art. F.2), or sanctioned (Art. F.3).

superseded. [Not defined] – used for a designation of a type that is not fol-lowed but is replaced by a subsequent designation of a different type under the provisions of Art. 9.15, 9.18, 9.19, 10.2, or 10.5.

suppressed name. See *nomen utique rejiciendum*.

suppressed works (opera utique oppressa). Works, ruled as suppressed, in which new names at specified ranks are not validly published and no nomenclatural act within the work associated with any name at the specified ranks is effective (Art. 34.1 and App. I).

synonym. [Not defined] – one of two or more names that apply to the same taxon. Also used in the sense of a name that applies to a taxon but is not the correct name for that taxon (see *heterotypic synonym, homotypic synonym*).

syntype. Any specimen cited in the protologue when there is no holotype, or any of two or more specimens simultaneously designated in the protologue as types (Art. 9.6). Illustrations cannot be syntypes.

tautonym. A binary designation in which the specific epithet exactly repeats the generic name (Art. 23.4).

taxon (taxa). A taxonomic group at any rank (Art. 1.1).

taxonomic equivalence. A taxonomic approach based on a noted morphological similarity or identicality between a fossil-taxon and a part of, or life-history stage of, a non-fossil taxon at the same rank, when the names of these two taxa are not considered to be synonyms (Art. 11.8) (see also *dual nomenclature*).

taxonomic synonym. See *heterotypic synonym.*

teleomorph. A meiotic sexual morph in pleomorphic fungi (Art. F.8 Notes 1 and 2).

type. See *nomenclatural type.*

type citation. [Not defined] – an explicit statement that specifies the type of a name but does not necessarily establish that type (see *type designation, type indication*).

type designation. [Not defined] – an explicit statement that establishes a type of a name; either (1) a holotype (Art. 9.1), syntype(s) (Art. 9.6), or type of a name of a genus or subdivision of a genus (Art. 10.2(a)) designated in the protologue or (2) a lectotype, neotype, or epitype subsequently designated under the provisions of Art. 9–10 and in accordance with Art. 7.8–7.11 and F.5.4 (see also *type indication*).

type indication. [Not defined] – a statement that provides only indirect evidence that establishes the type of a name (Art. 40; see also Art. 10.8) (see also *automatic typification, type designation*).

typify. [Not defined] – to attach a nomenclatural type to a name of a taxon; the process of which is *typification* (Art. 7–10 and F.5.4) (see also *automatic typification, nomenclatural type, type designation, type indication*).

typus conservandus (typ. cons.). See *conserved type.*

validate. [Not defined] – to make validly published; used in the context of a description or diagnosis, or illustration, effecting valid publication of a name (e.g. Art. 38 Ex. 23, 43.3, and 46 Ex. 7).

validly published. Effectively published and in accordance with the relevant provisions of Art. 32–45, F.4, F.5.2, F.5.3, and H.9 (Art. 6.2) (see *designation, name;* see also *effective publication*).

voted Example. An Example, denoted by an asterisk (*) in the *Code,* accepted by an International Botanical Congress in order to govern nomenclatural practice when the corresponding Article is open to divergent interpretation or does not adequately cover the matter. A voted Example is therefore comparable to a rule, as contrasted with other Examples provided by the Editorial Committee solely for illustrative or explanatory purposes (Art. 7 *Ex. 17 footnote).

APPENDICES I–VII

The *International Code of Nomenclature for algae, fungi, and plants* has seven Appendices, which together provide a permanent record of certain nomenclatural acts. The Appendices are as follows:

App. I. Suppressed works

App. IIA. Conserved, protected, and rejected names of families of algae, fungi, pteridophytes, and fossils

App. IIB. Conserved and rejected names of families of bryophytes and spermatophytes

App. III. Conserved, protected, and rejected names of genera and subdivisions of genera

App. IV. Conserved, protected, and rejected names of species and infraspecific taxa

App. V. Suppressed names

App. VI. Binding decisions on descriptive statements

App. VII. Binding decisions on confusability of names

The Appendices are available as an online searchable database at https://naturalhistory.si.edu/research/botany/codes-proposals.

INDEX OF SCIENTIFIC NAMES

This index includes the scientific names appearing in the Preamble and Division II of the *Code*. The references are not to pages but to the Articles, Examples, Notes, Preamble, and Recommendations, as follows: Ex. = Example; *Ex. = voted Example; F. = Chapter F (fungi); H. = Chapter H (hybrids); N. = Note; Pre. = Preamble. Arabic numerals indicate an Article (e.g. 60 or H.10); Arabic numerals immediately followed by an upper-case letter indicate a Recommendation (e.g. 19A or F.3A). Where more than one Article or Recommendation is cited, these are separated by a semicolon; in the few cases where a main paragraph in an Article is cited in addition to a Note or Example(s), these are separated by a comma (e.g. F.8.1, N.3 = Art. F.8.1 in addition to Art. F.8 Note 3). Sequences are indicated by a plus sign (e.g. 23.Ex.5+8). Double quotation marks indicate designations (i.e. not validly published names, e.g. *"Anema"* or *"Abies koreana* var. *yuanbaoshanensis"*). An incorrect spelling used in the protologue of a name at the rank of genus or below is cited in single quotation marks after the correct spelling (e.g. *Bougainvillea* Comm. ex Juss., *'Buginvillaea'*); an incorrect spelling of a name at the rank of genus or below used in a work later than the protologue is cited in single quotation marks as a separate index entry (e.g. *'Musenium'*). Pre-Linnaean designations are excluded, as are words or designations not to be regarded as names (Art. 20.4 and 23.7). Authors are cited after validly published names at the rank of genus or below, even when names are cited without authors in the main text of the *Code*. When author citations both with and without "ex" (Art. 46.5) are given for a name in the *Code,* only the alternative without "ex" appears in this Index. Under a generic name, names of subdivisions of genera precede species names. The multiplication sign (×) and single or double quotation marks are disregarded in the alphabetical sequence of names and epithets.

A sub-index of epithets appearing in the *Code* other than in combinations is provided in the Subject index, under Epithets (see p. 280).

SUBJECT INDEX

The references in this index are not to pages but to the Articles, Recommendations, etc. of the *Code,* as follows: Ex. = Example; *Ex. = voted Example; F. = Chapter F (fungi); fn. = footnote; Gl. = Glossary; H. = Chapter H (hybrids); N. = Note; Pre. = Preamble; Prin. = Principle; Prov. = Provision (Division III); R. = Recommendation (Div. III). Arabic numerals indicate an Article (e.g. 40); Arabic numerals immediately followed by an upper-case letter indicate a Recommendation in Div. II (e.g. 46A). Within an Article or Recommendation, the main paragraphs, including footnotes, are listed first, followed after commas by Notes, then Examples; then after a semicolon by the next relevant Article or Recommendation (e.g. 14.15, N.4, Ex.8; 34.2). Provisions and Recommendations in Div. III are treated similarly (e.g. Prov.4.13, R.1; Prov.8.1). Continuous sequences are indicated by a dash (e.g. 11.3–8 = Art. 11.3 to 11.8 inclusive; 60.8(a–b) = Art. 60.8(a) and (b)); interrupted sequences are indicated by a plus sign (e.g. Prov.1.1+4+fn. = Prov. 1.1, 1.4, and 1.4 footnote).

For ease of reference, a few sub-indices have been included under the following headings: Abbreviations and acronyms, Definitions, Epithets (other than in combinations), Publications, Transcriptions, and Word elements.

Scientific names appearing in the Preamble and Div. II are not included in this Subject index, but in the preceding Index of scientific names.

www.ingramcontent.com/pod-product-compliance
Lightning Source LLC
Chambersburg PA
CBHW022135020426
42334CB00015B/905